VOLUME NINETY THREE

# SEMICONDUCTORS AND SEMIMETALS

## Semiconductor Nanowires I: Growth and Theory

# SERIES EDITORS


## EICKE R. WEBER

*Director*
*Fraunhofer-Institut*
*für Solare Energiesysteme ISE*
*Vorsitzender, Fraunhofer-Allianz Energie*
*Heidenhofstr. 2, 79110*
*Freiburg, Germany*

## CHENNUPATI JAGADISH

*Australian Laureate Fellow*
*and Distinguished Professor*
*Department of Electronic*
*Materials Engineering*
*Research School of Physics*
*and Engineering*
*Australian National University*
*Canberra, ACT 0200*
*Australia*

VOLUME NINETY THREE

# SEMICONDUCTORS AND SEMIMETALS

## Semiconductor Nanowires I: Growth and Theory

Edited by


**ANNA FONTCUBERTA I MORRAL**
*Laboratory of Semiconductor Materials,*
*Institute of Materials,*
*Ecole Polytechnique Fédérale de Lausanne,*
*Lausanne, Switzerland*

**SHADI A. DAYEH**
*Department of Electrical and Computer Engineering,*
*University of California, San Diego, USA*

**CHENNUPATI JAGADISH**
*Department of Electronic Materials Engineering,*
*Research School of Physics and Engineering,*
*Australian National University,*
*Canberra, Australia*



ELSEVIER

AMSTERDAM • BOSTON • HEIDELBERG • LONDON
NEW YORK • OXFORD • PARIS • SAN DIEGO
SAN FRANCISCO • SINGAPORE • SYDNEY • TOKYO
Academic Press is an imprint of Elsevier

Academic Press is an imprint of Elsevier
225 Wyman Street, Waltham, MA 02451, USA
525 B Street, Suite 1800, San Diego, CA 92101-4495, USA
The Boulevard, Langford Lane, Kidlington, Oxford OX5 1GB, UK
125 London Wall, London, EC2Y 5AS, UK

First edition 2015



ISBN: 978-0-12-803027-1
ISSN: 0080-8784

For information on all Academic Press publications
visit our website at http://store.elsevier.com/

# CONTENTS

# CONTRIBUTORS

**Erik Bakkers**
Eindhoven University of Technology, Eindhoven, and Kavli Institute of Nanoscience, Delft University of Technology, Delft, The Netherlands. (ch5)

**Sonia Conesa-Boj**
Eindhoven University of Technology, Eindhoven, The Netherlands. (ch5)

**Vladimir G. Dubrovskii**
St. Petersburg Academic University, St. Petersburg, and Ioffe Physical Technical Institute RAS, St. Petersburg, Russia. (ch1)

**Frank Glas**
CNRS—Laboratoire de Photonique et de Nanostructures, Marcoussis, France. (ch2)

**Moïra Hocevar**
Institut Néel CNRS/UJF, Grenoble, France. (ch5)

**Young Joon Hong**
Faculty of Nanotechnology & Advanced Materials Engineering, Graphene Research Institute, Sejong University, Seoul, South Korea. (ch3)

**Nari Jeon**
Department of Materials Science and Engineering, Northwestern University, Evanston, Illinois, USA. (ch6)

**Lincoln J. Lauhon**
Department of Materials Science and Engineering, Northwestern University, Evanston, Illinois, USA. (ch6)

**Chul-Ho Lee**
KU-KIST Graduate School of Converging Science and Technology, Korea University, Seoul, South Korea. (ch3)

**Gyu-Chul Yi**
Department of Physics and Astronomy, Seoul National University, Seoul, South Korea. (ch4)

**Hosang Yoon**
Department of Physics and Astronomy, Seoul National University, Seoul, South Korea. (ch4)

# PREFACE

Prior to 1990 known as whiskers and nowadays as nanowires, filamentary crystals with a diameter in the nanoscaler range have recently inspired many scientists and engineers. Thanks to the small dimensions and particularly due to their special shape, nanowires have enabled novel applications and improved existing ones in a broad range of fields. These include the discovery of new phenomena in one-dimensional physics, next-generation solar cells, lasers, solid-state lighting, biosensing, and integrated circuits. Additionally, they have provided an excellent platform to study crystal growth in 1D and to synthesize materials in new phases, combinations, and architectures that are not possibly attained otherwise.

Obtaining nanowires with the tailored size, shape, crystal phase, orientation, and composition profile is the key requisite for enabling any application. Today, exquisite control on many of these properties has already been demonstrated, thanks to the focused and extensive efforts of the scientific community in the last decades which developed impressive understanding of their growth and of tailoring their properties. It is therefore timely to disseminate these results and to provide a comprehensive treatment of nanowire growth and theory, particularly for the III–V and II–VI compound semiconductor materials.

In this first volume, we review important aspects of nanowire growth. The first chapter concerns with the theory of VLS growth in compound semiconductors which is influenced by complex contributions of many growth parameters that are not typically present in elemental whisker or nanowire growth. The second chapter treats the role of strain in nanowires and related heterostructures, a key characteristic for tuning both the growth and properties of semiconductor nanowires. The third chapter discusses a recently emerging growth mechanism of 1D nanowires on 2D materials: van der Waals heteroepitaxy. The fourth chapter comprises a summary on the studies of position-controlled growth of ZnO nanostructures. The fifth chapter provides one archetypical example of what nanowires enable in contrast to thin-film technology: the quasi defect-free integration of group IV and III–V semiconductors in a single structure. The sixth and last chapter reviews atom probe tomography studies on nanowires, an important technique for the three-dimensional mapping of the chemical composition in nanowires. Our examples concern semiconductor materials. Nevertheless, the implications are in fact much more general and concern many more materials systems.

Anna Fontcuberta i Morral, Ecole Polytechnique Fédérale de Lausanne
Shadi A. Dayeh, University of California, San Diego
Chennupati Jagadish, Australian National University
Editors

CHAPTER ONE

# Theory of VLS Growth of Compound Semiconductors

**Vladimir G. Dubrovskii**[1]
St. Petersburg Academic University, St. Petersburg, Russia
Ioffe Physical Technical Institute RAS, St. Petersburg, Russia
[1]Corresponding author: e-mail address: dubrovskii@mail.ioffe.ru

## Contents



## 1. INTRODUCTION

This chapter deals with growth and structural modeling of vapor–liquid–solid (VLS) III–V nanowires catalyzed by either Au or group III metals. We have tried to balance a semi-quantitative description of the VLS growth process and presentation of the well-known results with some newly developed models. Consequently, the exposition of the chapter is the following. Section 2 contains an introduction to the VLS growth method in the case of III–V nanowires, which should be helpful to the beginners. Section 3 describes the driving force—the liquid chemical potential in ternary Au-III–V alloys. In Section 4, we briefly describe the VLS growth kinetics in the steady state where significant simplifications in modeling become possible. Section 5 presents a detailed analysis of the transport-limited axial

*Semiconductors and Semimetals*, Volume 93
ISSN 0080-8784
http://dx.doi.org/10.1016/bs.semsem.2015.09.002

and radial growth rates, with some refinements of the existing models. In Section 6, we consider our recently published results on the Zeldovich nucleation rate in Au-catalyzed III–V nanowires along with a self-consistency renormalization which appears very important in applications. Section 7 describes the position-dependent nucleation in nanowires. In Section 8, we present a simple self-consistent growth equation which allows one to describe the vast variety of experimental data.

The next sections contain mainly new results. In Section 9, we consider some new data on Ga-catalyzed growth of GaAs nanowires and in particular the effect of the radius self-equilibration. Section 10 deals with Au-catalyzed ternary InGaAs nanowires where we develop a new atomistic growth picture. Despite its simplicity, it helps to improve our understanding of complex growth phenomena in ternary nanowires. Finally, Section 11 presents a refined model of crystallographic polytypism in Au-catalyzed III–V nanowires. This model is capable of describing the thresholds between mononucleation, polynucleation, and atomistic growth, and explains very well the recently observed nonmonotonic behaviors of the preferred crystal structure versus different parameters such as the group V flux.

## 2. FUNDAMENTALS OF VLS GROWTH

In simple terms, the VLS method (Wagner and Ellis, 1964) makes use of catalytic effect of a liquid metal droplet (e.g., Au) to fabricate semiconductor nanowires that grow away from the substrate with the position and width determined by the initial location and size of this droplet, as shown in Fig. 1. Usually, nanowires grow in $\langle 111 \rangle$ direction perpendicular to (111) substrate. Parasitic growth on nonactivated parts of the substrate is either completely suppressed or proceeds at a lower rate than that of nanowires. Fast elongation of nanowires occurs because much more semiconductor material arrives to the droplet than to the substrate. In chemical epitaxies such as metal organic chemical vapor deposition (MOCVD) or hydride vapor phase epitaxy (HVPE), a liquid metal largely enhances the cracking efficiency of semiconductor precursors, while in molecular beam epitaxy (MBE) the droplet acts as a material collector that directs the diffusion fluxes to the nanowire top (Dubrovskii, 2014a). Whatever is the epitaxy technique used to produce nanowires, the VLS process necessarily involves two phase transitions of a semiconductor material: from vapor (e.g., Ga and $As_2$ beams in MBE) to liquid (Au–Ga–As liquid alloy in the droplet) and from liquid to solid (GaAs nanowire), which explains the term "vapor–liquid–solid."

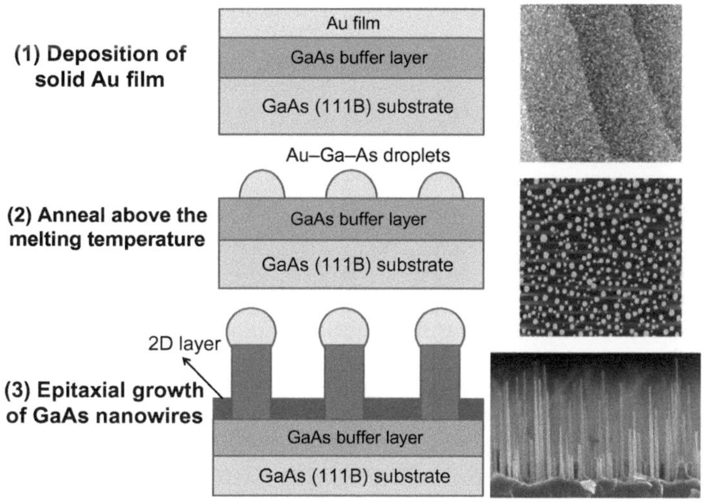


**Figure 1** Au-assisted VLS growth of GaAs nanowires on GaAs(111)B substrate by MBE with predeposition of ~1-nm-thick solid Au layer. Atomic force microscopy images show the surface after Au deposition (first stage) and 3D droplets after annealing (second stage). Scanning electron microscopy (SEM) image shows the vertically aligned GaAs nanowires (third stage). These nanowires have rather broad length and diameter distributions due to irregularity of the initial ensemble of droplets achieved by thermal dewetting.

The VLS growth usually consists of the following steps schematized in Fig. 1:

**(1)** Preparation of the initial catalyst layer or catalyst particles which can be organized into regular arrays by lithographical surface patterning.
**(2)** Anneal of the surface above the melting temperature of a given metal–semiconductor alloy to produce liquid droplets.
**(3)** Material deposition where the structural properties of nanowire ensembles (length, width, shape, and even the crystal phase) can be tuned by deposition conditions such as the growth time, temperature, group III and V fluxes.

Figure 2 presents the examples of regular arrays of InP nanowires grown by chemical beam epitaxy (CBE) from organized Au seeds in the openings of inert $SiN_x$ mask layer on InP(111)B (Kelrich et al., 2013, 2015). Since vertical and radial growth rates are determined by the droplet diameter, lateral spacing, and deposition conditions while fluctuations within the nanowire ensemble (e.g., randomness of starting times for growth or inhomogeneity of material influxes into different nanowires) are usually negligible, all nanowires can be made almost exactly identical. This feature of catalytic

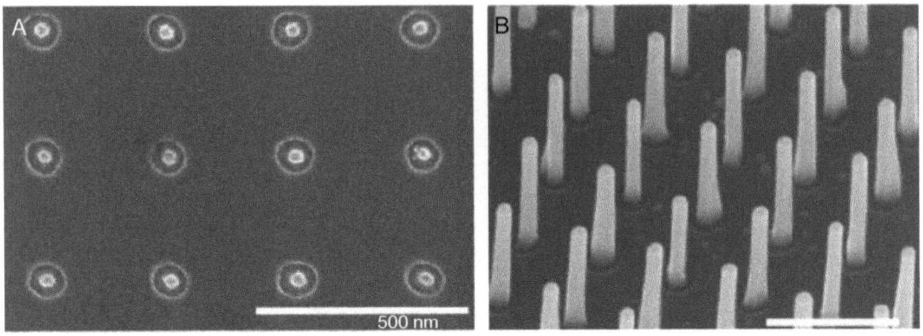


**Figure 2** (A) Plan view SEM image of a regular 2D array of Au nanoparticles inside the $SiN_x$ openings prior to CBE growth. (B) 45° tilted SEM image of a regular arrays of InP nanowires grown from 127-nm diameter openings at $T = 420°C$, TMI flow corresponding to 110 nm/h 2D growth rate on InP(100) at 500 °C, and $PH_3$ flow rate of 3 sccm (Kelrich et al., 2013). Tapering and crystal structure of these nanowires can be controlled by the growth temperature and phosphorus flux (Kelrich et al., 2015).

nanostructures is very advantageous for applications. Such a level of size uniformity cannot be achieved in self-induced approaches due to random character of nucleation. It should be noted, however, that perfectly regular nanowires can also be obtained by selective area (SA) methods (Noborisaka et al., 2005) so that the VLS technique is not exclusive in this respect. On the other hand, SA growth can acquire the VLS character under the excessive group III influx to the nanowire top.

The VLS mechanism applies only when the alloy of metal and semiconductor remains liquid during growth. The state of catalyst particles depends on many factors such as the growth temperature, droplet composition, and particle size. Alternatively, the so-called vapor–solid–solid (VSS) mechanism of III–V compounds introduced by Persson et al. (2004) and further developed by Dick et al. (2005) suggests that the catalyst particle at the nanowire tip is solid during growth. In Persson et al. (2004), through *in situ* transmission electron microscopy (TEM) analysis and energy-dispersive X-ray spectroscopy (EDXS) measurements of GaAs nanowires grown by CBE, the authors observed the crystallinity of the Au–Ga particle and a low Ga concentration below the eutectic melt. The analysis of Dick et al. (2005) refers to the case of Au-catalyzed InAs nanowires grown by MOCVD below the eutectic temperature of bulk Au–In alloy at 454.3 °C. Indeed, the VLS growth mechanism is always guaranteed for self-catalyzed III–V nanowires with group III metal catalysts, while in Au-catalyzed III–V nanowires the interplay between VLS and VSS modes requires close examination. In particular, the Au–Ga–As droplets should be liquid at typical MBE temperatures above 500 °C (Harmand et al., 2005), while Au-catalyzed MBE growth of InAs nanowires within the temperature window from 380 to 430 °C most probably yields partly solidified catalyst particles (Tchernycheva et al., 2007).

As any epitaxial growth, the VLS process should occur at a positive difference of the effective chemical potentials of a semiconductor material in the vapor (V) and solid (S) phases, $\Delta\mu_{VS} = \mu_V - \mu_S$. However, the VLS growth necessarily involves an intermediate liquid (L) state of this semiconductor at the chemical potential $\mu_L$. The latter should be larger than $\mu_S$ to ensure solidification but smaller than $\mu_V$ to drive the vapor-to-liquid transition. Therefore, the classical adsorption-induced VLS growth occurs when

$$\mu_V > \mu_L > \mu_S \quad \text{or} \quad \Delta\mu_{VS} > \mu_{LS} > 0. \tag{1}$$

For III–V semiconductors, the definitions of these chemical potentials require some care. First, we need to consider atomic chemical potentials of III and V elements in vapor such that $\mu_V = \mu_3^V + \mu_5^V$ (here and below, the subscript "3" stands for group III, "5" for group V, and "35" for a solid III–V pair). These are given by the atomic fluxes of group III and V elements onto the droplet surface—$I_3$ and $I_5$, respectively, MBE growth geometry or the corresponding cracking efficiencies in chemical epitaxies (see Fig. 3 for illustration of the main parameters in VLS III–V nanowires). For Au-catalyzed VLS growth, the liquid chemical potential at the surface temperature $T$ is generally defined as (Glas, 2010a): $\mu_L(c_3, c_5, T) = \mu_3^L(c_3, c_5, T) + \mu_5^L(c_3, c_5, T)$. Here, $c_3$ and $c_5$ are the relative atomic concentrations of group III and V atoms dissolved in gold such that the atomic concentration of Au equals $c_{Au} = 1 - c_3 - c_5$. The solid chemical potential refers to a bound III–V pair in stoichiometric zinc blende (ZB) crystal: $\mu_S = \mu_{35}^S(T)$ and generally cannot be decoupled to the sum of the group III and V contributions. If the nanowire phase is wurtzite (WZ), the solid chemical potential changes to $\mu_S + \psi$, with $\psi$ as the bulk energy difference between the WZ and ZB phases (Dubrovskii, 2014a) (for conventional III–V materials, $\psi$ is positive). Due to the known low solubility of most group V elements such as As in liquid metals, the VLS process usually proceeds at $c_5 \ll c_3$. In this case, we can neglect the dependence of $\mu_3^L$ on $c_5$ and write approximately (Dubrovskii, 2014b)

$$\Delta\mu_{LS}(c_3, c_5, T) = \mu_3^L(c_3, T) + \mu_5^L(c_3, c_5, T) - \mu_{35}^S(T). \tag{2}$$

Therefore, the group V chemical potential in liquid is the most difficult one since it depends on concentrations of both growth species.

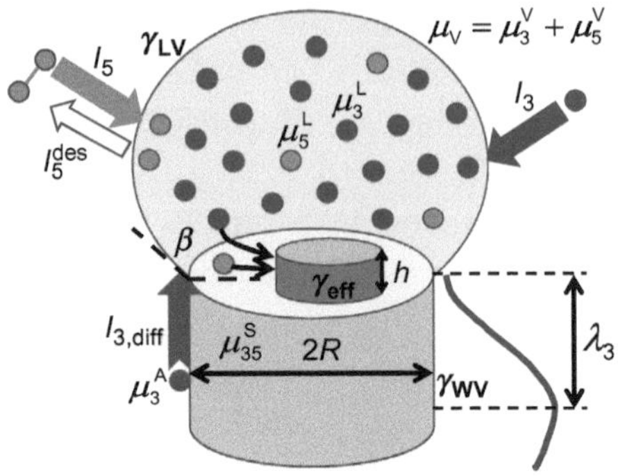


**Figure 3** Illustration of Au-catalyzed VLS growth of a cylindrical III–V nanowire of radius $R$ from vapor fluxes $I_3$ and $I_5$ via 2D nucleation. The droplet is a ternary Au–III–V alloy having the chemical potential $\Delta\mu_{LS} = \mu_3^L + \mu_5^L - \mu_{35}^S$ with respect to ZB solid. The $\Delta\mu_{LS}$ value depends on $c_3$, $c_5$, and $T$. At usual growth temperatures, group III atoms are not expected to desorb either from the droplet or nanowire sidewalls, while desorption of group V element from the droplet occurs at a temperature-dependent rate $I_5^{des}$. Diffusion flux of group III adatoms from the sidewalls $I_{3,diff}$ depends on the effective diffusion length $\lambda_3$ which is determined by the point of maximum of the adatom concentration $n_3(z)$. The latter determines the adatom chemical potential $\mu_3^A$. Surface energetics of the system is set by the wire–vapor ($\gamma_{WV}$), liquid–vapor ($\gamma_{LV}$), and island ($\gamma_{eff}$) interface energies as well as the contact angle $\beta$. The $\gamma_{WV}$ value depends on the facet type and crystal structure (ZB or WZ), $\gamma_{LV}$ is a function of $c_3$, while $\gamma_{eff}$ depends on $c_3$, the island position, and crystal structure. In the case shown in the figure, the island is entirely surrounded by liquid and thus $\gamma_{eff} = \gamma_{SL}(c_3)$ is the surface energy of lateral solid–liquid interface.

Semiconductor material can also be collected by the nanowire sidewalls. This portion of the material may contribute to radial nanowire growth but more usually under optimized conditions, group III adatoms migrate to the nanowire top and thus enhance its elongation rate (Dubrovskii, 2014a). Surface diffusivity of group V atoms is usually considered negligible, which is supported by the results and analysis of Ramdani et al. (2013). Diffusion transport of a group III element to the top occurs only when the chemical potential of group III adatoms (A) on the sidewalls is larger than that in the droplet:

$$\mu_3^A > \mu_3^L. \tag{3}$$

The two conditions given by Eqs. (1) and (3) define the VLS growth with positive adsorption and diffusion fluxes. Of course, the droplet must receive enough group V species from vapor to ensure that each group III adatom arriving to the droplet will subsequently find its pair to contribute to growth. In this case, the VLS growth is said to proceed under As-rich conditions and is limited by the kinetics of Ga species, where surface diffusion of Ga can increase the elongation rate by several times with respect to 2D equivalent growth rate (Dubrovskii et al., 2005). Chemical potential of Ga adatoms is given by (Dubrovskii, 2014a)

$$\mu_3^{\mathrm{A}} = k_{\mathrm{B}} T \ln(\sigma n_3) + \phi(T), \tag{4}$$

where $\sigma$ is the elementary surface area of relevant plane, $\phi(T)$ is a function of the surface temperature, $n_3$ is a coordinate-dependent adatom concentration, and $k_{\mathrm{B}}$ is the Boltzmann constant.

Usually, the adatom concentration on the sidewalls $n_3(z)$ has a maximum at a certain distance from the top (Dubrovskii et al., 2009a; Harmand et al., 2010), as shown in Fig. 3. This distance can be associated with the effective diffusion length of Ga adatoms $\lambda_3$, i.e., the feeding length from which material is collected by the droplet in a more phenomenological approach (Plante and LaPierre, 2009; Tchernycheva et al., 2007). Generally, the group III diffusion flux to the nanowire top is given by

$$I_{3,\,\mathrm{diff}} = -cRD_3 \frac{\partial n_3}{\partial z}\bigg|_{z=L}, \tag{5}$$

where $D_3$ is the temperature-dependent diffusion coefficient of group III adatoms on the nanowire sidewalls, $L$ is the nanowire length, and $c$ is the geometrical factor, e.g., $c = 2\pi$ for cylindrical nanowire of radius $R$ and $c = 6$ for regular hexahedral nanowire with side $R$. This is reduced to the simplified expression $I_{3,\,\mathrm{diff}} = cR\lambda_3 \omega I_3 \sin\alpha_3$ (with $\alpha_3$ as the incident angle of Ga beam and $\omega = 1/\pi$ for circular and $1/3$ for hexagonal cross section as the geometrical factor of MBE growth) only if $n_3 = n_{\max}(L - z)/\lambda_3$ and $n_{\max} = \left(\lambda_3^2/D_3\right)\omega I_3 \sin\alpha_3$. In other words, the adatom concentration at distance $\lambda_3$ from the top should be set by the balance of Ga adsorption and linear sink, with $\tau_3 = \lambda_3^2/D_3$ as the effective adatom lifetime. In self-catalyzed VLS procedures, Ga-rich droplet serves as a Ga reservoir. Therefore, Ga diffusion influences only the droplet volume, while the vertical growth should be a linear function of As flux (Colombo et al., 2008; Ramdani et al., 2013). These differences will be considered in detail in Sections 4, 6, and 9.

Long ago, Wagner and Ellis (1965) and Givargizov (1987) studied the minimum size of the wire which can be grown via the VLS mechanism at a given degree of vapor supersaturation. This size-dependent limitation on the VLS growth arises due to the Gibbs–Thomson (GT) effect of elevation of the liquid chemical potential due to the droplet curvature (Dubrovskii et al., 2009a). Consider unstable equilibrium between a semiconductor material in an infinite vapor phase at the chemical potential $\mu_V$ and in the liquid droplet at the effective chemical potential $\mu_L^{eff} = \mu_L + 2\gamma_{LV}\Omega_L/R_d$. Here, $\gamma_{LV}$ is the liquid–vapor surface energy of the droplet, $\Omega_L$ is the elementary volume per III–V pair in the liquid phase, $R_d = R/\sin\beta$ is the droplet radius, $R$ is the nanowire radius, and $\beta$ is the contact angle of the droplet. At very small $R_d$, we can write approximately $\mu_L^{eff} \cong \mu_S + 2\gamma_{LV}\Omega_L/R_d$, because the difference between $\mu_L$ and $\mu_S$ is small compared to the GT term. Then, from the condition $\mu_V = \mu_L^{eff}$ we get the minimum nanowire radius in the Wagner–Ellis form (Wagner and Ellis, 1965)

$$R_{min} \cong \frac{2\Omega_L\gamma_{LV}\sin\beta}{\Delta\mu_{VS}}. \qquad (6)$$

This formula can be more rigorously justified in the case of adsorption-induced VLS growth whose elongation rate is limited by nucleation at the liquid–solid interface (Dubrovskii, 2014a), as originally considered by Givargizov (1987).

Consider, for example, the Au-catalyzed growth of GaAs nanowires with $\Omega_L \cong \Omega_S \equiv \Omega_{35} = 0.0452\,\text{nm}^3$ (the elementary volume of a GaAs pair in the solid state), $\beta = \pi/2$, $\gamma_{LV} \cong 1\,\text{J/m}^2$ at $\Delta\mu_{VS} \cong 800\,\text{meV}$. These parameters yield a very small value of $R_{min} \cong 1.8\,\text{nm}$. Therefore, the GT effect should not be important for nanowires wider than ~10 nm. Several works previously reported on the purely increasing radius dependences of the nanowire elongation rate, for example, Dayeh and Picraux (2010) for Au-catalyzed Ge nanowires and Moewe et al. (2008) for Au-catalyzed InP nanowires. Such dependences are sometimes modeled by phenomenological Givargizov–Chernov expression (Givargizov, 1987)

$$\frac{dL}{dt} = K\Delta\mu^2\left[1 - \frac{R_{min}}{R}\right]^2 \qquad (7)$$

which assumes that the elongation rate is proportional to the squared effective chemical potential (according to Eq. (6), $\Delta\mu$ here should be put to $\Delta\mu_{VS}$) with the GT correction. Kashchiev (2006) argued that the suppression of the elongation rate at small $R$ might be due to another size-dependent effect, namely, the mononuclear growth mechanism of nanowires (Dubrovskii and Sibirev, 2004) whose rate must tend to zero at $R \to 0$

simply due to the lack of surface area available for nucleation. This can be incorporated into a self-consistent growth model considered below in Section 8.

Assuming a planar liquid–solid interface under the droplet, the VLS growth should proceed by two-dimensional (2D) nucleation of crystal islands from a supersaturated liquid alloy at chemical potential $\Delta\mu_{LS}$. Macroscopic approximation for the dimensionless formation energy of regular triangle island of height $h=1$ monolayer (ML) composed of $i$ III–V pairs (monomers) writes down as (Dubrovskii and Grechenkov, 2015)

$$F(i) = -i\Delta\overline{\mu} + ai^{1/2}, \quad a = 2\times 3^{3/4}\frac{\gamma_{eff}}{k_B T}(\Omega_{35}h)^{1/2}, \tag{8}$$

where $\Delta\overline{\mu}\equiv\Delta\mu_{LS}/k_B T$ and $\gamma_{eff}$ is the effective surface energy of the island lateral interface (see Fig. 3). For a circular island, the coefficient in Eq. (8) for $a$ changes to $\pi$. The first, negative term in Eq. (8) describes thermodynamically favorable lowering of the system energy by the monomer assembling into a nucleus, while the second, positive term stands for the surface energy of the nucleus lateral facets having the surface area $3hr$, with $r=\left(2/3^{1/4}\right)(\Omega_{35}/h)^{1/2}i^{1/2}$. Maximizing this $F(i)$ readily yields the necessary parameters of classical nucleation theory (Dubrovskii, 2014a; Kashchiev, 2000; Markov, 2003): the critical size $i_c$, the nucleation barrier $F_c\equiv F(i_c)$, and the second derivative of the formation energy at $i_c$ which defines the half-width $\Delta i_c$ of $F(i)$ near the critical size according to $|F''(i_c)|\Delta i_c^2/2=1$:

$$i_c = \frac{a^2}{4\Delta\overline{\mu}^2}; \quad F_c = \frac{a^2}{4\Delta\overline{\mu}}; \quad |F''(i_c)| = \frac{2\Delta\overline{\mu}^3}{a^2}. \tag{9}$$

The surface energy constant $a$ is determined by the ratio of the characteristic surface energy per elementary area to thermal energy and therefore is expected to be larger or even much larger than unity for most systems of interest—otherwise, 2D islands could not exist because their boundaries would be destroyed by thermal fluctuations.

According to classical nucleation theory (Dubrovskii, 2014a; Kashchiev, 2000; Markov, 2003), the nucleation rate $J$, i.e., the number of stable nuclei which emerge per unit time per unit surface area at a given supersaturation of a metastable mother phase (in our case—a liquid alloy at chemical potential $\Delta\mu_{LS}$), is driven by thermodynamic fluctuations. Unstable nuclei surpass the nucleation barrier of a given height and enter the stable growth region where the surface curvature effect becomes negligible. Therefore, $J$ must be proportional to the inverse exponent of the nucleation barrier:

$$J \propto \exp(-F_c). \tag{10}$$

Kinetic approach pioneered by Zeldovich (1942) allows one to compute the prefactor of the nucleation rate which depends on $i_c$, $|F''(i_c)|$, and some other kinetic parameters. The Zeldovich nucleation rate in Au-catalyzed III–V nanowires will be considered in Section 6. Now we note that the strong inequality $a \gg 1$ leads to a very steep dependence of $J$ on $\Delta\bar{\mu}$ such that rising chemical potential by only a few percent increases the nucleation rate by the order of magnitude.

Due to large values of liquid chemical potentials typically employed in VLS growth conditions, we can really reach the limit of the macroscopic theory. For example, at $T = 550$ °C, $\Omega_{35} = 0.0452$ nm$^3$, $h = 0.326$ nm, and $\gamma_{\text{eff}} = 0.5$ J/m$^2$, the $a$ value equals 24.3. Dimensionless formation energies with these parameters and variable $\Delta\mu_{\text{LS}}$ are shown in Fig. 4. It is seen that the macroscopic model is well justified below 400 meV where the critical size of classical nucleation theory is of the order of ten. When the chemical potential rises to above 500 meV, the critical nucleus rapidly approaches one meaning that the 2D islands grow almost irreversibly. Such situations may completely change the nucleation rate as will be discussed later on.

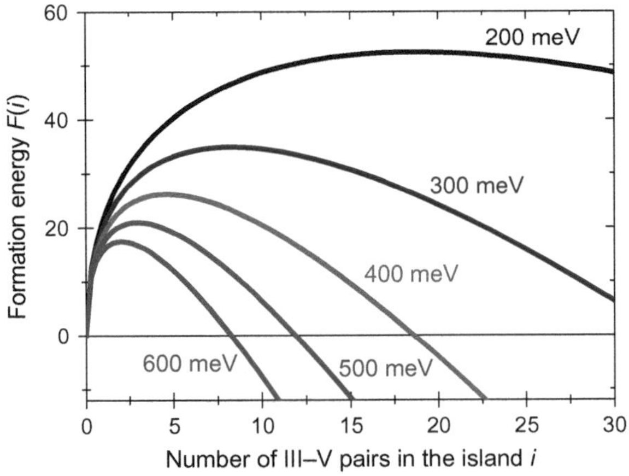


**Figure 4** Dimensionless formation energy given by Eq. (8) for GaAs islands at $T = 550$°C, $\gamma_{\text{eff}} = 0.5$J/m$^2$, $a = 24.3$, and different $\Delta\mu_{\text{LS}}$. The macroscopic nucleation theory works well at $\Delta\mu_{\text{LS}} = 400$meV and below where the critical size is of the order of ten but reaches its limit at $\Delta\mu_{\text{LS}} = 500$meV and above where the critical island nucleus consists of only a few GaAs pairs.

The probability to nucleate a 2D island on the top facet of cylindrical nanowire of radius $R$ per unit time equals $\pi R^2 J$. On the other hand, the island will fill the complete ML slice of a nanowire after the time period $R/v$, where $v = dr/dt$ is the growth rate of a linear size of the island. The ratio of these characteristic times naturally yields the control parameter $\varepsilon = \pi R^3 J/v$. Analyses of Dubrovskii (2013), Dubrovskii and Sibirev (2004), Dubrovskii et al. (2006, 2008a), Gamalski et al. (2011), Glas (2014), Glas et al. (2007, 2010), and Wen et al. (2011) show that nanowires often grow at $\varepsilon \ll 1$, i.e., in the mononuclear regime whereby only one 2D island succeeds in nucleation in each nanowire layer and then almost instantaneously spreads to fill the complete ML.

Such a character of growth has far-reaching implications from many perspectives. First, the position (or orientation) of a tiny island (whose size is larger but comparable with the critical size) with respect to the preceding layer will dictate the preferred crystal structure of the whole nanowire (Glas et al., 2007). Second, formation of a ML removes a substantial fraction of growth constituencies from the droplet. This effect is particularly important for group V atoms due to their low concentration in the droplet. Immediately after the ML formation, the droplet appears nearly emptied with its group V atoms and should be refilled until the next ML can nucleate. This effect leads to a sub-Poissonian character of the nucleation statistics that manifests through temporal anticorrelations between successive nucleation events (Dubrovskii, 2013; Dubrovskii et al., 2008a; Glas, 2014; Glas et al., 2010). For a small enough nanowire radius, the timescale hierarchy within the growth cycle implies that the island growth time is much smaller than the duration of the nucleation step, which is in turn much shorter than the refill time Third, the sawtooth time dependence of supersaturation controls the periodically changing geometry of the growth interface in the presence of a stable truncated facet (Gamalski et al., 2011; Sibirev et al., 2014).

## 3. CHEMICAL POTENTIALS FOR AU-CATALYZED VLS GROWTH OF III–V NANOWIRES

Determination of chemical potential of a liquid alloy is of paramount importance for modeling and controlling the growth rates, morphologies, and crystal structures of VLS nanowires (Dubrovskii, 2014a,b; Dubrovskii and Grechenkov, 2015; Dubrovskii et al., 2008a; Gil et al., 2014; Glas et al., 2007, 2013; Johansson et al., 2009; Kelrich et al., 2015). According to general theory (Landau and Lifshitz, 1980), chemical potential of a ternary alloy is given by the expansion

$$\Delta\mu_{LS}(c_3, c_5, T) = k_B T \ln c_3 + k_B T \ln c_5 + f(T) + \sum_{j,k} \phi_{jk}(T) c_3^j c_5^k, \quad (11)$$

where the first three terms correspond to diluted perfect alloy and the sum of the terms $c_3^j c_5^k$ with $j + k \geq 1$ describes interatomic interactions with certain temperature-dependent parameters $\phi_{ij}(T)$. These interactions become more and more important as the concentrations (or at least one of the two concentrations) increase. In the simplest approximation of perfect alloy, one can write: $\Delta\mu \cong k_B T \ln[c_3 c_5/(c_3 c_5)^{eq}]$, where $(c_3 c_5)^{eq}$ stands for the activity of the ground solid state. The latter is expected to depend on temperature as $(c_3 c_5)^{eq} = \exp(-\Lambda_{35}/k_B T)$ with the activation energy $\Lambda_{35}$.

This simple expression already shows the following important trends:

**(i)** Chemical potential increases with both $c_3$ and $c_5$, as it should be due to the condition of thermodynamic stability $(\partial\Delta\mu_{LS}/\partial c_x)_T > 0$. The $\Delta\mu_{LS}$ value is more sensitive to the smallest concentration due to the steep logarithmic behavior at $c_x \to 0$.

**(ii)** At a given temperature, there is no simple way to decouple which part of the $(c_3 c_5)^{eq}$ relates to group III or group V element. This value sets a certain threshold for the product of the two concentrations below which the VLS growth cannot proceed.

**(iii)** Different group III and V concentrations can relate to the same value of the chemical potential. In particular, exceeding the threshold to start the VLS growth can be achieved by increasing $c_3$ or $c_5$ alone, or both concentrations.

**(iv)** Due to $\Delta\mu_{LS} \cong k_B T \ln(c_3 c_5) + \Lambda_{35}$ and $c_3 c_5 < 1$, chemical potential decreases with temperature for given concentrations.

**(v)** While the elongation rate in Au-catalyzed VLS growth of III–V nanowires is usually limited by the kinetics of group III species (see the next section for the detail), group V concentration strongly affects the chemical potential value (which can be very different for the same $dL/dt$) and therefore the preferred crystal phase.

In a more quantitative approach, Glas (2010a) provided an explicit scheme for calculation of the liquid–solid chemical potential $\Delta\mu_{LS}(c_3, c_5, T)$ of ternary Au-III–V alloys, based on the refined Stringfellow model for the interaction parameters of couples of different atoms (Stringfellow, 1972). Several groups have analyzed the composition of the catalyst droplets after growth by MBE

(Dheeraj et al., 2008; Glas et al., 2010; Harmand et al., 2005; Tchernycheva et al., 2007) or MOCVD (Burke et al., 2010; Johansson et al., 2009). Typical values of concentrations of the group III elements in gold during the VLS growth of III–V nanowires amount to several tens of percent. On the contrary, for arsenide and phosphide nanowires, the EDXS analyses reveal no significant amount of As and P in the droplets after growth. This sets an upper limit of the order of 1% to the postgrowth group V concentration. Hence, the concentration of group V atoms should be of a few percents at most. This should be due to the known low solubilities of group V atoms in gold and group III melts (Massalski, 1986). The only one exception from this rule is antimony which can be found in the droplets in a significant fraction (Burke et al., 2010).

Therefore, Fig. 5 (Glas, 2010a) shows the calculated liquid chemical potentials for the Au–Ga–As system as functions of atomic concentration of As $c_5 = c_{As}$ between 0.001 and 0.06 at three different temperatures within the typical growth domain and at three different Ga concentrations of 20%, 35%, and 70%. Figure 6 shows the liquid chemical potential of the Au–Ga–As alloy depending on $c_3 = c_{Ga}$, for three different $c_{As} = 0.5\%$, 2%, and 5% and at two different temperatures. It is seen $\Delta\mu_{LS}$ is most sensitive to the group V concentration when the latter is less than ~1%, whereas

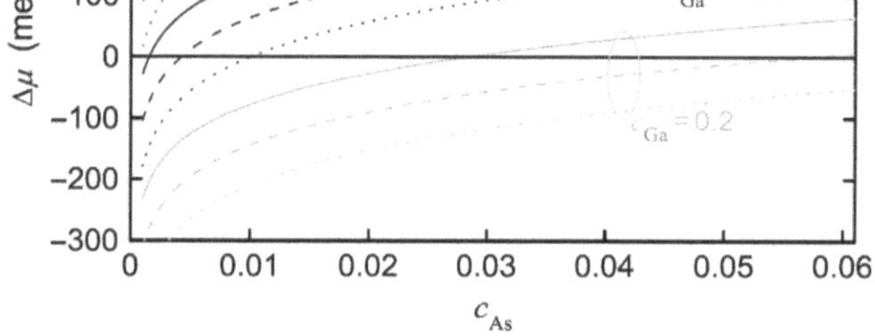


**Figure 5** Variation with the atomic concentration of As in the liquid of the difference of chemical potential per Ga–As pair between a ternary Au–Ga–As liquid and solid ZB GaAs $\Delta\mu \equiv \Delta\mu_{LS}$, calculated for a set of temperatures and Ga concentrations in the liquid (Glas, 2010a).

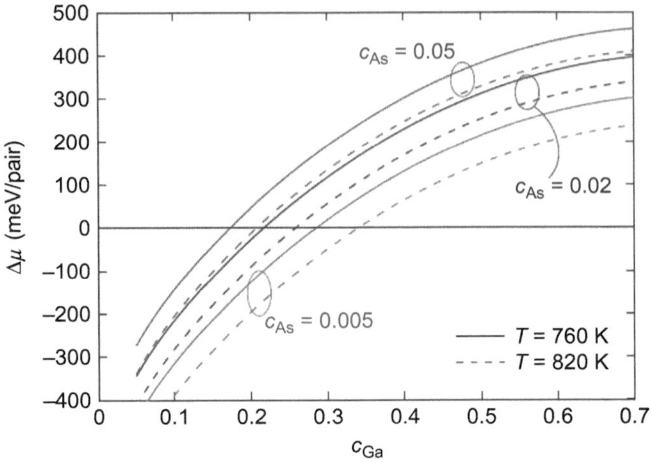


**Figure 6** Variation with the atomic concentration of Ga in the liquid of the difference of chemical potential per Ga–As pair between a ternary Ga–As–Au liquid and solid ZB GaAs $\Delta\mu \equiv \Delta\mu_{LS}$, calculated for a set of temperatures and As concentrations in the liquid (Glas, 2010a).

the variations with group III concentrations are more uniform in the range considered. This trend remains the same for other Au–III–V systems such as Au–In–As, Au–In–P, and Au–Ga–P.

Furthermore, Glas (2010a) found that the chemical potential curves can be well fitted by the expressions that split $\Delta\mu_{LS}$ into the terms depending logarithmically on the concentrations, and the complementary polynomials of $c_3$, $c_5$, and $T$:

$$\Delta\mu_{LS}(c_3, c_5, T) = k_B T \ln c_3 + k_B T \ln c_5 + \sum_{i=0}^{2} p_i(c_3, T) c_5^i. \tag{12}$$

Here,

$$p_i(c_3, T) = \sum_{j=0}^{m_i} A_{ij}(T) c_3^j, \tag{13}$$

with $m_0 = 3$, $m_1 = m_2 = 4$, and

$$A_{ij}(T) = \sum_{k=0}^{2} B_{ijk} T^k. \tag{14}$$

The dependence of $\Delta\mu_{LS}$ on three independent variables thus requires a sizable number of 42 parameters $B_{ijk}$, which are tabulated in Glas (2010a) for different binaries.

## 4. GROWTH KINETICS OF III–V NANOWIRES

Most growth models of Au-catalyzed III–V nanowires (see, e.g., Dubrovskii, 2014a for the review), with only few exceptions (Chou et al., 2014; Dubrovskii, 2014b; Krogstrup et al., 2013), treat the VLS process as a group III limited, as in standard theories of planar growth under As-rich conditions. According to the data of the previous sections, considering only the kinetics of group III species is applicable only if the growth equation does not contain $\mu_5^L$. Dubrovskii (2014b) considered this issue theoretically by using the following simple equations for the total number of group V ($N_5$) and III ($N_3$) atoms in an Au–III–V droplet:

$$\frac{dN_5}{dt} = \chi_5 I_5 (1+\varepsilon)\pi R^2 - I_5^{des}\frac{2}{(1+\cos\beta)}\pi R^2 - \frac{\pi R^2}{\Omega_{35}}\frac{dL}{dt}, \tag{15}$$

$$\frac{dN_3}{dt} = \chi_3 I_3 \pi R^2 + 2\pi R j_{3,\,diff} - \frac{\pi R^2}{\Omega_{35}}\frac{dL}{dt}. \tag{16}$$

The first equation accounts for the direct atomic flux of group V ($I_5$) and the reemitted flux ($\varepsilon I_5$) (Glas et al., 2013; Ramdani et al., 2013) which are arriving at a droplet seated on top of a nanowire having radius $R$ (with $\chi_5$ reflecting the beam geometry in MBE (Glas, 2010b) and the cracking efficiency in MOCVD); desorption flux $I_5^{des}$ from the surface area of a spherical cap droplet with the contact angle $\beta$; and the sink of group V atoms due to nanowire elongation $dL/dt$. As discussed, we neglect surface diffusion of group V species. The influx of group III atoms in the second equation is the sum of the impinging flux $\chi_3 I_3$ (with the same meaning of the parameters as for group V) and the diffusion current per unit length of the triple phase line (TPL) of cylindrical nanowire $j_{3,\,diff} = -D_3(\partial n_3/\partial z)|_{z=L}$ such that the total diffusion flux $I_{3,\,diff} = 2\pi R j_{3,\,diff}$, as in Eq. (5). Desorption of group III atoms from the droplet is neglected. The sink is exactly the same as for group V. A very similar model has been considered in Chou et al. (2014).

In the steady state ($dN_5/dt = dN_3/dt = 0$), Eq. (16) gives the elongation rate

$$\frac{dL}{dt} = \chi_3 \Omega_{35} I_3 + \frac{2\Omega_{35} j_{3,\,diff}}{R}. \tag{17}$$

As discussed briefly in Section 2 and will be considered in more detail in Section 5, $j_{3,\text{diff}}$ should depend on the chemical potentials of group III atoms in the droplet ($\mu_3^L$) and group III sidewall adatoms at large enough distance from the droplet ($\mu_3^A$). If $\mu_3^L$ is insensitive to $c_5$, as in Eq. (2), the diffusion flux of group III adatoms and therefore the entire elongation rate are also independent of group V kinetics. In particular, the simplest approximation $j_3 = \phi_3 \lambda_3 I_3$ (where $\phi_3$ summarizes the geometrical effect in MBE or the cracking efficiency at the sidewalls in MOCVD and $\lambda_3$ can depend on the group V flux) yields the typical inverse radius dependence of the elongation rate (Dubrovskii and Sibirev, 2007; Dubrovskii et al., 2005, 2007; Johansson et al., 2005; Plante and LaPierre, 2006, 2009; Seifert et al., 2004)

$$\frac{dL}{dt} = \Omega_{35} I_3 \left[ \chi_3 + \phi_3 \frac{2\lambda_3}{R} \right]. \tag{18}$$

In this case, the elongation rate does not even depend on the state of the catalyst particle. The dependence of $dL/dt$ on $\mu_3^L$ appears only when we consider the reverse diffusion flux from the droplet (Dubrovskii et al., 2009b). In such regimes, the $\mu_3^L$ value serves as an external parameter that influences the diffusion flux and cannot be determined within the transport equations alone.

The group III-limited elongation rate given by Eq. (17) can alternatively be obtained from Eq. (15) at $dN_5/dt = 0$, i.e., based on the kinetics of group V species:

$$\frac{dL}{dt} = \chi_5 \Omega_{35} I_5 (1 + \varepsilon) - \frac{2}{1 + \cos\beta} \Omega_{35} I_5^{\text{des}}. \tag{19}$$

Desorption flux from the droplet can be computed by considering an imaginary group V vapor which is at equilibrium with group V atoms in the liquid at the chemical potential $\mu_5^L$, as in Glas et al. (2013). We first equate the chemical potential of group V molecules (X)V$_m$ (incorporating $m$ group V atoms) in this imaginary vapor phase to the chemical potential of $m$ group V atoms in the liquid: $\mu_{5(m)}^V = m\mu_5^l$. Then, we use the perfect gas approximation for the vapor chemical potential, $\mu_{5(m)}^v = k_B T \ln\left(P_{5(m)}/P_5^0\right)$ (with $P_{5(m)}$ being the partial pressure of group V species in the imaginary vapor and $P_5^0$ a certain temperature-dependent function) along with the known formula for the vapor flux: $I_5^{\text{des}} = mk_5 P_5/\sqrt{2\pi M_5 k_B T}$. Here, $k_5$ is the Knudsen evaporation coefficient and $M_5$ is the mass of a (X)V$_m$ molecule. This yields the group V desorption flux in the form

$$I_5^{\text{des}} = \frac{mk_5 P_5^0}{\sqrt{2\pi M_5 k_B T}} \exp\left(\frac{m\mu_5^L}{k_B T}\right) \tag{20}$$

Using Eqs. (18)–(20), we are able to express the chemical potential of a group V element in the droplet through the vapor flux of group V species and the nanowire elongation rate, the quantities that can be directly measured (Dubrovskii, 2014b):

$$\mu_5^L(G, T) = k_B T\left(G - \frac{\ln \nu_5^0(T)}{m}\right). \tag{21}$$

Here, the parameters are defined as

$$G = \frac{1}{m} \ln\left[\chi_5 \Omega_{35} I_5(1+\varepsilon) - \frac{dL}{dt}\right]; \ \nu_5^0 = \frac{2}{(1+\cos\beta)} \frac{m\Omega_{35} k_5 P_5^0}{\sqrt{2\pi M_5 k_B T}}. \tag{22}$$

These equations circumvent the uncertainty in the group V contribution into the overall liquid chemical potential in the steady state. According to Eq. (2), the $\Delta\mu_{LS}$ is now given by

$$\Delta\mu_{LS} = \mu_3^L(c_3, T) + k_B T\left(G - \frac{\ln \nu_5^0(T)}{m}\right) - \mu_{35}^S(T). \tag{23}$$

With the known $G$ (which requires determination of $\nu_5(1+\varepsilon)$ and $dL/dt$), the only unknown that remains in Eq. (23) is the group III concentration $c_3$.

Very importantly, the group V chemical potential depends on the VLS growth conditions only via their combination $G$ which is proportional to the logarithm of the effective group V to III influx imbalance. Indeed, the term under the logarithm in Eq. (22) for $G$ gives the difference between the total influx of As atoms due to direct and reemitted fluxes and the As sink due to nanowire growth whose rate is determined by the total influx of Ga atoms via direct impingement and surface diffusion. When this influx imbalance is negative, no VLS growth can proceed under the steady-state conditions. Figure 7 shows how the $G$ value and consequently the overall liquid chemical potential can be tuned in differently sized nanowires by the V/III atomic flux ratio $f_{53} = I_5/I_3$ in the MBE case with $dL/dt$ given by Eq. (18), $m = 2$, $\Omega_{35} I_3 = 1\,\text{ML/s}$, $\chi_3 = \chi_5 = 1/\sin^2\beta$, $\beta = 110°$, $\varepsilon = 2$, $\phi_3 = \sin\alpha_3/\pi$, $\alpha_3 = 17°$, and $\lambda_3 = 1000\,\text{nm}$.

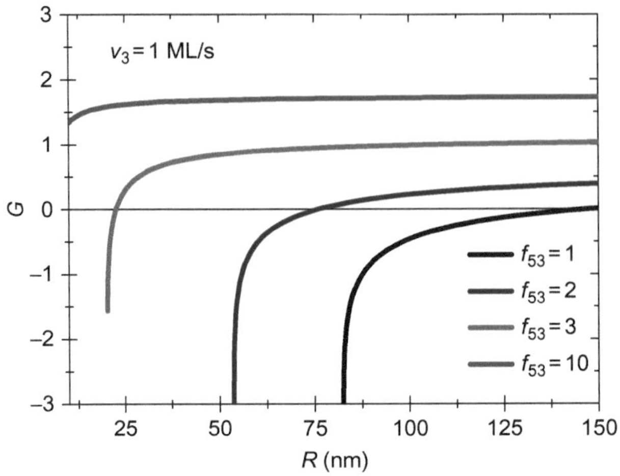


**Figure 7** Radius dependencies of $G$ at $v_3 = 1$ ML/s and different effective V/III flux ratios for the parameters of GaAs nanowires grown by Au-assisted MBE.

In the case of Ga-catalyzed GaAs nanowires and other III–V nanowires growing from group III metal particles, the nanowire elongation rate (under the assumption of a time-independent nanowire radius) is entirely determined by Eqs. (19) and (20), i.e., by the adsorption–desorption balance of As. Of course, the liquid chemical potential in this case is a function of only two variables, the As concentration in Ga–As droplet and the growth temperature. This simplification allows for the predictive modeling of Ga-catalyzed growth given by Glas et al. (2013), which will be described in Section 9. However, the kinetics of Ga species should also be considered in order to understand the establishment of a stationary nanowire radius or nonstationary droplet inflation and nanowire radial extension (Matteini et al., 2015; Priante et al., 2013).

## 5. TRANSPORT-LIMITED GROWTH OF NANOWIRES

As mentioned in Section 2, the early-days VLS growth was performed from chemical precursors that required high surface temperatures for their decomposition (Givargizov, 1987; Wagner and Ellis, 1964). In this case, a liquid droplet on the wire top facilitates chemical reactions at the vapor–liquid surface, while the pyrolysis at the solid surfaces can be almost completely suppressed at low temperatures. Consequently, the diffusion-induced contributions to the nanowire elongation rate are negligible. Without surface diffusion and the GT enhancement of desorption from small droplets, the sole rate-limiting step for nanowire growth is the thermally activated metal-catalyzed dissociative adsorption directly on the catalyst droplet and the vertical growth rate is radius-independent (Kodambaka et al., 2006). However, the first diffusion-induced growth models of one-dimensional structures were developed long ago in connection with non-VLS formation mechanisms, e.g., driven by screw dislocations (Blakely and Jackson, 1962; Ruth and Hirth, 1964).

Au-catalyzed growth of Si and III–V nanowires by MBE and related epitaxy techniques simply cannot be understood without considering surface diffusion (Dubrovskii et al., 2005; Schubert et al., 2004). In modern MOCVD technologies, pyrolysis also takes place at the solid surfaces producing adatoms which can subsequently diffuse to the nanowire top and contribute to its elongation (Seifert et al., 2004). Consequently, several models have been proposed that are capable of describing the major effects of surface diffusion (Dubrovskii and Hervieu, 2014; Dubrovskii et al., 2005, 2006, 2008a, 2009a, 2012a; Fröberg et al., 2007; Harmand et al., 2010; Johansson et al., 2005; Lu et al., 2012; Plante and LaPierre, 2009; Seifert et al., 2004; Yee et al., 2012). Since Au-catalyzed diffusion-induced growth of III–V nanowires is driven by the kinetics of group III adatoms, similar single-component approach applies to nanowires of III–V compounds and elemental semiconductors, and is not sensitive to the state of the catalyst particle (i.e., VLS or VSS growth). In particular, it has been shown that a purely diffusion growth yields the decreasing radius dependence of the elongation rate in the form $1/R^p$, where the exponent $p$ depends on the growth conditions. More advanced models account for surface diffusion and the GT effect simultaneously, leading to the experimentally observed non-monotonous length–radius correlation with a maximum (Dubrovskii et al., 2009a,b; Johansson et al., 2005). Diffusion growth models originally developed for VLS semiconductor nanowires are now widely used in studies of other highly anisotropic objects such as self-induced GaN nanowires (Consonni et al., 2012).

From Eq. (17), the diffusion-induced nanowire elongation rate equals

$$\left(\frac{\mathrm{d}L}{\mathrm{d}t}\right)_{\mathrm{diff}} = -\frac{2\Omega_{35}}{R} D_{\mathrm{f}} \frac{\mathrm{d}n_{\mathrm{f}}}{\mathrm{d}z}\bigg|_{z=L}. \tag{24}$$

Here and below in this section, we use new notations “f” and “s” instead of “3” for group III adatoms on the sidewall facets (f) and substrate surface (s) in order to distinguish between these two adatom subsystems. To calculate the diffusion flux, one should consider two steady-state diffusion equations for the substrate ($n_s$) and sidewall ($n_f$) adatoms:

$$D_s \Delta n_s + I_3 \cos\alpha_3 - \frac{n_s}{\tau_s} = 0, \tag{25}$$

$$D_f \frac{d^2 n_f}{dz^2} + I_3 \frac{\sin\alpha_3}{\pi} - \frac{n_f}{\tau_f} = 0. \tag{26}$$

The first terms here describe the diffusion with the corresponding diffusion coefficients $D_f$ and $D_s$, the second stand for the adsorption from the group III atomic flux $I_3$ directed at an angle $\alpha_3$ to the substrate normal in the case of MBE, and the third give the adatom sinks. Parameters $\tau_s$, $\tau_f$ are the effective lifetimes on the substrate and sidewall surfaces, respectively. These lifetimes are usually limited by incorporation to surface steps. In MOCVD case, the geometrical functions will be changed to the pyrolysis efficiencies of a Ga precursor at the substrate and sidewall planes. More precisely, Eqs. (25) and (26) should be written for the differences $n_{s,f} = n'_{s,f} - n^{eq}_{s,f}$ between the actual adatom concentration $n'$ and its equilibrium concentration $n^{eq}$ on a given surface. To get Eqs. (25) and (26), we then require $I_3 \gg n^{eq}_{s,f}/\tau_{s,f}$ meaning that the incoming flux of Ga is much larger than the desorption rate at the equilibrium adatom concentration.

General solutions to Eqs. (25) and (26) are given by

$$n_s(r) = I_3 \tau_s \cos\alpha_3 + c_1 I_0(r/\lambda_s) + c_2 K_0(r/\lambda_s), \tag{27}$$

$$n_f(z) = \frac{I_3 \tau_f \sin\alpha_3}{\pi} + a_1 \cosh(z/\lambda_f) + a_2 \sinh(z/\lambda_f). \tag{28}$$

Here, $r$ is the distance from the NW axis in the substrate plane, $\lambda_s = \sqrt{D_s \tau_s}$ is the effective diffusion length of group III adatoms on the substrate surface and $\lambda_f = \sqrt{D_f \tau_f}$ is that on the sidewalls. The functions $I_q(y)$ and $K_q(y)$ here and below denote the modified Bessel functions of the order $q$ in standard notations, the $I$ increasing and $K$ decreasing at $y \to \infty$. The physics of the growth process is now determined by the four boundary conditions that are required for finding the four coefficients $c_1$, $c_2$, $a_1$, and $a_2$.

For a dilute nanowire ensemble such that the average spacing between nanowires is much larger than $\lambda_s$, relevant boundary conditions are given by (Dubrovskii and Hervieu, 2014)

$$n_s(r \to \infty) = I_3 \tau_s \cos\alpha_3, \tag{29}$$

$$D_s \frac{dn_s}{dr}\bigg|_{r=R} = k_{sf} n_s(R) - k_{fs} n_f(0), \tag{30}$$

$$D_f \frac{dn_f}{dz}\bigg|_{z=0} = k_{fs} n_f(0) - k_{sf} n_s(R), \tag{31}$$

$$k_B T \ln[\sigma_f n_f(L)] = \mu_3^L. \tag{32}$$

The first boundary condition for the substrate adatoms means simply that the adatom concentration is constant far away from the nanowire. This yields $c_1 = 0$ in Eq. (27) to eliminate the increasing Bessel function $I_0$. The second and the third boundary conditions at the substrate–nanowire boundary equate the incoming and outgoing diffusion fluxes at the base to the difference of the direct and reverse transition rates between the substrate and sidewalls (Consonni et al., 2012). Here, $k_{sf} = (\nu\sigma_s/a)\exp(-E_{sf}/k_B T)$ is the rate of the adatom transition from the substrate to the sidewalls, and $k_{fs} = (\nu\sigma_f/a)\exp(-E_{fs}/k_B T)$ is the rate of the reverse transition, with $E_{sf}$, $E_{fs}$ as the corresponding activation energies, $a$ as the lattice spacing, and $\nu$ as the transition frequency. Summing up Eqs. (30) and (31), we resume the continuity of the diffusion flux at the base $D_s(dn_s/dr)_{r=R} = -D_f(dn_f/dz)_{z=0}$, while the continuity of chemical potentials $\sigma_s n_s(R) = \sigma_f n_f(0)$ (Dubrovskii et al., 2009a) holds only when $E_{sf} = E_{fs}$. The fourth condition at the nanowire–droplet boundary requires the continuity of the group III chemical potential at the top (where we put the function $\phi(T)$ in Eq. (4) to zero). This approximation is valid in the diffusion regime of nanowire–droplet atomic exchange such that the barrier for the adatom transition from the nanowire sidewall to the droplet equals the diffusion barrier (Dubrovskii and Hervieu, 2014).

With these boundary conditions, the solution for the diffusion-induced nanowire elongation rate is obtained in the form (Dubrovskii et al., 2009a,b)

$$\frac{1}{V}\left(\frac{dL}{dt}\right)_{diff} = \frac{BU(L/\lambda_f) + C}{U'(L/\lambda_f)}, \tag{33}$$

where $V \equiv I_3 \Omega_{35} \cos\alpha_3$ is the 2D equivalent growth rate. The functions $U$ and $U'$ are defined as follows

$$U(L/\lambda_f) = \sinh\left(\frac{L}{\lambda_f}\right) + b\frac{\lambda_f}{\lambda_s}\delta\left(\frac{R}{\lambda_s}\right)\left[\cosh\left(\frac{L}{\lambda_f}\right) - 1\right], \tag{34}$$

$$U'(L/\lambda_{\rm f})\equiv\frac{{\rm d}U}{{\rm d}(L/\lambda_{\rm f})}=\cosh\left(\frac{L}{\lambda_{\rm f}}\right)+b\frac{\lambda_{\rm f}}{\lambda_{\rm s}}\delta\left(\frac{R}{\lambda_{\rm s}}\right)\sinh\left(\frac{L}{\lambda_{\rm f}}\right), \qquad (35)$$

with $\delta(R/\lambda_s)=K_1(R/\lambda_s)/K_0(R/\lambda_s)$, $b=(D_s\sigma_f/D_f\sigma_s)\exp(-\Delta\varepsilon_{sf}/k_BT)$, and the parameter $\Delta\varepsilon_{sf}\equiv E_{fs}-E_{sf}$ describing the nonequivalence of the substrate and sidewall surfaces. The kinetic coefficients $B$ and $C$ stand for the diffusion fluxes from the sidewalls and the substrate, respectively:

$$B=\frac{2\lambda_{\rm f}\tan\alpha_3}{\pi R}g_{\rm f}, \qquad (36)$$

$$C=\frac{2\lambda_{\rm s}}{R}\delta\left(\frac{R}{\lambda_{\rm s}}\right)g_{\rm s}. \qquad (37)$$

The functions $g_f$ and $g_s$ describe the differences between the activities of sidewalls and substrate adatoms and the liquid phase

$$g_{\rm f}=1-\frac{\theta_{\rm L}}{\theta_{\rm f}}=1-\exp\left(-\frac{\Delta\mu_{\rm fL}}{k_BT}\right);\ g_{\rm s}=1-\frac{\theta_{\rm L}}{\theta_{\rm s}}=1-\exp\left(-\frac{\Delta\mu_{\rm sL}}{k_BT}\right). \qquad (38)$$

Here, $\Delta\mu_{fL}=\mu_f-\mu_3^L$, $\Delta\mu_{sL}=\mu_s-\mu_3^L$, and the activities of sidewall and substrate adatoms are given by

$$\theta_{\rm f}=\frac{I_3\tau_{\rm f}\sigma_{\rm f}\sin\alpha_3}{\pi};\ \theta_{\rm s}=I_3\tau_{\rm s}\sigma_{\rm s}\cos\alpha_3\exp(\Delta\varepsilon_{\rm sf}/k_BT), \qquad (39)$$

with the obvious relationships between the adatom chemical potentials $\mu_f$, $\mu_s$ and the activities. The diffusion fluxes of surface and sidewall adatoms are positive at $\theta_s>\theta_L$, $\theta_f>\theta_L$ and negative otherwise. Using Eq. (39), the coefficient $b(\lambda_f/\lambda_s)$ in Eqs. (34) and (35) can be alternatively presented as

$$b\frac{\lambda_{\rm f}}{\lambda_{\rm s}}=\frac{\lambda_{\rm s}\theta_{\rm f}}{\lambda_{\rm f}\theta_{\rm s}}\frac{\pi}{\tan\alpha_3}=\frac{\lambda_{\rm s}}{\lambda_{\rm f}}\left(\frac{1-g_{\rm s}}{1-g_{\rm f}}\right)\frac{\pi}{\tan\alpha_3}. \qquad (40)$$

The total group III-limited elongation rate of a nanowire is obtained by summing up Eq. (33) with the direct impingement of group III atoms, as in Eq. (17). Additionally, we should subtract the 2D growth rate of parasitic surface layer between the nanowires $V_{2D}$ (Harmand et al., 2010) if this process is not suppressed, e.g., by an inert mask layer. Without desorption, all group III atoms deposited should be distributed in surface layer and nanowires. Therefore, we get

$$V_{2D} \cong V - \pi R^2 N_W \frac{dL}{dt}, \tag{41}$$

with $N_W$ as the surface density of nanowires. The final result for the elongation rate thus takes the form

$$\frac{1}{V}\frac{dL}{dt} = \left(A + \frac{BU(L/\lambda_f) + C}{U'(L/\lambda_f)}\right)\left(1 + \pi R^2 N\right) - 1. \tag{42}$$

The adsorption term in the MBE case is given by

$$A = \frac{\chi_3(\alpha_3, \beta)}{\cos\alpha_3}, \tag{43}$$

with the geometrical function $\chi_3(\alpha_3, \pi/2) = (1 + \cos\alpha_3)/2$ at $\beta = \pi/2$ and $\chi_3 = 1/\sin^2\beta$ when $\beta \geq \alpha_3 + \pi/2$ (Glas, 2010b). Thus, the transport-limited elongation rate generally contains four unknown parameters: the two adatom diffusion lengths on the substrate and sidewalls ($\lambda_s$ and $\lambda_f$) and the two chemical potential differences between the surface and sidewall adatom subsystems and Ga atoms in the liquid state ($\Delta\mu_{sL}$ and $\Delta\mu_{fL}$, or $g_s$ and $g_f$). The group V flux can influence these transport equations only indirectly, e.g., through the dependence of the diffusion lengths or adatom activities on the As pressure. When $g_s \to 1$ and $g_f \to 1$, the group III diffusion flux is not sensitive to the state of the droplet and is regulated only by the two diffusion lengths.

Equation (42) was checked by comparing with the experimental dependences of the nanowire length on the growth time in Au-assisted MBE of InPAs nanowires with modulated composition at 420 °C (Harmand et al., 2010). Excellent agreement was obtained showing that the diffusion flux from the substrate surface [described by the $C$ term in Eq. (42)] is usually very low and even slightly negative with a small $\lambda_s$ of the order of a few nanometers. Such low value could be explained by a very rapid incorporation of In adatoms into 2D surface layer that grows concomitantly with nanowires. Therefore, the diffusion-induced contribution is governed by the diffusion of sidewall adatoms, i.e., the $B$ term, in which the diffusion length $\lambda_f$ was found to be at least 2000 nm.

Explicit solutions to Eqs. (25) and (26) yield the renormalized concentration of sidewall adatoms $\widetilde{n}_f = (2\Omega_{35} D_f / R\lambda_f V) n_f$ in the form

$$\widetilde{n}_f(x) = \widetilde{n}_f^0 - \left[\frac{I_{top} - I_{base}\cosh(l)}{\sinh(l)}\right]\cosh(x) - I_{base}\sinh(x), \tag{44}$$

with $x \equiv z/\lambda_f$, $l = L/\lambda_f$, and $\widetilde{n}_f^0 = (2\lambda_3 \tan\alpha_3)/(\pi R) = B/g_f$. The normalization is chosen in such a way that the flux to the top

$$I_{top} = -\frac{d\widetilde{n}_f}{dx}\bigg|_{x=l} = \frac{1}{V}\left(\frac{dL}{dt}\right)_{diff} = \frac{BU(l)+C}{U'(l)} \tag{45}$$

yields the diffusion-induced nanowire elongation rate and the flux to the base writes down as

$$I_{base} = -\frac{d\widetilde{n}_f}{dx}\bigg|_{x=0} = \frac{C - C_0[\cosh(l) - 1]}{U'(l)}, \tag{46}$$

with $C_0 = (2\lambda_s/R)\delta(R/\lambda_s)(\theta_f/\theta_s - 1)$. The adatom concentration defined in Eq. (44) reaches its maximum at the height $x_*$ given by

$$\tan(x_*) = \frac{\sinh(l)}{\cosh(l) - I_{top}/I_{base}}, \tag{47}$$

which is reduced to $z_* \cong L/(1 - I_{top}/I_{base})$ when $L/\lambda_f \ll 1$. The last expression shows that the concentration maximum lies between 0 and $L$ when $I_{top} > 0$ and $I_{base} < 0$, as in Harmand et al. (2010). This property is demonstrated in Fig. 8 for the parameters of Harmand et al. (2010): $\Delta\mu_{sL} = -145\,\text{meV}$, $\Delta\mu_{fL} = 160\,\text{meV}$ (corresponding to $g_s = -10.1$ and $g_f = 0.93$ at 420 °C), $\lambda_s = 2\,\text{nm}$, and $\lambda_f = 2000\,\text{nm}$. At $\alpha_3 = 30°$ and $\beta = 120°$,

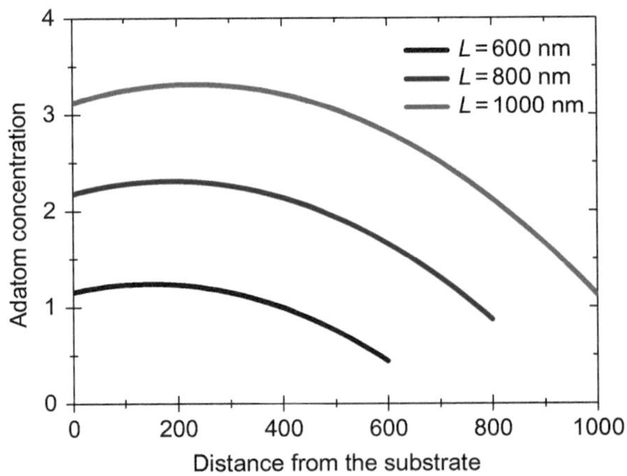


**Figure 8** Normalized adatom concentration $\widetilde{n}_f$ as a function of distance from the substrate $z$ for the parameters given in the text and three different nanowire lengths.

these parameters yield for $A=1.54$, $B=30.4$, $C=-1.80$, and $C_0=28.1$ for a 45-nm diameter nanowire.

Equation (42) leads to qualitatively different behaviors of the nanowire length for different $A$, $B$, and $C$ describing the contributions from direct impingement, surface diffusion of sidewall, and surface adatoms, respectively. Assuming $\pi R^2 N_W \to 0$, we obtain

$$\frac{1}{V}\frac{dL}{dt} = A_1 + \frac{BU + C}{U'}, \tag{48}$$

with $A_1 = A - 1$. It is seen that nanowires can start growing from the substrate (at $L/\lambda_f \ll 1$) when $A_1 + C > 0$ and at small $C \cong 0$ this requires $A_1 > 0$. In the MBE case, the $A_1$ value is positive due to magnifying effect of the droplet surface intercepting a larger portion of the group III beam compared to the substrate surface (Chatillon et al., 2009). The initial stage of nanowire growth can be thus described by the equation

$$\frac{1}{V}\frac{dL}{dt} = A_1 + \frac{2\lambda_s}{R}\delta\left(\frac{R}{\lambda_s}\right)g_s, \tag{49}$$

where the diffusion-induced term is further reduced to $2\lambda_s g_s/R$ at $R/\lambda_s \gg 1$ and $2(\lambda_s/R)^2[g_s/\ln(\lambda_s/R)]$ at $R/\lambda_s \ll 1$. On the other hand, at $L/\lambda_f \gg 1$ the nanowires will continue growing when $A_1 + B > 0$, and the elongation rate is reduced to

$$\frac{1}{V}\frac{dL}{dt} = A_1 + \frac{2\lambda_f \tan\alpha}{\pi R} g_f. \tag{50}$$

Three possible growth scenarios are schematized in Fig. 9 showing the nanowire elongation rates $dL/dH \equiv dL/d(Vt)$ at fixed $A_1 = 0.54$, $b(\lambda_f/\lambda_s)\delta(R/\lambda_s) = 1$, and different $B$ and $C$. The usual growth at $B=5$ and $C=-0.5$ yields positive $dL/dH$ for any $L$. The continuing growth scenario at $B=5$ and $C=-3$ means that the nanowire cannot start from the substrate surface but can continue growing from a certain critical length $l_1$, say a nanowire stem obtained under different conditions such as temperature (Dubrovskii et al., 2009b) or V/III flux ratio (Harmand et al., 2010). Limited growth to a certain critical length $l_2$ at $B=-1$ and $C=2$ is observed when the nanowire emerges from the substrate but cannot exceed the critical length due to a negative diffusion flux $B$.

Let us now consider an interesting situation at $\lambda_f \to \infty$ which holds when all adatoms collected from the substrate can reach the droplet regardless of the nanowire length (Dubrovskii et al., 2012a). Assuming that $b$ in Eqs. (34) and

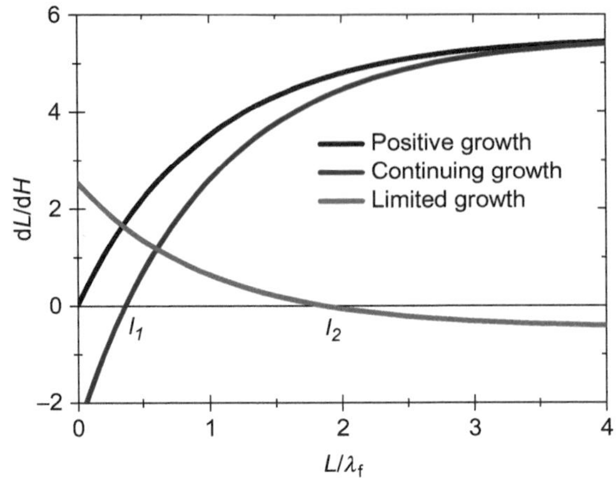


**Figure 9** Three scenarios of nanowire growth at a positive $A_1$ for different $B$ and $C$.

(35) is of the order of one and expanding the $U$ and $U'$ in $L/\lambda_f$, we obtain the universal, $\lambda_f$ independent asymptote of the elongation rate in the form

$$\frac{1}{V}\frac{\mathrm{d}L}{\mathrm{d}t} = A_1 + \frac{(L + b\delta L^2/2\lambda_s)(2\tan\alpha_3/\pi R)g_f + (2\lambda_s/R)\delta g_s}{1 + b\delta L/\lambda_s}. \tag{51}$$

In the case $R/\lambda_s \gg 1$, we can neglect the $A_1$ term because the collection area on the substrate surface is much larger than the nanowire radius and also assume that $b\delta L/\lambda_s \gg 1$. This yields the explicit representation of the nanowire length in the form

$$L = \lambda_s\sqrt{\frac{2g_s}{abg_f}\left[\exp\left(\frac{2ag_f H}{R}\right) - 1\right]}, \tag{52}$$

with $H = V(t - t_0)$ as the effective deposition thickness, $t_0$ as the incubation time for growth, and $a = \tan\alpha_3/\pi$ as the geometrical factor.

At the initial growth stage, Eq. (52) is reduced to

$$L \cong 2\lambda_s\sqrt{\frac{g_s H}{bR}}, \tag{53}$$

which is valid as long as $2ag_f Vt/R \ll 1$ and requires $g_s > 0$. This gives an unusual, square root time dependence of the nanowire length and the $1/\sqrt{R}$ radius correlation. The initial stage is followed by the exponential growth law

$$L \cong \lambda_s \sqrt{\frac{2g_s}{abg_f}} \exp\left(\frac{ag_f H}{R}\right), \quad (54)$$

which pertains as long as $L \ll \lambda_f$ and requires $g_f > 0$. In the case $g_f < 0$, the nanowires will tend to a finite length

$$L \to \lambda_s \sqrt{\frac{2g_s}{ab|g_f|}} \quad (55)$$

rather than growing infinitely.

Figure 10 shows the fits obtained from Eq. (54) to the experimental dependences of the length of differently sized ⟨110⟩-oriented Ge nanowires on the growth time (Dubrovskii et al., 2012a). These inclined nanowires were produced by Au-catalyzed MBE on Si(111) substrates with the Ge beam perpendicular to the substrate, and show an interesting focusing effect on the length distribution. It is seen that the narrowest 60-nm-thick nanowires grow faster at the initial stage, as described by Eq. (53), but then stop growing. Conversely, the thickest 130-nm diameter wires that have evolved slowly during the first 30 min overtake the length of the 60-nm-thick wires to become more than four times longer after 70 min of growth.

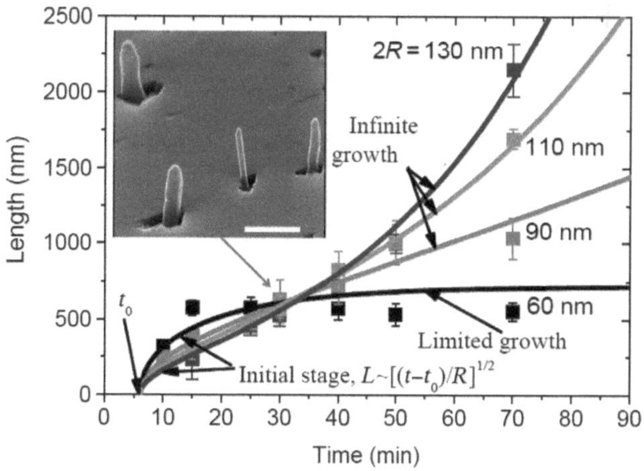


**Figure 10** Experimental length–time dependences of ⟨110⟩-oriented Ge nanowires for four different diameters $2R$ (symbols), and theoretical fits (lines) obtained from Eq. (52) at $t_0 = 6$ min, $g_f = -0.30$ for 60-nm, 0.13 for 90-nm, 0.45 for 110-nm, and 0.75 for 130-nm diameter nanowires. Inset shows SEM image of ⟨110⟩-oriented Ge nanowires grown for 30 min. The scale bar is 600 nm. Different growth regimes are indicated by arrows (Dubrovskii et al., 2012a).

Remarkably, a growth stage exists of around 30 min of growth where the length distribution of $\langle 110 \rangle$ Ge nanowires is quite narrow, and this effect is well supported by the theoretical curves. In particular, 60-nm-thick nanowires clearly exhibit the limited growth to a finite length at $g_f < 0$, while thicker nanowires follow the regime of exponential growth in the large time limit at $g_f > 0$.

The diffusion transport equations can be simplified in some cases such as the VLS growth on patterned surfaces where the diffusion flux from the substrate is defined by lateral confinement of the feeding zones. Consider, for example, the InP nanowires obtained by Au-catalyzed selective area CBE form TMIn and precracked phosphine on InP(111)B substrates covered with a $SiN_x$ mask, as shown in Fig. 2 (Kelrich et al., 2015). During selective area VLS growth, the InP nanowire formation has a diffusion-induced character, while 2D growth on a mask layer is suppressed (Dalacu et al., 2009; Dubrovskii et al., 2015a; Kelrich et al., 2013; Poole et al., 2012). The model of indium-limited growth is schematized in Fig. 11, for an idealized case of nontapered nanowires. Indium is deposited from the beam which is perpendicular to the substrate, with the equivalent deposition rate $V = \chi_3 I_3 \Omega_{35}$. This rate accounts for the cracking efficiency of TMI at a given substrate temperature and V/III flux ratio $\chi_3$ (Kelrich et al., 2015). A spherical droplet with the contact angle $\beta$ collects the direct flux $I_3/\sin^2\beta$. Indium also scatters from the inert SiN layer and reemits onto the nanowire sidewalls with the probability $\omega_r$. The latter equals ½ (Dalacu et al., 2009) with neglect of a fraction of the entire indium flux which is not reflected from the openings (Kelrich et al., 2013). These indium particles impinge the exposed annular area of the opening, $\pi\left(R_{op}^2 - R_b^2\right)$ (with $R_{op}$ as the opening radius and $R_b$ as

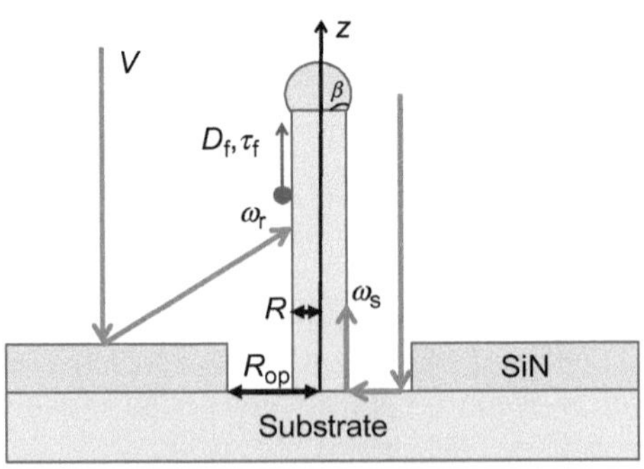


**Figure 11** Illustration of the growth model for a constant radius nanowire ($R_b = R_d = R$) with the parameters described in the text.

the nanowire base radius) and produce indium adatoms which subsequently migrate from the substrate to the nanowire sidewalls with the probability $\omega_s$.

Stationary diffusion equation describing the excessive concentration of sidewall adatoms $n_f \equiv n$ in this case writes down as

$$D\frac{d^2 n}{dz^2} + \frac{I_3}{2} - \frac{n}{\tau} = 0. \tag{56}$$

The linear sink $-n/\tau$ is expected to be due to the adatom incorporation to the growing shells around the nanowires. The boundary conditions are chosen in the form (Dubrovskii et al., 2015a)

$$\begin{gathered} -2\pi R_b D\frac{dn}{dz}\Big|_{z=0} = \omega_s I_3 \pi \left(R_{op}^2 - R_b^2\right), \\ n(z=L) = 0. \end{gathered} \tag{57}$$

The first condition determines the diffusion flux to the nanowire base from the openings of a given size, with $R_{op}$ as the radius of the opening, $R_b$ as the base radius, and $\pi\left(R_{op}^2 - R_b^2\right)$ as the exposed annular area. The second boundary condition at the nanowire top is relevant when the activity of sidewall In adatoms is much larger than that in the liquid droplet.

Solution to Eq. (56) with boundary conditions defined by Eq. (57) is given by

$$n(z) = \frac{I_3 \tau}{2}\left[1 - \frac{\cosh(z/\lambda)}{\cosh(L/\lambda)}\right] + \frac{\lambda}{D}\frac{\omega_s I_3\left(R_{op}^2 - R_b^2\right)}{2R_b}\frac{\sinh[(L-z)/\lambda]}{\cosh(L/\lambda)}, \tag{58}$$

where $\lambda \equiv \lambda_f = \sqrt{D\tau}$ is the migration length of In adatoms which is limited by surface incorporation. Its temperature dependence should be mainly governed by the Arrhenius-like behavior of $D$: $D \propto \exp(-E_D/k_B T)$; hence $\lambda$ is expected to increase for higher temperatures. Using Eq. (24) and adding a constant direct impingement term, we find the nanowire elongation rate in the form (Dubrovskii et al., 2015a)

$$\frac{dL}{dH} = \frac{1}{\sin^2\beta} + \frac{\lambda}{R_d}\tanh(L/\lambda) + \omega_s\frac{\left(R_{op}^2 - R_b^2\right)}{R_b R_d}\frac{1}{\cosh(L/\lambda)}, \tag{59}$$

where $R_d$ is the radius of the nanowire top. If we additionally take into account the fact that In atoms do not scatter from the mask openings, the growth equation modifies to (Kelrich et al., 2015)

$$\frac{dL}{dH} = \frac{1}{\sin^2\beta} + \frac{\lambda}{R_d}\tanh(L/\lambda)\left[1 - \frac{2}{\pi}\arctan\left(\frac{R_{op}}{\lambda\tanh(L/\lambda)}\right)\right] + \omega_s\frac{\left(R_{op}^2 - R_b^2\right)}{R_b R_d}\frac{1}{\cosh(L/\lambda)}. \tag{60}$$

Figure 12 shows InP nanowires grown at 420 and 480 °C for 50 min at $PH_3$ flow rates of 1, 3, 5, 7, and 9 sccm. The TMIn flow rate was kept constant and the corresponding planar growth rates for this TMIn flow rate on (100) substrates are given in Table 1 as functions of the $PH_3$ flow rate and temperature. Significant reduction of the nanowire length with increasing phosphine flow is evident with the only exception of $PH_3$ flow rate of 1 sccm at 480 °C (Fig. 12B), where radial growth took place at the expense of vertical elongation. It is important to note that no tapering is observed in nanowires grown at 480 °C at the phosphine flow rates that were larger than 1 sccm.

Within the model given by Eq. (60), the influence of the $P_2$ flux on the nanowire growth rate can be introduced by variation of the diffusion length $\lambda$ and/or the equivalent planar growth rate $V = \chi_3 I_3 \Omega_{35}$. The variation of the planar growth rate is attributed to desorption of TMIn molecules due to bond saturation by $P_2$. Figure 13 shows the experimental length–diameter dependence of the nanowires and their fits by Eq. (60) at $\beta = \pi/2$ (as revealed by measurements after growth), $R_b = R_d \equiv R$, $\omega_s = 1$, and with the measured values of $R_{op}$ for each sample. Since the SEM pictures in Fig. 12 indicate that tapering is not affected by the phosphine flow rate, it seems reasonable to assume a constant diffusion length at a given growth temperature to fit the data. The values of the planar growth rate and the diffusion length were obtained by the curve fitting. The obtained values of $\lambda$ were 650 and 2500 nm for the growth temperature of 420 and 480 °C, respectively. A comparison between the values of $V$ obtained by the fitting procedure and the measured growth rate on (001) surfaces is given in Table 1. A fairly close correspondence is obtained, indicating that the assumption of a constant diffusion length is reasonable. The only exception is the case of the lowest $PH_3$ flow rate of 1 sccm at 480 °C. At these growth conditions, an extremely high stacking fault density is observed. As a result, the diffusion length along the nanowire sidewalls is very short (~300 nm) inducing significant radial growth.

Since indium desorption is ineffective at our low temperatures, the linear sink $-n/\tau$ describes the radial growth under the assumption of an atomically rough sidewall surface (e.g., with a high density of atomic steps trapping the

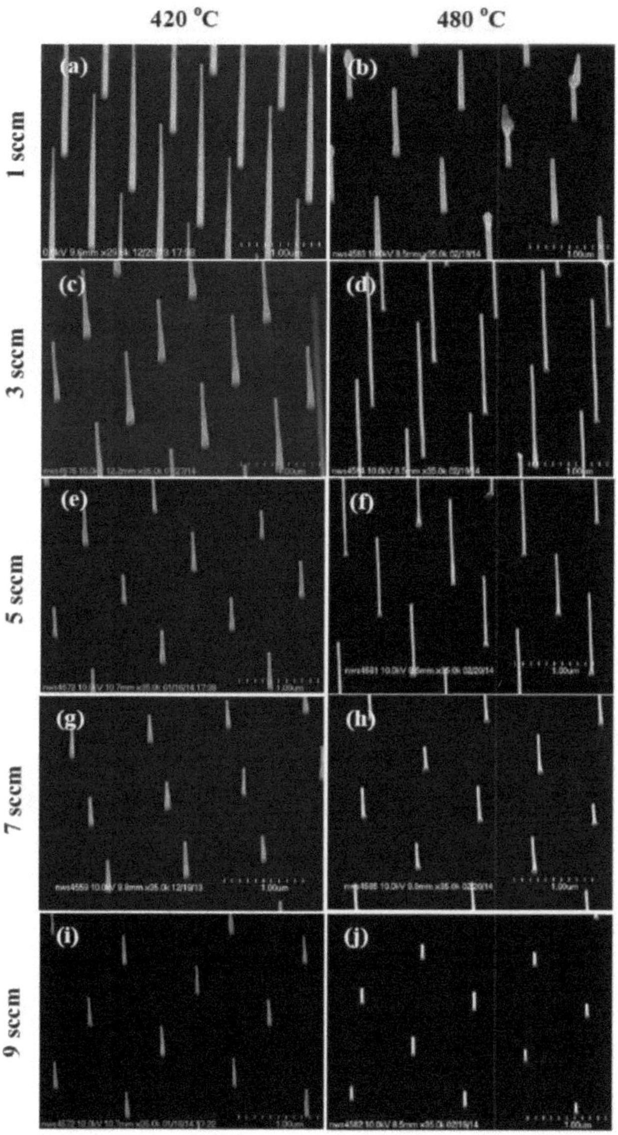


**Figure 12** SEM images of InP nanowires grown at different $PH_3$ flux rates and temperatures, indicated on the left side and top of the figure, correspondingly. The SEM images were taken at a tilt of 30°. All scale bars correspond to 1 μm (Kelrich et al., 2015).

In adatoms) (Dubrovskii et al., 2012b). When radial nanowire growth is mediated by 2D nucleation at the sidewalls, it should be described by an essentially nonlinear Zeldovich nucleation rate (Dubrovskii et al., 2008b), the case which will be considered later on. In the linear model, the radial

**Table 1** Measured and Fitted Planar Growth Rates of InP

| $PH_3$ Flow Rate (sccm) | Measured Growth Rate at 420 °C on (001) (nm/h) | Fitted Value of $V$ at 420 °C (nm/h) | Measured Growth Rate at 480 °C on (001) (nm/h) | Fitted Value of $V$ at 480 °C (nm/h) |
|---|---|---|---|---|
| 1 | 113 | 144 | 165 | 168 |
| 3 | 89 | 100 | 100 | 94 |
| 5 | 60 | 66 | 93 | 70 |
| 7 | 53 | 56 | 75 | 55 |
| 9 | 36 | 47 | 67 | 42 |

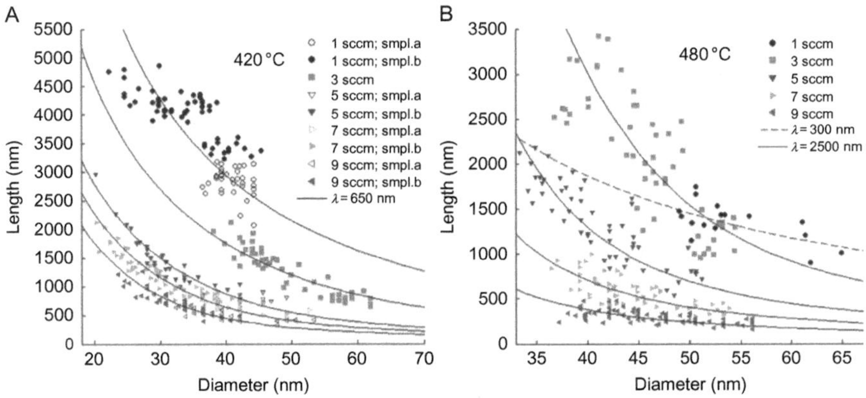


**Figure 13** Measured (symbols) and fitted by Eq. 60 (lines) length–diameter correlations of InP nanowires after 50 min of CBE growth at different $PH_3$ flow rates: (A) $T=420$ °C and (B) $T=480$ °C. The fitting values of the equivalent planar growth rates are summarized in Table 1 and the values of the diffusion length are shown in the insets (Kelrich et al., 2015).

growth rate is simply proportional to the local adatom concentration at a given distance from the substrate (Dubrovskii et al., 2012b):

$$\frac{dR(z,t)}{dt} = \frac{\Omega_{35}}{\tau} n(z, L(t)). \tag{61}$$

Here, $R(z,t)$ is the nanowire radius at height $z$ at time $t$, and the time dependence of the right hand side goes through $L(t)$. From Eq. (58), we obtain the $dR(z,L)/dH$ in the form

$$\frac{\mathrm{d}R(z,L)}{\mathrm{d}H}=\frac{1}{2}\left[1-\frac{\cosh(z/\lambda)}{\cosh(L/\lambda)}\right]+\frac{\omega_s}{2}\frac{\left(R_{\mathrm{op}}^2-R_{\mathrm{b}}^2\right)}{\lambda R_{\mathrm{b}}}\frac{\sinh[(L-z)/\lambda]}{\cosh(L/\lambda)}. \quad (62)$$

As in Dubrovskii et al. (2008b), we can now present $\mathrm{d}R(z,L)/\mathrm{d}L=[\mathrm{d}R(z,L)/\mathrm{d}H][\mathrm{d}L/\mathrm{d}H]^{-1}$, use Eq. (59) (where we neglect the direct impingement term) and (62), and integrate the result over $L'$ from $z$ to $L$, assuming $R_{\mathrm{b}}\cong R_{\mathrm{d}}$. This yields the nanowire shape $R(z,L)$ at a given distance from the substrate $z$ and total nanowire length $L$:

$$R(z,L)=R_{\mathrm{d}}+\frac{R_{\mathrm{d}}}{2}f\left(\frac{z}{\lambda},\frac{L}{\lambda}\right). \quad (63)$$

Here, the function $f(x,l)$ is given by

$$f(x,l)=[1-a\sinh(x)]\ln\left(\frac{\sinh(l)+a}{\sinh(x)+a}\right)+a\cosh(x)(l-x)$$
$$-\sqrt{1+a^2}\cosh(x)\ln\left(\frac{(\mathrm{e}^l+c_1)(\mathrm{e}^x+c_2)}{(\mathrm{e}^l+c_2)(\mathrm{e}^x+c_1)}\right) \quad (64)$$

with the constants $c_1=a-\sqrt{1+a^2}$ and $c_2=a+\sqrt{1+a^2}$, and $a=\omega_s\left(R_{\mathrm{op}}^2-R_{\mathrm{d}}^2\right)/(\lambda R_{\mathrm{d}})=\mathrm{const}$ under the assumption $R_{\mathrm{b}}\cong R_{\mathrm{d}}$ (otherwise, no analytical solution can be obtained). When $a\to 0$, Eq. (64) simplifies to

$$f(x,l)=\ln\left(\frac{\sinh(l)}{\sinh(x)}\right)-\cosh(x)\ln\left(\frac{\tanh(l/2)}{\tanh(x/2)}\right), \quad (65)$$

the expression which completely neglects diffusion from the opening.

Figure 14 shows how the nanowire shape changes with increasing the migration length when the nanowire profiles are calculated from the simplified Eqs. (63) and (65). It is seen that this simple model already qualitatively explains the experimental data regarding tapering of InP nanowires (Kelrich et al., 2015): the nanowire morphology appears tapered for $\lambda=400$–$800$ nm (at low temperatures), while at $\lambda=2200$ nm or more one obtains straight nanowires with uniform diameter from base to top.

## 6. NUCLEATION RATE IN VLS NANOWIRES

We now turn to discussion of the Zeldovich nucleation rate in VLS nanowires starting from Au-catalyzed growth where the liquid droplet is a ternary Au–III–V alloy, as shown in Fig. 15. When nuclei are single-

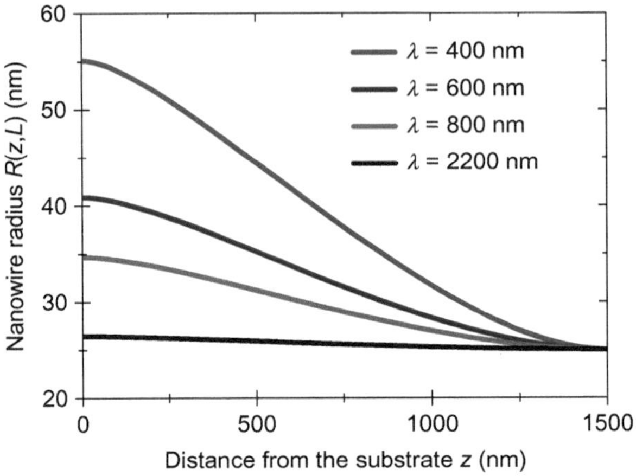


**Figure 14** Theoretical shapes of a 1500-nm-long nanowire with a top radius of 25 nm at different $\lambda$, obtained from Eq. (65).

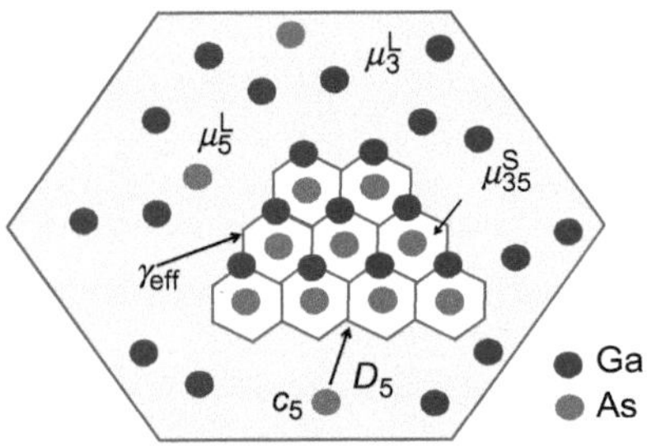


**Figure 15** 2D projection of the VLS growth from a liquid Au–Ga–As droplet showing macroscopic ZB GaAs island (surrounded by liquid) having the lateral surface energy $\gamma_{\mathrm{eff}} = \gamma_{\mathrm{SL}}$ and chemical potential $\mu_{35}^{\mathrm{S}}$, and nonstoichiometric alloy of Ga and As atoms dissolved in gold. The Ga and As chemical potentials in liquid equal $\mu_3^{\mathrm{L}}$ and $\mu_5^{\mathrm{L}}$, respectively, while the As concentration equals $c_5$.

component 2D surface islands fed from a diluted "sea" of surface adatoms, the macroscopic Zeldovich expression for the nucleation rate is given by (Dubrovskii, 2009, 2014a; Dubrovskii and Grechenkov, 2015; Dubrovskii and Nazarenko, 2010; Kashchiev, 2000; Kukushkin and Osipov, 1996; Markov, 2003)

$$J = \sqrt{\frac{|F''(i_c)|}{2\pi}} n_1 W^+(i_c) \mathrm{e}^{-F_c}. \tag{66}$$

Here, $F_c \equiv F(i_c)$ is the nucleation barrier in $k_B T$ units, $i_c$ is the number of adatoms in the critical nucleus, $n_1$ is the adatom surface concentration, $W^+(i_c)$ is the adatom attachment rate to the critical nucleus, and $F''(i_c)$ is the second derivative of the formation energy $F(i)$ at $i = i_c$. Equation (66) contains both thermodynamic $[F, i_c, F''(i_c)]$ and kinetic $[n_1, W^+(i_c)]$ parameters which correspond to a given supersaturation $\zeta$. The latter is generally defined as $\Delta\bar{\mu} = \ln(\zeta + 1)$, where $\Delta\bar{\mu}$ is the difference of chemical potentials between a metastable adatom phase and a 2D solid layer, expressed in $k_B T$ units. For a dilute adatom sea, supersaturation equals $\zeta = n_1/n_{1e} - 1$ in view of $\Delta\bar{\mu} = \ln(n_1/n_{1e})$ for a perfect gas, with $n_{1e}$ as the equilibrium adatom concentration at the surface temperature $T$.

As discussed in Sections 2 and 3, the chemical potential $\Delta\bar{\mu}$ should now be replaced to $\Delta\mu_{LS}/k_B T$ and the reformulation of $i_c$, $F_c$, and $F''(i_c)$ for VLS growth from a ternary alloy by means of Eq. (9) is rather straightforward. In Dubrovskii and Grechenkov (2015), we proposed a method for determining the analogues of $n_1$ and $W^+(i_c)$, based on the following considerations. We assume that a droplet is so small that the concentrations of both group III and V atoms dissolved in Au are spatially uniform in the entire liquid volume. The lateral island surface is considered as a linear sink for III–V pairs which should arrive to the island through its perimeter in stoichiometric one-to-one proportion. Due to the known low solubility of most group V elements in liquid metals, the relative atomic concentrations of group III and group V atoms in liquid should obey the strong inequality $c_5 \ll c_3$, as discussed in Sections 2 and 3. Therefore, the incorporation of III–V pairs into the island is expected to be limited by much fewer group V atoms in liquid, suggesting the substitution of the $n_1$ in Eq. (66) to the surface concentration of group V atoms:

$$n_1 \equiv c_5 \frac{h}{\Omega_{35}}. \tag{67}$$

As for the attachment rate $W^+(i_c)$, we first write down the general expression for the island growth rate (Dubrovskii, 2014a)

$$\frac{di}{dt} = W^+(i)\left[1 - e^{dF/di}\right] \cong W^+(i)\left[1 - e^{-\Delta\bar{\mu}}\right], \tag{68}$$

where the last expression holds for supercritical islands with $i \gg i_c$ where $dF/di \cong -\Delta\bar{\mu}$. On the other hand, in the diffusion transport model (Dubrovskii, 2014a), the growth rate of a supercritical triangle island with side $r$ would be

$$\frac{\mathrm{d}i}{\mathrm{d}t} = 3rD\frac{\mathrm{d}n_1}{\mathrm{d}x} = 3rD\frac{h}{\Omega_{35}}\frac{\mathrm{d}c_1}{\mathrm{d}x}. \tag{69}$$

This would pertain if islands were growing from a dilute adatom sea having the diffusion coefficient $D$, the surface coverage $c_1$, and on the surface with the elementary site area $\Omega_{35}/h$. Here, $\mathrm{d}c_1/\mathrm{d}x$ denotes the concentration gradient at the island boundary. In the perfect gas approximation, we have simply $\Delta\bar{\mu} = \ln(c_1/c_{1e})$ and hence $\mathrm{d}c_1/\mathrm{d}x = c_{1e}\mathrm{d}[\exp(\Delta\bar{\mu})]/\mathrm{d}x$. For III–V growth, we use the same expression with $\Delta\bar{\mu} = \bar{\mu}_3^{\mathrm{L}} + \bar{\mu}_5^{\mathrm{L}} - \bar{\mu}_{35}^{\mathrm{S}}$—indeed, no growth should occur when $\Delta\bar{\mu} = 0$. The equilibrium coverage $c_{1e}$ should be changed to $\exp\left(\bar{\mu}_{35}^{\mathrm{S}}\right)$, which is the equilibrium activity of a III–V solid. Overall, these considerations yield the substitution of $\mathrm{d}c_1/\mathrm{d}x$ to $\mathrm{d}\left[\exp\left(\bar{\mu}_3^{\mathrm{L}} + \bar{\mu}_5^{\mathrm{L}}\right)\right]/\mathrm{d}x$ in Eq. (69). Finally, within this "lattice gas" model, the chemical potential changes abruptly from $\bar{\mu}_3^{\mathrm{L}} + \bar{\mu}_5^{\mathrm{L}}$ in a spatially uniform liquid to $\bar{\mu}_{35}^{\mathrm{S}}$ in the reference solid state, over the elementary distance $(\Omega_{35}/h)^{1/2}$. This allows us to write approximately

$$\frac{\mathrm{d}}{\mathrm{d}x}\mathrm{e}^{\bar{\mu}_3^{\mathrm{L}} + \bar{\mu}_5^{\mathrm{L}}} \cong \left(\frac{h}{\Omega_{\mathrm{s}}}\right)^{1/2}\left(\mathrm{e}^{\bar{\mu}_3^{\mathrm{L}} + \bar{\mu}_5^{\mathrm{L}}} - \mathrm{e}^{\bar{\mu}_{35}^{\mathrm{S}}}\right). \tag{70}$$

We now use this in Eq. (69) instead of $\mathrm{d}c_1/\mathrm{d}x$. It is also clear that the diffusion transport from liquid to solid should be limited by group V atoms, which are a thinner and slower diffuser, so the diffusion coefficient $D$ should be put to $D_5$. Thus, we arrive at

$$\frac{\mathrm{d}i}{\mathrm{d}t} = 2 \times 3^{3/4} D_5 \frac{h}{\Omega_{35}} \mathrm{e}^{\bar{\mu}_3^{\mathrm{L}} + \bar{\mu}_5^{\mathrm{L}}} i^{1/2}\left(1 - \mathrm{e}^{-\Delta\bar{\mu}}\right), \tag{71}$$

where the $i^{1/2}$ size dependence shows that the growth rate is proportional to the island perimeter $3r$, as usual in the ballistic growth regime (Dubrovskii, 2009, 2014a; Dubrovskii and Nazarenko, 2010; Kukushkin and Osipov, 1996). Comparing Eqs. (71) and (68) and then using Eq. (9) for $i_c$, we find:

$$\begin{aligned} W^+(i) &= 2 \times 3^{3/4} D_5 \frac{h}{\Omega_{35}} \mathrm{e}^{\bar{\mu}_3^{\mathrm{L}} + \bar{\mu}_5^{\mathrm{L}}} i^{1/2}, \\ W^+(i_c) &= 3^{3/4} D_5 \frac{h}{\Omega_{35}} \mathrm{e}^{\bar{\mu}_3^{\mathrm{L}} + \bar{\mu}_5^{\mathrm{L}}} \frac{a}{\Delta\bar{\mu}}, \end{aligned} \tag{72}$$

where the last expression is what is required in Eq. (66).

Using Eqs. (67) and (72) in Eq. (66), we obtain the Zeldovich nucleation rate in Au-catalyzed III–V nanowires in the form

$$J=\frac{3^{3/4}}{\sqrt{\pi}}D_5c_5\left(\frac{h}{\Omega_{35}}\right)^2 e^{\overline{\mu}_{35}^{S}+\Delta\overline{\mu}}\Delta\overline{\mu}^{1/2}\exp\left(-\frac{a^2}{4\Delta\overline{\mu}}\right). \tag{73}$$

This expression is valid when the nucleus formation energy is given by Eq. (8). We now discuss the so-called self-consistency renormalization of the nucleation rate which arises due to reformulation of the formation energy based on the following considerations. Despite its clear physical meaning and simplicity, Eq. (8) has one important drawback as $F(i)$ does not equal zero at $i=1$. This makes the whole theory self-inconsistent—indeed, if we first consider a simplified case of elemental transition from adatoms to 2D islands with $\Delta\overline{\mu}=\ln(\zeta+1)$, the quasi-equilibrium size distribution $n_{eq}(i)=n_1\exp[-F(i)]$ [relating to a given adatom concentration $n_1=n_{1e}(\zeta+1)$] does not give correctly this adatom concentration at $i=1$. Therefore, it has been proposed (see the discussion by Kashchiev, 2000, Ch. 7 and references therein, or Eq. (3) of Kashchiev, 2008) to use the renormalization

$$\widetilde{F}(i)=F(i)-F(1)=-i\Delta\overline{\mu}+ai^{1/2}+\Delta\overline{\mu}-a, \tag{74}$$

which corrects the macroscopic formation energy to acquire the property $\widetilde{F}(1)=0$. Obviously, this renormalization does not affect $i_c$ and $F''(i_c)$ in Eq. (9), while the effective nucleation barrier is changed to

$$\widetilde{F}_c=\frac{a^2}{4\Delta\overline{\mu}}+\Delta\overline{\mu}-a. \tag{75}$$

Since $a\gg1$ as discussed earlier, the additional factor $\exp(a)$, appearing in Eq. (73) after the renormalization, increases the corrected Zeldovich nucleation rate by several orders of magnitude.

It seems logical to introduce the same renormalization for the VLS growth of III–V compounds. Reformulation of the self-consistency criterion in this case is rather straightforward—we request that the renormalized macroscopic free energy of forming a monomer is zero and thus the condition $\widetilde{F}(1)=0$ applies also for III–V pairs. We note that our liquid monomers are not identical to bound Ga–As pairs on the solid substrate which are nuclei of size one (because the difference of their chemical potentials equals $\Delta\overline{\mu}$), while in elemental growth from the adatom sea nuclei of size one are the adatoms themselves. Thus, the chemical potential difference $\Delta\overline{\mu}$ appearing in Eq. (74) can be further discussed. However, the shift by the surface energy constant $a$ seems to be justified as it eliminates the unphysical surface energy of two spatially separated atoms in the droplet. Therefore, we propose to use the self-consistent formula for the free energy of forming a 2D III–V island from liquid group III and V atoms given by

$$\widetilde{F}_c = \frac{a^2}{4\Delta\overline{\mu}} + (1-\nu)\Delta\overline{\mu} - a. \tag{76}$$

Here, $\nu = 0$ or 1 depending on whether the chemical potential difference between two Ga and As atoms in the liquid and a bound GaAs dimer is disregarded or included.

This changes the Zeldovich nucleation rate in VLS nanowires to

$$J = \frac{3^{3/4}}{\sqrt{\pi}} D_5 c_5 \left(\frac{h}{\Omega_{35}}\right)^2 e^{\overline{\mu}_{35}^S + \nu\Delta\overline{\mu}} \Delta\overline{\mu}^{1/2} \exp\left(a - \frac{a^2}{4\Delta\overline{\mu}}\right) \tag{77}$$

Comparing Eqs. (73) and (77), the nucleation rate with self-consistency renormalization contains the additional exponential factor $\exp(a) \gg 1$ and does not contain $\exp(\Delta\overline{\mu})$ when $\nu = 0$. Both Eqs. (73) and (77) yield the nucleation rate which increases with $\Delta\overline{\mu}$ and decreases with the surface energy $a$. In Eq. (77), the latter property holds due to $\partial\left(a - a^2/4\Delta\overline{\mu}\right)/\partial a = 1 - i_c^{1/2} < 0$ in the physical region where $i_c > 1$.

In the case of Ga-catalyzed GaAs nanowires and other III–V nanowires catalyzed by the respective group III metals, the chemical potential $\Delta\mu_{LS}$ should be a function of the group V atomic concentration in the droplet and temperature, thus $\Delta\overline{\mu} = \Delta\overline{\mu}(c_5, T)$. This chemical potential can be obtained as described in Glas et al. (2013). The equation for the Zeldovich nucleation rate used in Glas et al. (2013) is

$$J = A(T) c_5 \Delta\overline{\mu}^{1/2} \exp\left(-\frac{a^2}{4\Delta\overline{\mu}}\right). \tag{78}$$

The authors assumed that the prefactor $A$ depends on temperature but neither on $\Delta\mu$ nor $c_5$. The nucleation rate given by Eq. (78) is shown in Fig. 16 as a function of chemical potential, under the assumption of a constant $Ac_3 =$ and at different $a$.

Comparing Eq. (78) to Eqs. (73) and (77), we find

$$A = \frac{3^{3/4}}{\sqrt{\pi}} D_5 \left(\frac{h}{\Omega_{35}}\right)^2 e^{\varepsilon a + \mu_{35}^S + \nu\Delta\overline{\mu}}, \tag{79}$$

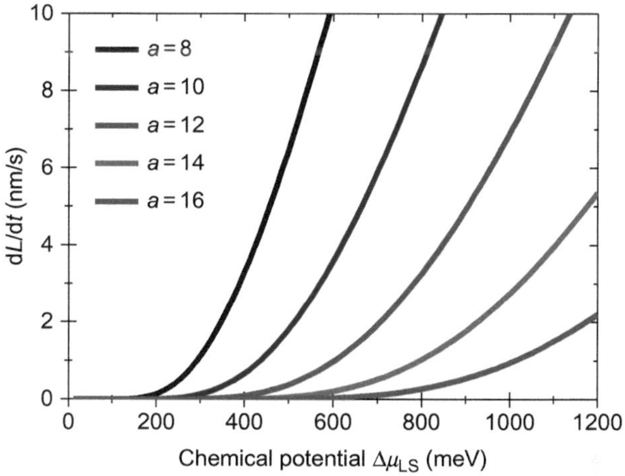


**Figure 16** Nucleation-mediated elongation rate of GaAs nanowires with 20 nm radius versus chemical potential at different *a*.

where $\varepsilon = 0$ if we use Eq. (73) and 1 in Eq. (77). For a binary Ga–As liquid and at a low As concentration, the surface energy $\gamma_{\text{eff}}$ entering Eq. (8) for $a$ and therefore the $a$ value itself are expected to relate to a pure Ga droplet. All other physical constants entering Eq. (79) except for $\Delta\overline{\mu}$ depend only on temperature. Thus, the assumption of $A = A(T)$ and all the results of Glas et al. (2013) correlate with the renormalized nucleation rate given by Eq. (77) when $\nu = 0$. Equations (73) and (77) at $\nu = 1$ contain the additional exponential factor $\exp(\Delta\overline{\mu})$ because these nucleation rates are proportional to the nonequilibrium liquid activity rather than the equilibrium activity of the reference solid state. This property is seen already from Eq. (66) which is proportional to the nonequilibrium vapor concentration $n_1 = \sigma^{-1}\exp(\overline{\mu}_1^{\text{eq}} + \Delta\overline{\mu}_1)$ in the perfect gas approximation.

As mentioned in Section 2, the Zeldovich nucleation rate can only be used in macroscopic theory where the critical nucleus consists of at least several III–V pairs. The Zeldovich approach is widely used in modeling the nanowire growth and crystal structure and has been proven adequate for describing the major effects (Dubrovskii, 2014a), although we really hit the limit of macroscopic models at high chemical potentials and low surface energies. In theories of one-component irreversible growth (at $i_c = 1$) of 2D surface islands, the nucleation rate is given by $J = \sigma_1 n_1^2$, i.e., proportional to the squared adatom concentration with a certain dimerization rate $\sigma_1$ (Dubrovskii and Sibirev, 2014; Einax et al., 2013). This expression shows simply that a stable nucleus emerges whenever two mobile adatoms meet and are assembled into an immobile dimer. The situation is much more complex in our case of VLS nanowires growing from liquid. Relevant expressions to consider would be $J \propto c_5^2$ for Ga-rich droplets or $J \propto c_3 c_5$ for Au-rich droplets. In particular, in the case of Au-catalyzed III–V nanowires the nucleation rate at high supersaturations should be given by $J = (h/\Omega_{35})\sigma c_3 c_5$ for binaries and $J = (h/\Omega_{35})(\sigma_{Ga} c_{Ga} c_{As} + \sigma_{In} c_{In} c_{As})$ for ternaries such as InGaAs. These expressions and their impact on the axial growth rate and nanowire composition will be studied in more detail in Section 10.

## 7. POSITION-DEPENDENT NUCLEATION IN NANOWIRES

We now consider a very important question of the position-dependent nucleation rate in VLS nanowires. Glas et al. (2007) raised this issue in connection with the preferred crystal structure of Au-catalyzed III–V nanowires as will be discussed in Section 11. However, the position of the nucleus also affects the effective surface energy of the nucleus, the nanowire elongation rate, and the self-consistent determination of the group III concentration during growth. We mentioned in Section 2 that the energetics of a particular VLS system is set by the wire–vapor ($\gamma_{WV}$), liquid–vapor ($\gamma_{LV}$), and lateral solid–liquid ($\gamma_{SL}$) interface energies, and the contact angle of the droplet $\beta$. Let us study the stability of a TPL (i.e., the line where vapor, liquid, and solid phases meet) at the top of a cylindrical nanowire with straight sidewalls and a planar liquid–solid interface. When the TPL is randomly shifted downward (Fig. 17A), the vertical capillary force $F = \gamma_{LV}\sin\beta + \gamma_{SL} - \gamma_{WV}$ is directed upward provided that

$$\gamma_{WV} < \gamma_{LV}\sin\beta + \gamma_{SL} \tag{80}$$

This inequality is exactly identical to the Nebol'sin–Shchetinin condition (Nebol'sin and Shchetinin, 2003) for the stability of nanowire growth.

Whenever the inequality given by Eq. (80) is reversed, the capillary force is directed downward (Fig. 17B) and will only increase as the nanowire cylinder penetrates into the liquid until it hits the droplet surface (Dubrovskii et al., 2011). After that, the droplet should acquire a nonspherical shape relating to the minimum surface energy (Dubrovskii et al., 2011). One possibility to preserve the droplet on top at $\gamma_{WV} > \gamma_{LV}\sin\beta + \gamma_{SL}$ would be to introduce a nonplanar growth interface with a truncated facet.

These simple considerations can approximately be reproduced for 2D nucleus (Dubrovskii et al., 2008a; Glas et al., 2007) showing that the

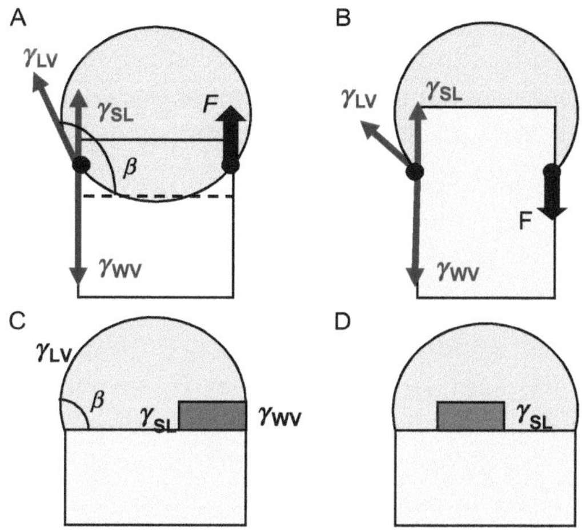


**Figure 17** Stable (A) and unstable (B) position of the triple phase line at the nanowire top depending on whether the inequality given by Eq. (80) is satisfied or reversed, favoring either triple phase line nucleation with the effective surface energy $\gamma_{eff}$ (C) or center nucleation with the surface energy $\gamma_{SL}$ (D).

inequality given by Eq. (80) favors the TPL nucleation where a portion of the island boundary $x$ is in contact with vapor (Fig. 17C). In this case, the island eliminates some part of preexisting surface area of the droplet which reduces its effective surface energy by $x\gamma_{LV}\sin\beta$, resulting in the effective surface energy (Glas et al., 2007)

$$\gamma_{eff} = (\gamma_{WV} - \gamma_{LV}\sin\beta)x + \gamma_{SL}(1-x). \quad (81)$$

If the sidewall planes are inclined with respect to the vertical direction by angles $\theta_j$ and their surface energies are $\gamma_j$, i.e., composed of alternating (111)A and (111)B planes of ML height (Glas et al., 2007) ($\theta_j = \pm 19.5°$), the effective surface energy of a TPL island is given by

$$\gamma_{eff} = \left[\frac{\gamma_j}{\cos\theta_j} - \left(\gamma_{SL}^0 + \gamma_{LV}\cos\beta\right)\tan\theta_j - \gamma_{LV}\sin\beta\right]x + \gamma_{SL}(1-x). \quad (82)$$

Such surface energies have similar expression as the energies calculated by Ross et al. (2005) in a 2D model of facetted nanowires. Equations (81) and (82) can be proved rigorously for MLs under the assumption of instantaneous nucleation under the constraint of a fixed liquid volume (Dubrovskii, 2014a). Whenever the inequality (80) is reversed, nucleation is expected to occur away from the TPL, in the center (C) position at the liquid–solid interface under the droplet. Of course, in this case the island is entirely surrounded by liquid and its surface energy simply equals $\gamma_{SL}$.

It is noteworthy that, for a given nanowire material, the stability condition given by Eq. (80) and therefore the TPL nucleation scenario become less probable for smaller $\gamma_{LV}$. Hence, we can tune the nucleation position by changing the catalyst composition. In particular, liquid group III metals are known to have a much smaller surface energy (0.87 J/m$^2$ for Al, 0.71 J/m$^2$ for Ga, and 0.57 J/m$^2$ for In at their respective melting temperatures) compared to liquid gold (1.14 J/m$^2$) (Jasper, 1972). Therefore, for GaAs nanowires and III–V nanowires in general, 2D nucleation is expected to occur at the TPL position if catalyzed by Au and at the C position if catalyzed by group III metals (Cirlin et al., 2010). Furthermore, suppression of the TPL nucleation may be achieved in Ga-rich AuGa alloys in Au-seeded VLS growth. This feature has an interesting impact on the crystal structure as will be discussed in Section 11.

The Zeldovich nucleation rate gives the probability of nucleation per unit surface area. Therefore, the C nucleation occurs on the whole nanowire facet without a ring of width $2r_c$, while the TPL nucleation occurs only inside such ring. For regular triangle islands, the side of the critical nucleus is given by

$$r_c = \frac{1}{3^{1/4}}\left(\frac{\Omega_{35}}{h}\right)^{1/2}\frac{a}{\Delta\bar{\mu}}. \tag{83}$$

Due to this purely geometrical effect of the reduction in the available nucleation area for the TPL islands, the nucleation-mediated nanowire elongation rate in mononuclear regime should be written as (Dubrovskii et al., 2008a)

$$\frac{dL}{dt} = \frac{3\sqrt{3}}{2}R^2 h\eta J, \tag{84}$$

with $\eta = 1$ for the C and $4r_c/R$ for the TPL growth scenario (at $r_c \ll R$). Therefore, the elongation rate is proportional to $R^2$ for the center nucleation and only to $R$ for the TPL nucleation.

## 8. SELF-CONSISTENT GROWTH EQUATION

We saw earlier in Section 5 that the transport-limited elongation rate usually depends only on $c_3$ (or $\mu_3^L$), while the effect of As pressure can be accounted for in the external parameters such as the Ga diffusion lengths or effective deposition thickness. The influence of As on the entire chemical potential $\Delta\mu_{LS}$ can be expressed through the group V flux and $dL/dt$ by Eqs. (21) and (22), so the remaining unknown value is $\mu_3^L(c_3)$. Therefore, the nucleation-mediated elongation rate given by Eq. (84) depends on $\mu_3^L(c_3)$ and the island surface energy. Whatever is the energetics of the VLS system considered, the effective surface energies of the island in different positions are not expected to depend strongly on the As concentration in the droplet simply because of its low As content. In particular, the liquid–vapor surface energy $\gamma_{LV}$ can be taken as linear approximation between the two pure metals. These considerations show that the unknown $c_3$ can be obtained theoretically by equating the nucleation-mediated elongation rate given by Eq. (84) to the transport-limited elongation rate whose different representations were considered in Section 5. With the measured $dL/dt$ under the given growth conditions, Eq. (84) alone allows one to obtain the $c_3$ value and to get the particular $\Delta\mu_{SL}$ in these conditions.

Let us now consider a simplified situation where we study the nanowire elongation rate as a function of the growth parameters and the nanowire radius and the liquid chemical potential does not strongly depend on $dL/dt$. This happens, for example, when Au-catalyzed growth proceeds at high group V input and the $G$ in Eq. (22) is insensitive to the elongation rate. We saw in Section 5 that $dL/dt$ in the diffusion-induced growth is often a power law function of the nanowire radius, and is proportional to $g = 1 - \theta_L/\theta_A$ for an adatom subsystem ("s" or "f") which mostly contributes to the diffusion transport to the nanowire top. Writing $\theta_L/\theta_A = (\zeta + 1)/(\Phi + 1)$, with $\zeta$ and $\Phi$ as the effective supersaturations of the liquid and adatom phases with respect to the reference solid, we can present the transport-limited elongation rate in the form

$$\left(\frac{dL}{dt}\right)_{\text{transport}} = V\left(\frac{\Lambda_a}{R}\right)^p \left(\frac{\Phi - \zeta}{\Phi + 1}\right). \tag{85}$$

with $\Lambda_a$ as the effective diffusion length of group III adatoms. The power index $p$ depends on the mechanisms of the diffusion transport and equals ½, 1, and 2 in the typical cases considered in Section 5. Since the transport-limited growth rate contains the difference $\Phi - \zeta$, nothing prevents us from adding a constant As contribution to both supersaturations and rearranging the $\Lambda_a$ to preserve the form of Eq. (85).

On the other hand, presenting the liquid chemical potential in Eq. (84) as $\Delta\bar{\mu} = \ln(\zeta+1)$ and using the linear expansion under the exponent in $\Phi-\zeta$, one obtains (Dubrovskii, 2014a; Yee et al., 2012)

$$\begin{aligned}\left(\frac{\mathrm{d}L}{\mathrm{d}t}\right)_{\mathrm{nucl}} &= \frac{3\sqrt{3}}{2}R^2 h\eta J_0(\zeta)\exp\left[-\frac{a^2}{4\ln(\zeta+1)}\right] \\ &\cong \frac{3\sqrt{3}}{2}R^2 h\eta J_0(\Phi)\exp\left[-i_c(\Phi)\left(\frac{\Phi-\zeta}{\Phi+1}\right)\right],\end{aligned} \tag{86}$$

with $i_c(\Phi) = a^2/4\ln^2(\Phi+1)$. All the preexponential $\zeta$-dependent terms are combined in $J_0(\zeta)$ and are taken at $\zeta=\Phi$ in the approximate expression. Such method is known in general nucleation theory and is justified due to an extremely steep dependence of the nucleation rate on supersaturation through the Zeldovich exponent (Dubrovskii, 2014a). Of course, Eq. (86) assumes that the island surface energy is a weak function of $\zeta$. Comparing Eqs. (85) and (86), we get the solution for the elongation rate (Dubrovskii and Hervieu, 2014)

$$\frac{\mathrm{d}L}{\mathrm{d}t} = V\left(\frac{\Lambda}{R}\right)^p W\left[\left(\frac{R}{R_*}\right)^{p+q}\right], \tag{87}$$

where $W(x)$ is the Lambert function such that $W\exp(W) = x$. The index $q$ equals 2 for the C and 1 for the TPL nucleation position. The effective collection length $\Lambda$ and the characteristic radius $R_*$ are defined as

$$\Lambda = \frac{\Lambda_a}{i_c^{1/p}(\Phi)};\quad \frac{1}{R_*^{p+q}} = \frac{3\sqrt{3}}{2}\frac{h}{V}\frac{J(\Phi)}{\Lambda^p}\begin{cases}1, q=2\\ 2r_c(\Phi), q=1\end{cases}. \tag{88}$$

Obviously, the elongation rate given by Eq. (87) is controlled by two parameters $\Lambda$ and $R_*$ associated with surface diffusion of adatoms and the characteristics of 2D nucleation on a facet of size $R$. Both values may depend on the As pressure and temperature. When the nucleation process is fast ($R/R_* \gg 1$), $\mathrm{d}L/\mathrm{d}t$ is limited by the diffusion transport:

$$\frac{\mathrm{d}L}{\mathrm{d}t} \cong \mathrm{const}\times V\left(\frac{\Lambda}{R}\right)^p, \tag{89}$$

and decreases with the radius as $1/R^p$. When the nucleation is slow ($R/R_* \ll 1$), the liquid supersaturation gets higher and approaches that of adatoms. This quickly decreases the diffusion flux, and the VLS growth becomes nucleation limited. In this case, the elongation rate increases with $R$ as

$$\frac{dL}{dt} \cong V \frac{\Lambda^p R^q}{R_*^{p+q}}, \quad R_* \gg R, \tag{90}$$

i.e., as $R$ for the TPL and as $R^2$ for the C nucleation positions. Generally, Eq. (87) gives a nonmonotonic length–radius correlation starting from zero and reaching its maximum at $R \sim R_*$, as shown in Fig. 18.

In particular, Figs. 19 and 20 show the images and experimental length–diameter dependences of InP and InAs–InP heterostructured nanowires at different concentrations of Be dopant, fitted by Eq. (87) with the parameters summarized in Table 2 (Yee et al., 2012). The fits were obtained at $q=1$, because all these nanowires have the WZ crystal structure and therefore nucleation is expected at the TPL (Glas et al., 2007). The best fits correspond to $p=2$, suggesting that the VLS growth is controlled by the collection of substrate adatoms under these conditions. We hypothesize that increasing Be doping level results in a decrease of the adatom supersaturation with respect to the droplet, a drastic decrease in the nucleation rate $J(\Phi)$, and a rapidly increasing $R_*$ in Eq. (87). The growth is then converted from the diffusion induced to the nucleation limited. More examples of using Eq. (87) for modeling different types of length–radius correlations of III–V and elemental VLS nanowires can be found in Dubrovskii and Hervieu (2014).

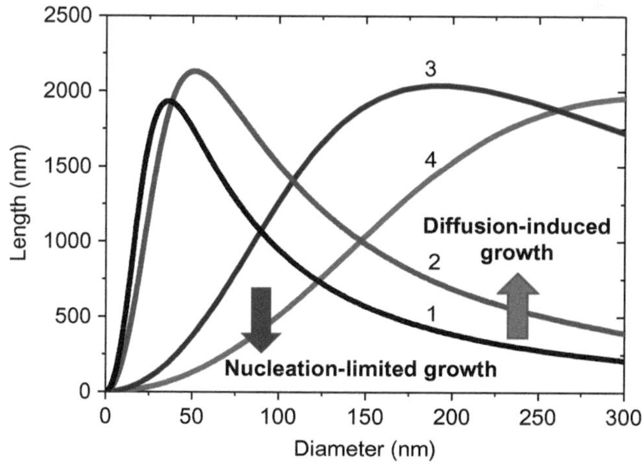


**Figure 18** Transition from the nonmonotonic to purely increasing length–diameter correlation with increasing $R_*$, obtained from Eq. (87) at $p=q=2$: curve 1—$R_*=14$nm, $\Lambda=50$nm; curve 2—$R_*=40$nm, $\Lambda=75$nm; curve 3—$R_*=75$nm, $\Lambda=225$nm; curve 4—$R_*=125$nm, $\Lambda=450$nm.

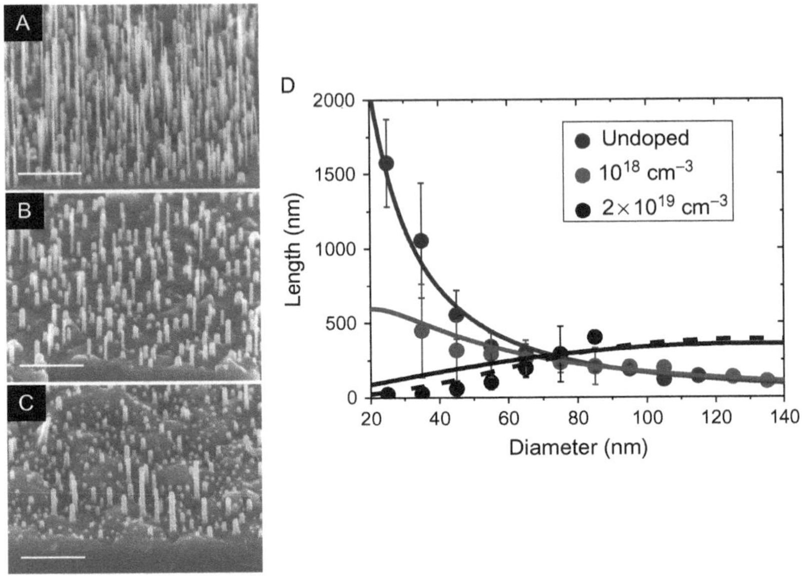


**Figure 19** Tilted view (45°) SEM images of (A) sample A1, undoped, (B) sample A2, Be doping $= 10^{18}$ cm$^{-3}$, and (C) sample A3, Be doping $= 2 \times 10^{19}$ cm$^{-3}$. Scale bars indicate 1 μm. (D) Length versus diameter for different Be-doped InP nanowires. Experimental data (symbols) are fitted by Eq. (87) with $q = 1$, $p = 2$, $H = Vt = 250$nm (lines), and other parameters summarized in Table 2. Dotted line in (D) shows the fit for sample A3 with $q = 2$.

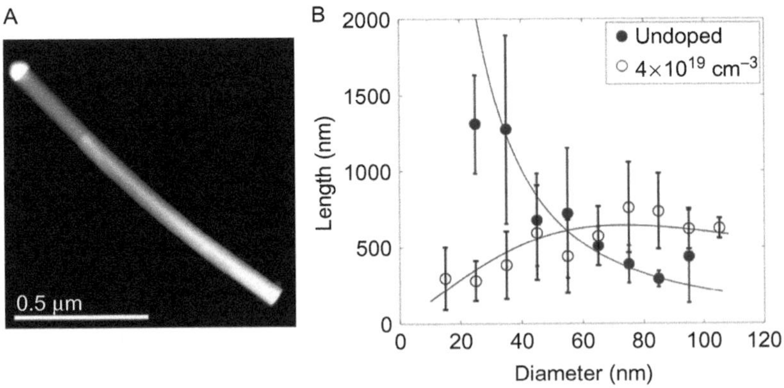


**Figure 20** (A) HAADF image of InAs–InP nanowire from sample B1; (B) InAs segment length versus diameter for different Be-doped InAs/InP nanowires (symbols), fitted by Eq. (87) with $q = 1$, $p = 2$, $H = 108$nm, and other parameters summarized in Table 2 (lines).

**Table 2** Fitting Parameters for Different Be-Doped InP and InAs–InP Nanowires

| Sample | Be Doping Level ($cm^{-3}$) | $R_*$ (nm) | $\Lambda$ (nm) |
|---|---|---|---|
| A1 | Undoped InP | 1.5 | 14 |
| A2 | InP $10^{18}$ | 11 | 22.5 |
| A3 | InP $2\times10^{19}$ | 72.5 | 115 |
| B1 | Undoped InAs | 1.5 | 25 |
| B2 | InAs $4\times10^{19}$ | 39 | 126.5 |

## 9. GA-CATALYZED GROWTH OF GaAs NANOWIRES

Triggered by the well-known incompatibility of gold with Si technology, great effort has been devoted to the development of Au-free growth protocols for nanowire synthesis. We mentioned earlier the self-catalyzed III–V nanowires whose VLS growth is promoted by group III metal rather than by a foreign metal catalyst. Effective results have been obtained, in case of GaAs, by self-assisted methods as reported, e.g., in Colombo et al. (2008) and Jabeen et al. (2008). By a suitable choice of either growth conditions or substrates, the formation of Ga droplets is promoted, and these droplets induce the epitaxial growth of GaAs nanowires either on GaAs (Fontcuberta i Morral et al., 2008; Priante et al., 2013) or on Si (Colombo et al., 2008; Matteini et al., 2015). In both cases, the substrate surface is covered with a thin layer of Si oxide. Concerning the growth process, an important difference between Au- and Ga-catalyzed growths is that the size of the metal droplet, which in case of Au does not change appreciably during the growth under the optimized conditions, is not necessarily constant in the case of Ga, and is highly sensitive to the V/III flux ratio.

When a balanced As/Ga ratio is used, the radius of Ga droplet stays constant. Under the assumption of a time-independent radius and steady-state conditions for As atoms in the droplet, Glas et al. (2013) presented a quantitative model for Ga-catalyzed VLS growth. The model is based on Eq. (19) for the As-limited elongation rate, Eq. (20) describing the temperature-dependent As desorption, Eq. (78) for the Zeldovich nucleation rate, and Eq. (84) for the nucleation-mediated elongation rate with $\eta = 1$. The latter assumption is based on the well-known fact of predominantly ZB crystal structure of Ga-catalyzed GaAs nanowires, and thus, the C nucleation should be the dominant growth scenario. Since chemical potential in a

Ga–As droplet is the known function of only two variables, the As concentration and the growth temperature, equalizing the transport-limited and nucleation-mediated elongation rates allows for self-consistent determination of the As concentration at a given temperature. Therefore, the model contains two unknown parameters—the surface (or edge) energy $\gamma_{\mathrm{eff}} = \gamma_{\mathrm{SL}}$ entering Eq. (8) for $a$ and the prefactor $A$ in the Zeldovich nucleation rate. We note, however, that calculations should be highly dependent on the value of the As reemission coefficient $\varepsilon$ in Eq. (19) which determines the effective influx of As atoms into the droplet and can even influence the very possibility of steady-state growth.

By carefully relating the atomic As flux to the measured beam equivalent pressure (BEP) of $As_4$ and fitting the experimental dependences of the $dL/dt$ on the growth temperature ($T = 575$–$630$ °C) and $As_4$ BEP ($1$–$5 \times 10^{-6}$ - Torr), the authors found the following recommended values of the two fitting parameters (corresponding to the best fits at $\varepsilon = 3.25$):

$$A = 7\,\mathrm{nm}^{-2}\mathrm{s}^{-1};\ \gamma_{\mathrm{SL}} = 0.123\,\mathrm{J/m}^2, \tag{91}$$

with $A$ independent of $T$ in this temperature window. The result is remarkable as it allows for predictive modeling of Ga-catalyzed VLS growth in different conditions and can be used as a limiting case of a Ga-rich droplet in the case of ternary Au–Ga–As alloys. This will be discussed in more detail in connection with the crystal phase modeling in Section 11. Here, we only note that reasonable fits to the data of Glas et al. (2013) can be obtained with different sets of $A$ and $\gamma_{\mathrm{SL}}$ (up to $\gamma_{\mathrm{SL}} \cong 0.2$) and that increasing $\gamma_{\mathrm{SL}}$ leads to the corresponding increase of the prefactor $A$ by the orders of magnitude. The obtained value for $\gamma_{\mathrm{SL}}$ in Ga droplets appears surprisingly low compared to $\gamma_{\mathrm{SL}} \sim 1\,\mathrm{J/m}^2$ used previously for Au–Ga–As droplets at ~20% Ga concentration (Dubrovskii et al., 2008a; Glas et al., 2007). If we use Eq. (79) for $A$ with the self-consistency renormalization included, the value of As diffusivity in Ga liquid would be $D_5 = 3 \times 10^{-12}\,\mathrm{m}^2/\mathrm{s}$. This is noticeably smaller than the values estimated for Ga-rich alloys (~$1$–$5 \times 10^{-10}\,\mathrm{m}^2/\mathrm{s}$ at temperatures around 600 °C) (Shawarz, 1980).

We now consider Ga-catalyzed growth of GaAs nanowires in a more general case where the amount of Ga in the droplet and therefore the droplet size changes with time. According to Eq. (19), the radius-independent elongation rate is still determined by the kinetics of As species: $dL/dt = C$, with $C$ independent of $R$. On the other hand, the total number of Ga atoms in the droplet $N_{\mathrm{Ga}}$ changes with time according to

$$\frac{dN_{Ga}}{dt} = \chi_{Ga} I_{Ga} \pi R^2 + 2 I_{Ga} \lambda_{Ga} R \sin\alpha_{Ga} - \frac{\pi R^2}{\Omega_{GaAs}} \frac{dL}{dt}. \tag{92}$$

Here, $I_{Ga}$ is the direct atomic flux of Ga atoms, $\chi_{Ga}$ is the geometrical function which depends on the contact angle of the droplet $\beta$ and the incident angle of the Ga beam $\alpha_{Ga}$ and equals $1/\sin^2\beta$ when $\beta \geq \alpha_{Ga} + \pi/2$ (Glas, 2010b), $R$ is the radius of cylindrical nanowire, $\lambda_{Ga}$ is the effective diffusion length of Ga adatoms on the nanowire sidewalls, and $\Omega_{GaAs} = 0.0452\,\mathrm{nm}^3$ is the volume per GaAs pair in the solid. The first term stands for the direct impingement of Ga, the second gives the diffusion-induced contribution of Ga adatoms from the length $\lambda_{Ga}$ beneath the droplet (in the simplest scenario of the diffusion-induced growth), and the third describes the sink due to nanowire elongation.

At a low As concentration, the droplet volume equals $V_{drop} = \Omega_{Ga} N_{Ga} = (\pi R^3/3) f(\beta)$, where $\Omega_{Ga} \cong 0.02\,\mathrm{nm}^3$ is the Ga atomic volume in liquid (Kodambaka et al., 2013) and $f(\beta) = (1 - \cos\beta)(2 + \cos\beta)/(1 + \cos\beta)\sin\beta$ is the geometrical function relating the volume of a spherical cap to the radius of its base. Assuming $\beta$ as being independent of radius and using Eq. (92), we arrive at

$$\frac{dR}{dt} = -A + \frac{B}{R}, \tag{93}$$

with

$$A = \frac{\Omega_{Ga}}{\Omega_{GaAs} f(\beta)} \left( \frac{dL}{dt} - \chi_{Ga} v_{Ga} \right); \quad B = \frac{2\Omega_{Ga}}{\pi \Omega_{GaAs} f(\beta)} v_{Ga} \lambda_{Ga} \sin\alpha \tag{94}$$

and $v_{Ga} = I_{Ga} \Omega_{GaAs}$ as the Ga influx in nm/s. Obviously, the behavior of the nanowire radius is very different at $A > 0$ and $A < 0$. When the effective Ga imbalance is positive ($A < 0$), i.e., more Ga atoms are brought from vapor than removed from the droplet to grow a nanowire, the droplet will inflate, possibly with the help of surface diffusion ($B$), and consequently, the nanowire will extend radially regardless of its initial dimension (the regime of radial growth). This situation corresponds to low As influxes which yield small elongation rates such that $dL/dt < \chi_{Ga} v_{Ga}$. Whenever $A > 0$, i.e., more Ga are removed from the droplet than brought from vapor, the situation becomes completely different. Additional diffusion flux of sidewall Ga adatoms will have a focusing effect on the diameter, i.e., small nanowires with $R < R_c$ will extend and large nanowires with $R > R_c$ shrink to reach the critical radius $R_c = B/A$ (the regime of radius self-equilibration).

Figure 21a shows the results of Priante et al. (2013) where the Ga-catalyzed VLS growth of GaAs nanowires was stopped and resumed at will by consuming the Ga droplet under As flux and then nucleating a secondary (or regrown) nanowire on top of the primary one under Ga-rich conditions. Primary GaAs nanowires were fabricated by standard solid-source MBE on Si-treated GaAs(111)B substrates. Self-catalyzed nanowire growth was carried out at 640 °C with a V/III BEP ratio of 5. The Ga flux was set to a BEP of $3.6 \times 10^{-7}$ Torr. This flux corresponds to a 2D growth rate of 16.8 nm/min on a GaAs(100) surface. The $As_4$ flux was therefore set to a BEP value of $1.8 \times 10^{-6}$ Torr. Growth was initiated by simultaneous exposure to Ga and $As_4$ fluxes. After 60 min, the Ga shutter was closed and the sample was kept for 10 min under As flux to obtain the complete consumption of the nanoparticle. After this step, the Ga shutter was opened again for different durations. Nanowire A in Fig. 21a represents the starting point of the secondary growth step: after the complete consumption of the Ga droplet, the tip appears flat and it is possible to notice a slight diameter reduction. This morphology is characteristic of the Ga droplet

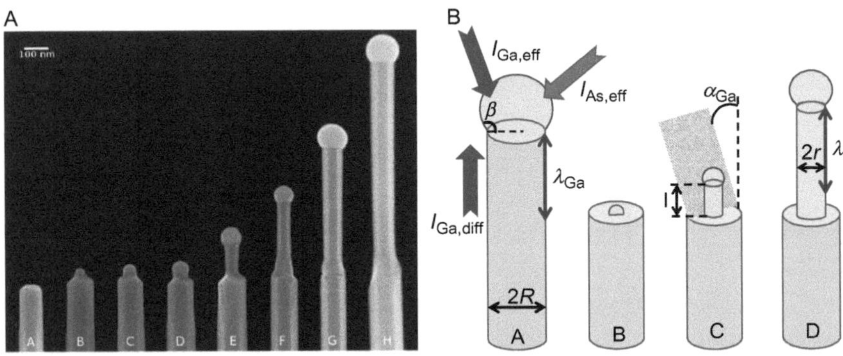


**Figure 21** (a) Collage of SEM images of the GaAs nanowires coming from different samples after the consumption of the Ga nanoparticle (A) and after exposure to Ga and As beams for 30 s (B–D), 60 s (E), 150 s (F), 300 s (G), and 420 s (H), respectively. The three different stages of nanoparticle formation coming from the same sample give an indication of the local fluctuations of the effective amount of material reaching the nanowire top. (b) Schematic of the nanowire growth process: regular growth stage with $R(t)$ increasing linearly with time due to a positive Ga imbalance arising from $I_{Ga,\,eff} + I_{Ga,\,diff} > I_{As,\,eff}$, with $I_{eff}$ denoting the effective incoming fluxes of Ga and As (A); nucleation of a Ga droplet on a flat top facet of the primary nanowire (B); fast radial growth of a short secondary nanowire (with the length $l < \lambda_{Ga}$) due to additional diffusion flux of Ga from the top surface of the primary nanowire (C); and regular growth of a long secondary nanowire with $l > \lambda_{Ga}$ (D).

consumption by exposure to As (Kim et al., 2012). After 30 s of exposure to Ga and As beams (wires B, C, and D), the morphology of the tip changes. A spherical cap droplet, with diameter smaller than that of the wire, becomes visible. This droplet is now sitting on top of a pyramid-like structure, which morphology resembles that of wires terminating with {110} facets. Wires B–D belong to the same sample, and the differences between them can be ascribed to local fluctuations of the effective beam fluxes. After 60 s (wire E), a new wire portion has grown and the droplet has already reached its equilibrium geometry, with a contact angle exceeding 110°. For increased deposition times (wire F after 150 s and G after 300 s), the wire continues to elongate. Nanowires B–F show two diameter discontinuities along their bodies: the first related to the consumption of the droplet, as in A, and the second due to the growth resumption with smaller diameter. By increasing the deposition time, the separation between the two discontinuities tends to decrease. After 420 s (wire H), only one discontinuity is observed, with an irregular shape.

The model schematized in Fig. 21b and described in more detail in Priante et al. (2013) is capable of explaining these trends by

**(i)** Description of the regular growth stage and in particular the observed linear length–time and radius–time dependences shown in Fig. 22. This stage is modeled by Eq. (93) with a negative $A$ and $B \rightarrow 0$. This is required to fit the radial extension at a time-independent rate; thus, the Ga influx imbalance should be positive and much larger than the diffusion-induced contribution.

**(ii)** Explanation of why Ga nanodroplet reappears on the wire top. The droplet formation under Ga-rich conditions is well understood intuitively; however, the preferential location of secondary droplets on the wire tops requires that the nucleation probability there is higher than on the sidewalls.

**(iii)** Modeling of the regrowth process and explanation of the observed nonlinear radius–time [$r(t)$] dependence shown in Fig. 22B, while the length–time correlation remains almost exactly identical to that at the regular growth stage (Fig. 22A). The observed nonlinearity of $r(t)$ is attributed to additional diffusion flux of Ga from the top surface of the primary nanowire which is effective as long as $l < \lambda_{Ga}$, as shown in Fig. 21b. After the length of secondary wire exceeds the Ga diffusion length, the radial growth retains to the regular conditions, i.e., a linear radius–time correlation.

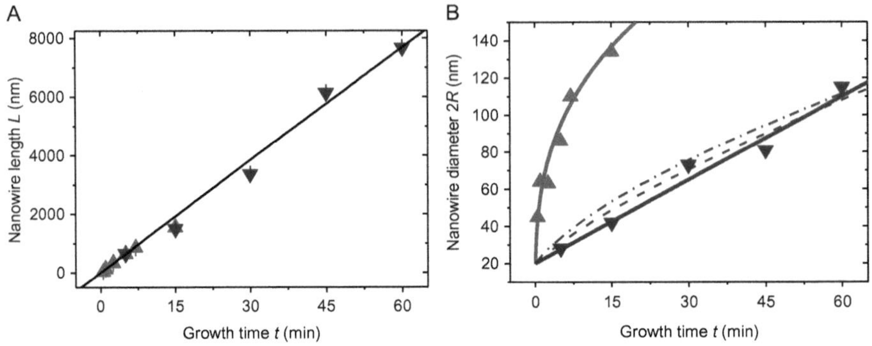


**Figure 22** (A) Regrowth length as measured from the interface (up triangles) and length of the regular Ga-catalyzed nanowires grown in the same conditions (down triangles) as a function of time. Solid line is the fit by $L=Ct$ with $C=130$ nm/min. (B) Diameters of the regrown nanowires (up triangles) and diameters of the Ga-catalyzed nanowires at the regular growth stage (down triangles) in the same conditions, as functions of time. Lines are theoretical fits: for regular nanowires, with use Eq. (93) at $A=-0.75$ nm/min, $B=0$ (bold solid line); $A=-0.43$ nm/min, $B=8.75$ nm$^2$/min (dashed line); and $A=-0.275$ nm/min, $B=15$ nm$^2$/min (dash-dotted line). It is seen how increasing $B$ leads to nonlinearity of the curves and worsens the correspondence. Regrown nanowire is modeled by a more complex equation that includes the diffusion flux of Ga from the top facet of the primary nanowire (Priante et al., 2013).

Let us now see what happens when $A>0$, the regime corresponding to the radius self-equilibration. The radius distribution of nanowires $f(R, t)$ at time $t$ obeys the following first-order equation (Dubrovskii, 2014a)

$$\frac{\partial f(R,t)}{\partial t}=-\frac{\partial}{\partial R}\left[\frac{\mathrm{d}R}{\mathrm{d}t}f(R,t)\right], \tag{95}$$

where the initial condition is determined by the size distribution of Ga droplets prior to growth. Introducing the scaled size $r=R/R_c$ and scaled time $x=t/\tau$, with $\tau=R_c/A=B/A^2$ as the characteristic relaxation time, solutions to Eqs. (93) and (95) can be put in the universal form with the minimized number of parameters. The time-dependent radius of an individual nanowire that has emerged from the droplet of radius $r_0=R_0/R_c$ at $t=0$ is given by

$$x=r_0-r+\ln\left(\frac{r_0-1}{r-1}\right), \tag{96}$$

In particular, this equation applies for the most representative nanowire radius $r_*(x)$ which corresponds to the most representative size of the initial droplets $r_*^0$ at $t=0$. The size distribution is obtained in the form

$$f(r, r_*) = \frac{cr}{|r-1|} g\left[r_* - r + \ln\left|\frac{r_*-1}{r-1}\right|\right], \tag{97}$$

where the time dependence $r_*(x)$ is determined by Eq. (96) with $r_0 = r_*^0$. The function $g(y)$ is determined by the initial size distribution of Ga droplets. The combination of variables in the argument of $g$ is the first integral of Eq. (95), while the prefactor $c$ ensures the correct normalization of the distribution.

Figure 23 shows the narrowing effect in the scaled variables for an ensemble of droplets described by the Gaussian distribution $g(y) = \left(1/\sqrt{\pi}\Delta r\right)\exp\left(-y^2/\Delta r^2\right)$ with the initial distribution width $\Delta r = 0.5$ and the initial mean size $r_*^0 = 1.5$. It is seen that this nonlinear system features quite unique behavior where the mean size self-equilibrates to $R_c$ and this process gradually decreases the effective width of the size distribution.

To observe the focusing effect on the size of Ga-catalyzed nanowires, one needs to fulfill the condition $dL/dt > \chi_{Ga}v_{Ga}$, which requires high enough As influx. Figure 24 shows the integrated histograms of diameters of the initial Ga droplets on patterned Si(111) substrates and the MBE-grown GaAs nanowires after 300 s of growth, where the narrowing of the diameter distributions is seen very clearly (Dubrovskii et al., 2015b).

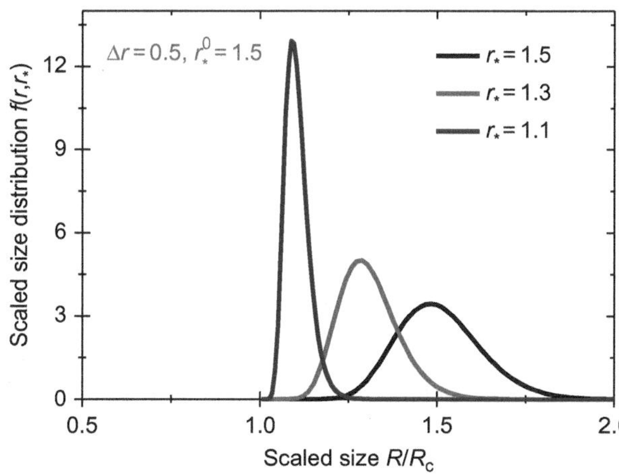


**Figure 23** Evolution of the initial Gaussian radius distribution around $r_*^0 = 1.5$ with the width $\Delta r = 1.5$ to a much narrower distribution as the mean size $r_*$ tends to one.

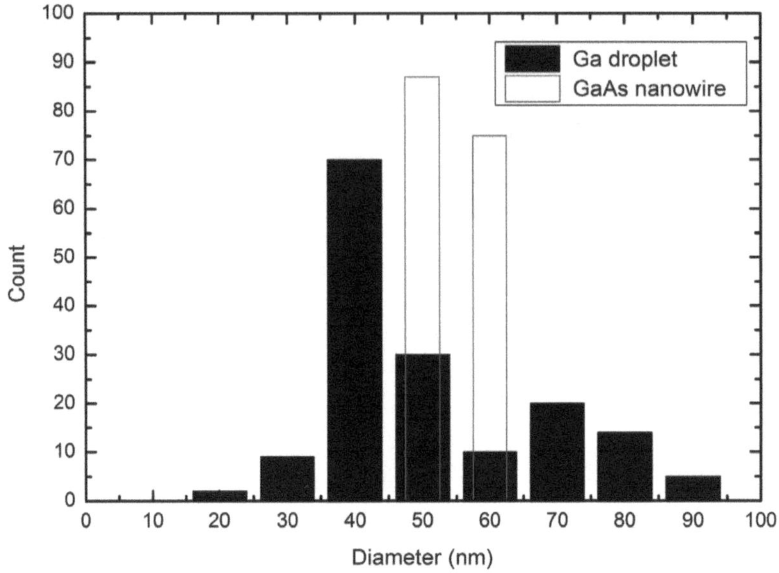


**Figure 24** Experimental histograms of Ga droplet diameters and GaAs nanowire top diameters after 300 s of MBE growth showing substantial narrowing effect with a stationary diameter between 50 and 60 nm (Dubrovskii et al., 2015b).

The growth parameters are described in detail in Plissard et al. (2011). In brief, these GaAs nanowires reach the mean length of 1 μm length after 300 s of growth (this value is the average height over all the data); therefore, the average elongation rate equals $dL/dt = 3.33\,\text{nm/s}$. The equivalent 2D growth rate $v \cos \alpha$ equals 0.326 nm/s, yielding $v_{Ga} = 0.360\,\text{nm/s}$ at $\alpha_{Ga} = 25°$. Taking the average value of the contact angle $\beta = 115°$, we obtain the Ga flux impinging the droplet at $\chi_{Ga} v_{Ga} = 0.438\,\text{nm/s}$. Therefore, the difference $dL/dt - \chi_{Ga} v_{Ga} = 2.89\,\text{nm/s}$ in this case is positive and large, which definitely favors the self-equilibration regime. With these plausible parameters, we obtain $A = 0.300$ nm/s. Using the value of $R_c = D_c/2$ of 25 nm (since the critical top diameter $D_c$ is close to 50 nm from Fig. 24), we arrive at $B = 7.50$. From Eq. (94), this yields a reasonable estimate for the effective diffusion length of Ga adatoms at the nanowire sidewalls of $\lambda_{Ga} = 750\,\text{nm}$. Finally, the relaxation time $\tau$ equals 83 s, which is shorter than the total growth time of 300 s. The distribution of the nanowire top diameters is therefore expected to be substantially narrowed after stopping the growth.

Figure 25 shows the plots of diameters $D = 2R$ versus time obtained from Eq. (5) with the above parameters and at different initial diameters from

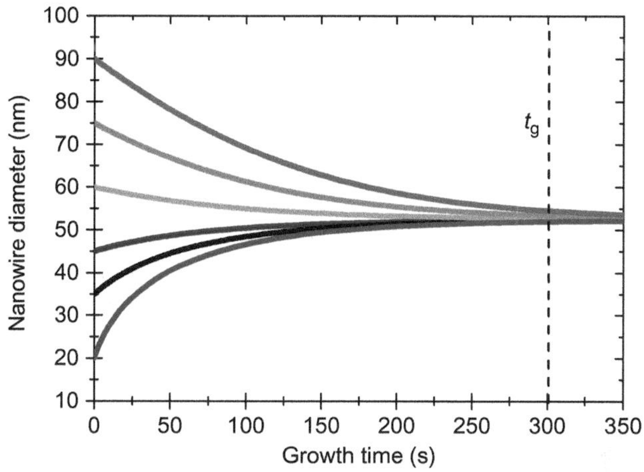


**Figure 25** Theoretical curves obtained from Eq. (96) for different initial diameters between 20 and 90 nm. Self-equilibration of the top nanowire diameters leads to narrowing the diameter range from 20–90 nm at the beginning of growth to only less than 50–60 nm after 300 s of growth, as observed in the experimental histograms presented in Fig. 24.

20 to 90 nm. The effect of self-equilibration is clearly seen and correlates with the experimental histogram shown in Fig. 24—at $t_g = 300$s, the size distribution is narrowed from 20–90 to ~50–65 nm, while thinner and thicker nanowires are no longer present in the histograms.

The effect of diameter self-equilibration will transform an arbitrary broad diameter distribution of the initial Ga droplets to the asymptotically uniform distribution provided that the growth time is long enough to allow for the complete size relaxation. After reaching this size-uniform state, the nanowires will continue to elongate with stationary diameter and at almost diameter-independent rate which yields also the uniform length. This explains why the ensembles of Ga-catalyzed GaAs nanowires usually have more uniform dimensions compared to Au-catalyzed ones (Cirlin et al., 2010; Colombo et al., 2008; Dubrovskii et al., 2011; Plissard et al., 2011; Ramdani et al., 2013). In this sense, the diameter self-equilibration in self-catalyzed III–V nanowires resembles the height equilibration of self-induced GaN nanowires (Sabelfeld et al., 2013). This effect is caused by completely different physical reasons but also leads to stable asymptotic size uniformity, while narrowing the length distribution of Ge nanowires described in Dubrovskii et al. (2012a) could be reached only at a certain moment of time.

## 10. FORMATION OF TERNARY AU-CATALYZED III–V NANOWIRES

To this end, the vast majority of device structures are based upon binary III–V nanowires. Ternary nanowires are interesting as they allow for the desired bandgap tunability by changing the growth conditions (Ameruddin, 2015). Improving our fundamental understanding of ternary nanowire growth and composition evolution as a function of directly accessible experimental parameters would therefore be crucial to achieve better control over the device structures. Despite this demand, true understanding and modeling of ternary III–V nanowires obtained by the VLS mechanism with Au catalysts is solely lacking in the literature. This is due to several issues among which we mention (i) unknown equilibrium phase diagrams and nonequilibrium chemical potentials of quaternary alloys such as the Au–Ga–In–As system considered below; (ii) uncertainties in determining the composition of the initial critical nucleus even if both droplet and solid compositions were exactly known; (iii) absence of any data on relevant surface energies of different interfaces; and (iv) lack of reliable data on the kinetic parameters (diffusion lengths, desorption rates of arsenic, etc.) as functions of the compositions. Therefore, the existing approach based on chemical potentials, surface energies, and the Zeldovich nucleation rate for macroscopic islands seems to be out of reach for ternary nanowires or requires some crude assumptions which cannot be easily checked against experimental data.

Recently, the growth rates and compositions of Au-seeded VLS InGaAs nanowires obtained by MOVPE under different conditions were systematically studied and modeled within a new approach that relies upon the irreversible atomistic growth picture (Ameruddin, 2015). InGaAs nanowires were fabricated with TMGa, TMIn, and $AsH_3$ as metal and arsenic precursors on GaAs (111)B substrates with size-uniform Au droplets. The diameters of the initial Au droplets were varied from 10 to 80 nm. The nanowire growths were conducted at growth temperatures ranging from 400 to 525 °C. V/III ratios were also varied ranging from 1.4 to 86.0. In different series of experiments, the main parameters which are expected to determine the nanowire growth rates and compositions are the droplet diameter, the V/III flow ratio in vapor, and the growth temperature. Consequently, Fig. 26 shows the nanowire axial growth rate $dL/dt$ (nm/min) and In percentage in the solid state $x_{In}$ (%) as functions of the droplet diameter $d = 2R_d$

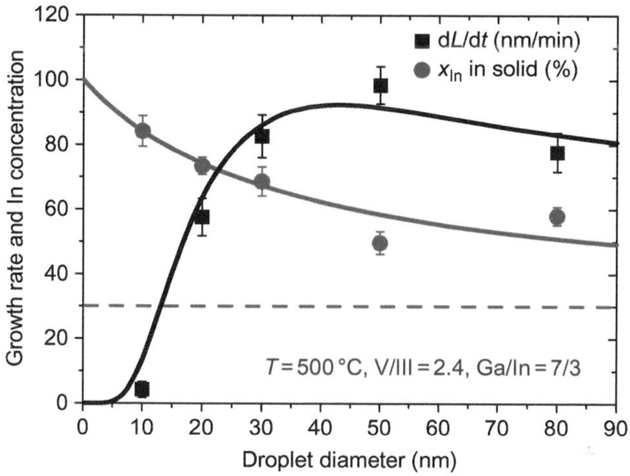


**Figure 26** Nanowire composition and axial growth rate versus droplet diameter at fixed $T = 500$ °C, V/III ratio $= 2.4$, and nominal Ga/In $= 7/3$ in vapor (symbols), fitted by the model expressions (lines).

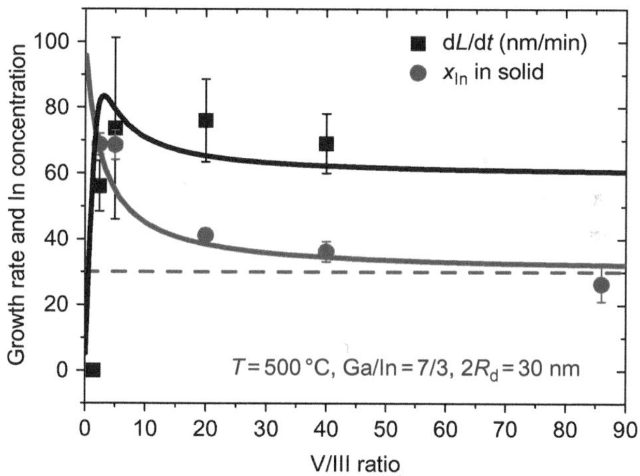


**Figure 27** Nanowire composition and axial growth rate versus droplet diameter at fixed $T = 500$ °C, $2R_d = 30$nm, and nominal Ga/In $= 7/3$ in vapor (symbols), fitted by the model expressions (lines).

(which equals the diameter of the NW top) at the fixed temperature $T = 500$ °C, V/III ratio $= 2.4$, and TMGa to TMIn flow ratio Ga/In $= 7/3$, along with the corresponding SEM images of differently sized nanowires and TEM/EDX images. Figure 27 shows the $dL/dt$ and $x_{In}$ versus V/III ratio at the fixed $T = 500$ °C, Ga/In $= 7/3$, and $d = 30$ nm. Figure 28 shows $dL/dt$

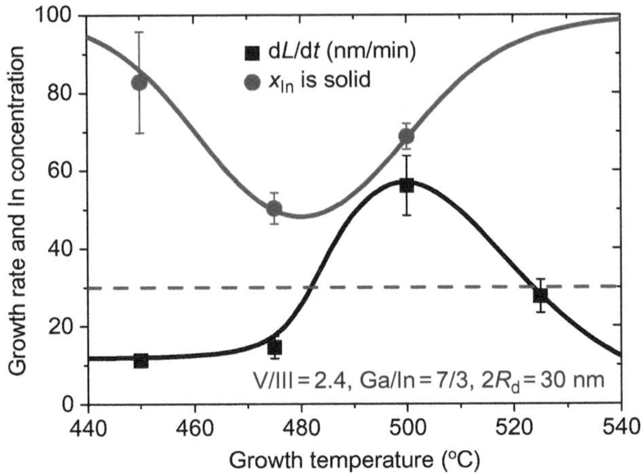


**Figure 28** Axial growth rate versus growth temperature at fixed $2R_d = 30$ nm, nominal Ga/In = 7/3 in vapor, and two different V/III flux ratios (symbols), fitted by the model expressions (lines).

and $x_{In}$ as a function of the growth temperature at the fixed V/III ratio = 2.4, Ga/In = 7/3, and $d = 30$ nm.

It is seen that the observed behaviors are complex and nonlinear, while the composition of the nanowire is distinctly different from that in vapor. In particular, the In concentration in solid gradually decreases with increasing the droplet size, while the axial growth rate is a nonmonotonic function of the droplet size with a maximum (Fig. 26). Figure 27 reveals a monotonically decreasing dependence of the In composition in solid on the V/III ratio, while the growth rate behavior is again nonmonotonic. Figure 28 shows nonmonotonic temperature dependences of both $x_{In}$ and $dL/dt$.

To understand these findings, we propose the following model of the ternary nanowire growth. First and most important, we assume that relative atomic concentrations of In and Ga are the same in the liquid and solid state, i.e., the nanowire composition is given by

$$x_{In} = \frac{c_{In}}{c_{In} + c_{Ga}}; \quad x_{Ga} = \frac{c_{Ga}}{c_{In} + c_{Ga}}, \tag{98}$$

where $c_{In}$ and $c_{Ga}$ are the atomic concentrations of In and Ga atoms in the droplet. This implies a transport-limited regime of nanowire growth rather than a nucleation-limited growth mode in which the liquid and solid composition can be very different for many reasons (preferential nucleation and/or 2D growth of one compound versus the other, nonstoichiometric nuclei, etc.) (Dick et al., 2008; Guo et al., 2013; Johansson et al., 2009). While the nucleation-limited regime should be described by the Zeldovich nucleation rate, some recent studies revealed that we really approach the limit of macroscopic approach in such modeling even for binaries. For instance, the critical size of Au-catalyzed GaAs nanowires grown by HVPE approaches 1 pair (Gil et al., 2014) and equals only 2 pairs in Ga-catalyzed GaAs nanowires (Glas et al., 2013).

In order to eliminate most of these uncertainties of macroscopic approach and in line with the first assumption on the uniform composition, we write the nucleation rate in the form $J = \sigma D_{\mathrm{As}} n_{\mathrm{As}}^{\mathrm{L}} \left(n_{\mathrm{In}}^{\mathrm{L}} + n_{\mathrm{Ga}}^{\mathrm{L}}\right)$, as in irreversible growth models (Dubrovskii and Sibirev, 2014). As above, $D_{\mathrm{As}}$ is the As diffusion coefficient in liquid, $n_{\mathrm{k}}^{\mathrm{L}}$ are the surface concentrations of atoms of species k (k = As, In, Ga) at the liquid–solid interface, and $\sigma$ is the capture rate which is assumed identical for k = In, Ga. Otherwise, the nanowire composition would differ from the liquid one due to different dimerization probabilities. Assuming also a spatially homogeneous composition in the droplet due to its small dimension and high diffusivities at the typical growth temperatures, we can write $n_{\mathrm{k}}^{\mathrm{L}} = (h/\Omega)c_{\mathrm{k}}$, with $h$ as the height of a ML and $\Omega$ as the volume per III–V pair in the solid state. More details regarding the transformation from macroscopic nucleation to atomistic irreversible growth in Au-catalyzed III–V nanowires are given in Section 11. Using our atomistic growth picture, we can thus write the second essential assumption of our model in the form

$$J = \sigma D_{\mathrm{As}} \left(\frac{h}{\Omega}\right)^2 c_{\mathrm{As}}(c_{\mathrm{In}} + c_{\mathrm{Ga}}). \tag{99}$$

This nucleation rate gives the probability of nucleation per unit surface area. Hence, the nanowire elongation rate is given by

$$\frac{\mathrm{d}L}{\mathrm{d}t} = \pi R_{\mathrm{d}}^2 hJ. \tag{100}$$

For the diffusion transport of both group III species, we adopt the following simple expressions:

$$j_{\mathrm{k}} = 2V_{\mathrm{k}}\left[1 + \frac{\lambda_{\mathrm{k}}}{R_{\mathrm{d}}}\left(1 - \frac{c_{\mathrm{k}}}{c_{\mathrm{k}}^{\mathrm{A}}}\right)\right] \tag{101}$$

for k = In, Ga, written for hemispherical droplet with the radius $R_{\mathrm{d}}$, which equals the top nanowire radius. Here, $V_{\mathrm{k}}$ are the atomic vapor influxes of group III species, $\lambda_k$ are the effective diffusion lengths of the group III adatoms on the sidewalls (both accounting for the temperature-dependent cracking efficiencies of TMIn and TMGa at different surfaces (Stringfellow, 2001) and influenced by As flux), $D_k$ are the diffusion coefficients of In and Ga adatoms, $\tau_k$ are their effective lifetimes (limited by radial growth), and $c_k^A = V_k \tau_k / h$ are the effective adatom activities on the sidewalls. The first terms in Eq. (101) describe the direct vapor fluxes and the second stand for surface diffusion of In and Ga adatoms. Desorption of both group III species are neglected.

The droplet can maintain a time-independent radius $R_d$ only when the material influxes of group III atoms into the droplet equal their sink due to nanowire growth, i.e.,

$$x_k \frac{dL}{dt} = j_k \tag{102}$$

for $k = \text{In}, \text{Ga}$. Summing up two Eq. (102) for In and Ga, we obtain: $dL/dt = j_{In} + j_{Ga}$, which gives the transport-limited elongation rate. Neglecting surface diffusion of As species and accounting for its desorption from the droplet, the As-limited elongation rate in the steady state is given by (Kelrich et al., 2015):

$$\frac{dL}{dt} = 2\left[V_{As}(1+\varepsilon) - \nu_{As}^{des} \exp\left(R_{GT}/R\right) c_{As}\right], \tag{103}$$

where $V_{As}$ is the direct atomic flux of As and $\varepsilon$ accounts for possible reemitted flux of As species. Desorption of As atoms from the droplet is assumed as being proportional to the As concentration with a certain temperature-dependent coefficient $\nu_{As}^{des}$. Additionally, we take into account the GT effect (Dubrovskii et al., 2009a) which exponentially increases arsenic desorption from small droplets due to the curvature effect, with a certain characteristic GT radius $R_{GT}$.

Solution to the above system can be obtained in simple analytical form if $D_{Ga}/\lambda_{Ga} \cong D_{In}/\lambda_{In}$. In this case, the composition is not influenced by the reverse diffusion fluxes from the droplet [the $-c_k/c_k^A$ terms in Eq. (101)] and equals

$$\frac{c_{Ga}}{c_{In}} = \frac{V_{Ga}\left(R_d + \lambda_{Ga}\right)}{V_{In}\left(R_d + \lambda_{In}\right)}. \tag{104}$$

The nanowire axial growth rate is obtained in the form

$$\frac{dL}{dt} = \frac{\gamma}{(2a+\gamma)} 2\tilde{V}_{As}; \ \gamma = \sqrt{(1+a-b)^2 + 4ab} - (1+a-b). \tag{105}$$

The coefficients are given by

$$a = \left(\frac{R_0}{R_d}\right)^3 \exp\left(\frac{R_{GT}}{R}\right); \ R_0^3 \equiv \frac{2v_{As}^{des}}{\tilde{V}_{As}} \frac{D_{In}}{\pi \lambda_{In} D_{As} \sigma} \left(\frac{\Omega}{h}\right)^2;$$
$$b = \frac{\tilde{V}_{In} + \tilde{V}_{Ga}}{\tilde{V}_{As}}, \tag{106}$$

where $\tilde{V}_{As} = V_{As}(1+\varepsilon)$, $\tilde{V}_{In} = V_{In}(1+\lambda_{In}/R_d)$, and $\tilde{V}_{Ga} = V_{Ga}(1+\lambda_{Ga}/R_d)$ are the total influxes of As, In, and Ga atoms into the droplet, respectively. Same model can also help to access the effect of nanowire tapering and shape, but this is not elaborated here.

We now analyze the obtained expressions and compare them with the experimental data. From Eq. (104), the nanowire composition in this model can be different from the vapor composition for two reasons: (i) different diffusion lengths of In and Ga adatoms and (ii) different cracking efficiencies of TMIn and TMGa at a given temperature. We remind that $V_{Ga}/V_{In}$ gives the ratio of group III influxes entering the droplet but not of the precursor fluxes impinging approaching the droplet. Higher In composition in nanowires than in vapor requires either $\lambda_{Ga} < \lambda_{In}$ or $V_{Ga} < V_{In}$. The latter inequality is supported by the data of Stringfellow (2001) showing that TMIn is a low-temperature precursor, while the decomposition of TMGa at the solid surface starts only at above 400 °C and might be very sensitive to the group V flux. From Eq. (104), the In composition is expected to gradually decrease with the droplet size due to lower diffusion flux, and reach the influx ratio at $R \to \infty$. From Eqs. (105) and (106), the axial growth rate must first increase with $R$ as $R^2$, then reach a maximum, and further decrease to a constant, as in most growth models for binary nanowires. From Eq. (104), the In content decreases for higher V/III ratios due to a lower diffusivities of adatoms, while the growth rate should feature a nonmonotonic behavior with the III/V ratio.

Eventually, rising the V/III ratio has the same effect on the $dL/dt$ as increasing the droplet size. For low V/III ratios and small $R_d$, growth and composition are limited by the As transport in the excess of group III atoms (where In should reach very high concentrations as a faster diffuser). As the droplet size or V/III ratio increases, the VLS growth is transformed to a group-III limited transport mode such that the growth rate decreases with $R_d$ and V/III and saturates to a constant.

Finally, the temperature dependences of the $dL/dt$ and $x_{In}$ should be nonmonotonic. At low temperatures, the cracking efficiency of TMGa and the diffusion length of In adatoms tend to zero, which suppresses the axial growth rate, while at high temperatures it is limited by the As desorption. The In content in solid should tend to one at low temperatures due to a low cracking efficiency of TMGa and at high temperatures due to enhanced thermally activated diffusion of In adatoms.

These trends correspond to the experimental data, with Figs. 26–28 showing a good quantitative correlation. The fits in Figs. 26 and 27 are obtained at $\lambda_{Ga}=0$ even at the lowest V/III ratio of 2.4. Since the adatom diffusivity is expected to decrease toward higher group V fluxes, the Ga diffusion is considered ineffective in all fits. The best fits are obtained at $\lambda_{In}=57$ nm, $R_0=7$ nm, and $R_{GT}=9$ nm under the conditions of 500 °C and V/III ratio = 2.4. We also assume that the cracking efficiencies for TMIn and TMGa equal one at 500 °C, and thus, the value of $V_{Ga}/V_{In}=7/3$ is given by the vapor composition. No reemission of As is required to obtain the fits ($\varepsilon=0$) and thus $\widetilde{V}_{As}=V_{As}$, $V_{Ga}=0.292\times V_{As}$, and $V_{In}=0.125\times V_{As}$. We use the value of $V_{As}=71$ nm/min for the overall normalization of the growth rates. To fit the V/III ratio dependences of $x_{In}$ and $dL/dt$, we assume that the diffusion length of In scales with the As flux as $\lambda_{In}\propto 1/V_{As}$, i.e., decreases inversely proportional to the V/III flux ratio.

Experimental temperature behaviors of $dL/dt$ and $x_{In}$ as well as their fits shown in Fig. 28 are more complex and require some additional parameters. We use the Arrhenius temperature dependences of the In diffusion length and the coefficient $a$ in Eq. (106) in the form $\lambda_{In}\propto \exp(-T_\lambda/T)$ and $a\propto \exp(-T_{des}/T)$ (assuming that the temperature dependence of $A$ is dominated by the enhanced desorption of As through $\nu_{As}^{des}$), with rather large $T_\lambda=60{,}000$K and $T_{des}=105{,}000$K. The observed decrease of the growth rate toward lower temperatures can be attributed only to low cracking efficiencies of TMGa at low temperatures. For the latter, we use the temperature dependence of Bouravleuv et al. (2010): $V_{Ga}\propto (1/2)[1+\tanh((T-T_{Ga})/\Delta T_{Ga})]$ with $T_{Ga}$ around 480 °C and very narrow $\Delta T_{Ga}$ of the order of 10 °C.

## 11. IMPACT OF GROWTH CONDITIONS ON THE CRYSTAL STRUCTURE OF III–V NANOWIRES

It is generally accepted that the surprising effect of the WZ phase formation in III–V nanowires of cubic ZB materials is closely connected with a lower surface energy of relevant WZ sidewalls (compared to their ZB counterparts) which can more than compensate the bulk energy difference between the WZ and ZB structures. This assumption is supported by different calculations of the surface energies (see Dubrovskii, 2014a for a review). Even more interestingly, the crystal structure of Au-catalyzed III–V nanowires can be tuned by growth conditions such as the growth temperature and V/III ratio. Recently, well-controlled formation of the WZ layers in III–V nanowires resulted in fabrication of WZ/ZB heterostructures with extremely abrupt interfaces (Akopian et al., 2010; Dick et al., 2010; Spirkoska et al., 2009).

Theoretical understanding of ZB–WZ polytypism in Au-catalyzed III–V nanowires as well as developing the methods to control it started from the work of Glas et al. (2007). In this paper, based on the experimental findings on MBE-grown, Au-catalyzed III–V nanowires, the WZ phase formation was explained by the TPL mononucleation of WZ 2D nuclei at high chemical potentials in the droplet, while low chemical potentials were related to the ZB structure. As discussed above, high chemical potentials correspond to low growth temperatures and high As fluxes; therefore, the WZ phase should be expected at low temperatures and high V/III ratios, while the ZB structure forms at high temperatures and low V/III ratios. However, this tendency is observed only in MBE and related growth techniques. Furthermore, several groups have reported pure ZB structure of Au-catalyzed GaAs nanowires obtained by MOCVD or HVPE at rapid elongation rates and high group V fluxes (Gil et al., 2014; Joyce et al., 2009; Ren et al., 2011), definitely relating to high chemical potentials. Finally, the crystal structure of Ga-catalyzed GaAs nanowires is almost pure ZB (Cirlin et al., 2010; Dubrovskii et al., 2011; Priante et al., 2013).

The initial structural models of III–V nanowires are well known and have been used for a long time for controlling the crystal phase by the growth parameter tuning (Dubrovskii, 2014a). Consequently, here we present a development of this approach which allows one to describe nonmonotonic behaviors of the crystal structure versus different parameters as well as to understand the threshold between the macroscopic mono- or polynucleation and atomistic VLS growths.

As mentioned above, one useful method for switching between the two phases is modulation of the group V flow, but here different authors report contradictory results. In the case of MBE or CBE with modest material inputs, high group V fluxes usually favor the WZ structure (Dheeraj et al., 2013; Fakhr et al., 2010; Husanu et al., 2014; Kelrich et al., 2015; Krogstrup et al., 2012). In MOCVD, this tendency is inverted and higher group V flows yield the ZB structure (Assali et al., 2013; Lehmann et al., 2012; Lehmann et al., 2013), suggesting a nonmonotonic behavior over a wider range of conditions. Switching from the WZ to ZB phase at high group V flows is not within the initial structural model (Glas et al., 2007). This approach is based on macroscopic description of the position-dependent nucleation of 2D islands with well-defined boundaries. Modification of the nucleation approach to include the known effect of suppression of the TPL nucleation in Ga-rich droplets suggests the existence of the so-called WZ corner where the WZ structure is preferred only within a range of Ga concentrations (Dubrovskii and Grechenkov, 2015). However, in these considerations we often reach the limit of macroscopic model because the critical size of emerging 2D islands approaches one. Of course, the macroscopic nucleation theory can hardly work in such cases. Rather, the VLS growth should acquire an irreversible atomistic character in which meeting of any two Ga and As atoms at the liquid–solid interface produces a stable nucleus.

Furthermore, it is commonly believed that the VLS growth proceeds in the so-called mononuclear regime with only one 2D island emerging in each NW monolayer and then meeting no other islands until it fills the whole top facet, as discussed in Section 7. This view is supported, e.g., by a self-regulated oscillatory behavior of supersaturation and the growth interface (Glas et al., 2010; Wen et al., 2011), which is not expected to persist without mononucleation. On the other hand, purely atomistic growth is polynuclear by its definition. Therefore, there is a gap in our understanding and modeling the VLS growth of III–V nanowires and especially the crystal structure when going beyond the standard assumptions such as macroscopic islands or mononuclear growth.

To fill this gap, we consider in more detail the self-consistency renormalization of the Zeldovich nucleation rate which was introduced in Section 6. The very first question which should be resolved in macroscopic modeling of 2D nucleation is the formation mechanism of a bound immobile Ga–As–Ga–As dimer. We argue that macroscopic nucleation requires such a dimer to be formed from two unbound atoms of Ga and As dissolved in liquid. This process should be limited by meeting two As atoms because arsenic is a much more diluted and slower diffuser (André et al., 2014). When instead it becomes more probable to form a stable Ga–As solid monomer, the next As atom will form the second Ga–As monomer rather than attach to the first dimer and hence we arrive at the atomistic growth picture. In the nucleation regime with $i_c \gg 1$, the macroscopic formation energy should therefore be renormalized as described by the modified Zeldovich equation:

$$J = \frac{3^{3/4}}{\sqrt{\pi}} \left( \frac{h}{\Omega_{35}} \right)^2 D_5 c_5 e^{\overline{\mu}^S_{35}} \Delta\overline{\mu}^{1/2} \exp\left( a - \frac{a^2}{4\Delta\overline{\mu}} \right). \tag{107}$$

On the other hand, Eq. (71) yields the following expression for the lateral growth rate in terms of a linear island size $r$:

$$v \equiv \frac{dr}{dt} = 2\sqrt{3} D_5 \left( \frac{h}{\Omega_{35}} \right)^{1/2} e^{\overline{\mu}^S_{35}} \left[ e^{\Delta\overline{\mu}} - 1 \right]. \tag{108}$$

For the C nucleation which is enabled on the whole facet, the mononuclear growth occurs when the time interval between the two successive nucleation events, $t_n = 1/\left(R^2 J_C\right)$, is much longer than the time required for an island to fill the whole facet of size $R$, $t_g = R/v$ (Kashchiev, 2000). This naturally leads to the dimensionless control parameter $\varepsilon_C = t_g/t_n = R^3 J_C/v$, with mononuclear growth occurring at $\varepsilon_C \ll 1$ and polynuclear at $\varepsilon_C \gg 1$. For arbitrary $\varepsilon_C$, the nanowire elongation rate is given by the Kashchiev formula (Kashchiev, 2006)

$$\left( \frac{dL}{dt} \right)_C = \frac{bhR^2 J_C}{1 + b\varepsilon_C^{2/3}}, \tag{109}$$

where $b = 3\sqrt{3}/2$ for hexahedral nanowire. This growth rate equals $bR^2 J_C$ at $\varepsilon_C \ll 1$ (mononuclear regime) and $(J_C v^2)^{1/3}$ at $\varepsilon_C \gg 1$ (polynuclear regime). The latter expression shows that the time $t_*$ required for filling the facet in polynuclear regime is limited by merging the islands with a time-linear surface density $Jt$ and hence the $t_*$ is obtained from $vt_* = 1/\left(Jt_*\right)^{1/2}$.

For the TPL nucleation, the situation is different because stable TPL islands can emerge only within the TPL ring of width $2r_c$. On the other hand, even if many islands emerge simultaneously within the TPL ring, the distance between them and the nanowire center remains $R$. Thus, we modify Eq. (109) to

$$\left( \frac{dL}{dt} \right)_{TPL} = \frac{bh\eta R^2 J_{TPL}}{1 + b\varepsilon_{TPL}}, \tag{110}$$

in which $\varepsilon_{TPL} = t_g/t_n = \eta R^3 J_{TPL}/v$ and $\eta = 4r_c/R$. This growth rate equals $b\eta R^2 J_{TPL} = 4bRr_c J_{TPL}$ at $\varepsilon_{TPL} \ll 1$ (mononuclear regime) and $v/R$ at $\varepsilon_{TPL} \gg 1$ (polynuclear regime).

From Eqs. (107) and (108), we get the expression for the polynucleation threshold

$$\varepsilon = \eta \frac{R^3 J}{v} \cong \frac{3^{1/4}}{2\sqrt{\pi}} c_5 \Delta\bar{\mu}^{1/2} e^{-F_c} \eta \left(\frac{R}{r_0}\right)^3, \qquad (111)$$

with $r_0 = (\Omega_{35}/h)^{1/2}$ as the elementary distance on the solid surface. With logarithmic accuracy, the main terms competing with each other in Eq. (111) are a very small $\exp(-F_c)$ and a very large $(R/r_0)^3$ ($r_0 = 0.37\,\mathrm{nm}$ for GaAs). The mono-to-polynuclear growth transformation occurs at $\varepsilon = 1$ corresponding to $F_c \cong \ln\left[0.37 c_5 \Delta\bar{\mu}^{1/2} \eta (R/r_0)^3\right] \gg 1$ where macroscopic theory is still valid.

We now consider how the present approach modifies our understanding of polytypism of Au-catalyzed III–V nanowires. It seems logical to assume that WZ nucleus can only emerge in the TPL position and in mononuclear regime or in the TPL polynuclear regime for the WZ structure but not for the ZB one. In other words, we imply that WZ nanowires form when the emerging islands all have the WZ orientation, while the ZB phase wins whenever there is a possibility for merging the islands with different structures (or orientations). This can be well understood when, for example, there is only one WZ island having emerged at the corner which is blocked by many ZB islands that nucleate concomitantly with it in the center. Such a suppression mechanism for the WZ structure is somewhat similar to the WZ-to-ZB phase transition induced by epitaxial burying where the WZ GaAs core retains the ZB structure of the AlGaAs shell (Patriarche et al., 2008). More complex scenarios such as crystal defects or formation of higher order polytypes are worth studying but not implemented here.

Figure 29 shows the WZ corner of Dubrovskii and Grechenkov (2015) (the TPL-TPL and TPL-C lines) for 30-nm radius Au-catalyzed GaAs nanowires grown at 550 °C from a Ga precursor and $As_2$ vapor, obtained from the modified Glas condition (Glas et al., 2007)

$$\Delta\mu > \Delta\mu_c = \frac{\psi}{1 - (a_{WZ}/a_{ZB})^2} = \frac{\psi}{1 - \left(\gamma_{\mathrm{eff}}^{WZ}/\gamma_{\mathrm{eff}}^{ZB}\right)^2}. \qquad (112)$$

The Glas condition shows that the WZ structure forms when the surface energy of WZ island is lower than that of ZB one, and requires a high enough chemical potential (larger than the critical value $\Delta\mu_c$) to overcome the bulk energy difference between the WZ and ZB structures $\psi$. The WZ

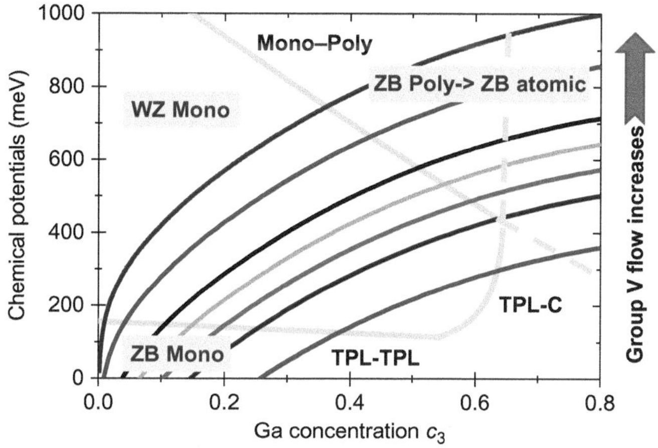


**Figure 29** "Wurtzite corner" (bold yellow (light gray in the print version) lines) for Au-catalyzed GaAs nanowires of Dubrovskii and Grechenkov (2015), restricted by the upper mono-to-polynuclear line for the C nucleation obtained from Eq. (111) at $\varepsilon = 1$ and $R = 30$ nm. Chemical potentials rising with increasing $G$ from −5 (green (gray in the print version) line) to 4 (blue (dark gray in the print version) line) correspond to increasing As flux at a fixed temperature of 550 °C. The WZ phase formation is possible only inside the reduced WZ corner via the mononuclear TPL nucleation. The WZ phase region is separated from the (ZB,TPL) mononuclear scenario by the Glas line at the bottom (low chemical potentials) (Glas et al., 2007), by the (WZ,TPL)-to-(ZB,C) line on the right (high Ga concentrations, hence low surface energies that suppress the TPL nucleation) (Dubrovskii and Grechenkov, 2015; Gil et al., 2014), and by the mono-to-polynucleation transition at the top. Very soon after acquiring polynuclear regime, the VLS growth becomes atomistic and no longer requires any nucleation, which is the most probable growth scenario of ZB nanowires at high chemical potentials.

corner is restricted by the upper Mono–Poly line relating to the mono-to-polynuclear transformation for C nucleation ($\varepsilon_C = 1$). For the TPL nucleation position, we use Eq. (81): $\gamma_{\mathrm{eff}}^{\mathrm{ZB}} = (\gamma_{\mathrm{ZB}} - \gamma_{\mathrm{LV}} \sin\beta)x + \gamma_{\mathrm{SL}}(1-x)$; $\gamma_{\mathrm{eff}}^{\mathrm{WZ}} = (\gamma_{\mathrm{WZ}} - \gamma_{\mathrm{LV}} \sin\beta)x + \gamma_{\mathrm{SL}}(1-x)$, while for the (ZB,C) nucleation scenario $\gamma_{\mathrm{eff}}^{\mathrm{ZB}} = \gamma_{\mathrm{SL}}$. The $\gamma_{\mathrm{ZB}}$, $\gamma_{\mathrm{WZ}}$ are the surface energies of relevant ZB and WZ nanowire sidewall facets, $\gamma_{\mathrm{LV}}$ is the liquid–vapor surface energy of the droplet, $\gamma_{\mathrm{SL}}$ is the surface energy of the lateral solid–liquid interface of a 2D island, $\beta$ is the contact angle of the droplet at the nanowire top, and $x = 1/3$ is the fraction of the island perimeter at the TPL.

The diagram in Fig. 29 was obtained for the following parameter values (Dubrovskii and Grechenkov, 2015): $\psi = 24\,\mathrm{meV/pair}$, $\gamma_{\mathrm{WZ}} = 1.30\,\mathrm{J/m^2}$ for the lowest energy $(1\bar{1}00)$ WZ side facets and $\gamma_{\mathrm{ZB}} = 1.543\,\mathrm{J/m^2}$ for the lowest energy (110) ZB side facets, and $\beta = 110°$. The $\gamma_{\mathrm{LV}}(c_3)$

dependence is taken as the linear approximation between pure liquids ($\gamma_{Ga}=0.675\text{J/m}^2$; $\gamma_{Au}=1.248\text{J/m}^2$ at 550 °C); however, the temperature dependences of these surface energies are almost negligible. The $\gamma_{SL}(c_3)$ is taken as the linear approximation between $1.0\,\text{J/m}^2$ at $c_3=0.2$ and $0.123\,\text{J/m}^2$ for a pure Ga droplet according to Eq. (91). The chemical potential of As atoms in liquid is expressed through the effective group V to III imbalance $G$, yielding the liquid–solid chemical potential difference as in Eq. (23). The chemical potentials of liquid Ga are obtained based on the Glas model described in Section 3.

More precisely, the actual Ga concentration during growth under a given As flux can be obtained from the following equations:

$$\frac{\mathrm{d}L}{\mathrm{d}t}=hbR^2\times K\Delta\overline{\mu}^{1/2}B, \tag{113}$$

$$B=\sum_{k=1}^{3}B_k=\sum_{k=1}^{3}\frac{\eta_k\exp\left[a_k-a_k^2/4(\Delta\overline{\mu}-\psi_k)\right]}{1+b\varepsilon_k^{q_k}}, \tag{114}$$

with a kinetic factor $K=\left(3^{3/4}/\sqrt{\pi}\right)c_5D_5r_0^{-4}\exp\left(\overline{\mu}_{35}^{S}\right)$. Here, we sum up the contributions to the overall growth rate from the three scenarios $k=$(WZ,TPL), (ZB,TPL), and (ZB,C), relating to different surface energies, critical sizes, and mono-to-poly nucleation thresholds. We introduce the Kolmogorov exponents (Dubrovskii, 2014a) which describe the WZ phase suppression by ZB nuclei [for scenarios $2=(\text{ZB},\text{TPL})$ and $3=(\text{ZB},\text{C})$] according to

$$p_{WZ}=\mathrm{e}^{-\varepsilon_2-\varepsilon_3}\frac{B_1}{B}, \tag{115}$$

assuming Poissonian random nucleation in polynuclear regimes.

Figure 30 show the WZ percentages, elongation rates, the $\varepsilon_k$, and critical sizes for the three nucleation regimes obtained from our model for the same parameters of 30-nm radius GaAs nanowires with linear fits for the surface energies versus Ga concentration, as in Fig. 29. However, we use a larger value for the solid–liquid energy of a GaAs island in pure Ga droplet of $\gamma_{SL}=0.208\text{J/m}^2$ compared to the value of $0.123\,\text{J/m}^2$ in Eq. (91). This requires simultaneous increase of the preexponent in the nucleation rate (the As diffusion coefficient $D_5$ in our model) from $3\times10^{-12}$ to $\sim 4\times10^{-10}\,\text{m}^2/\text{s}$. This value is closer to the values estimated for Ga-rich alloys (Shawarz, 1980). The increased surface energy also allows for a better fit of the known results, in particular, predominantly WZ phase of MBE-grown VLS GaAs nanowires (at typical V/III ratios ~2). Extrapolating our model to a limiting case of Ga-catalyzed GaAs nanowires, our estimates show that the VLS growth at $\gamma_{SL} = 0.123 J/m^2$ is polynuclear, consistent with a very small critical size of the order of 2.

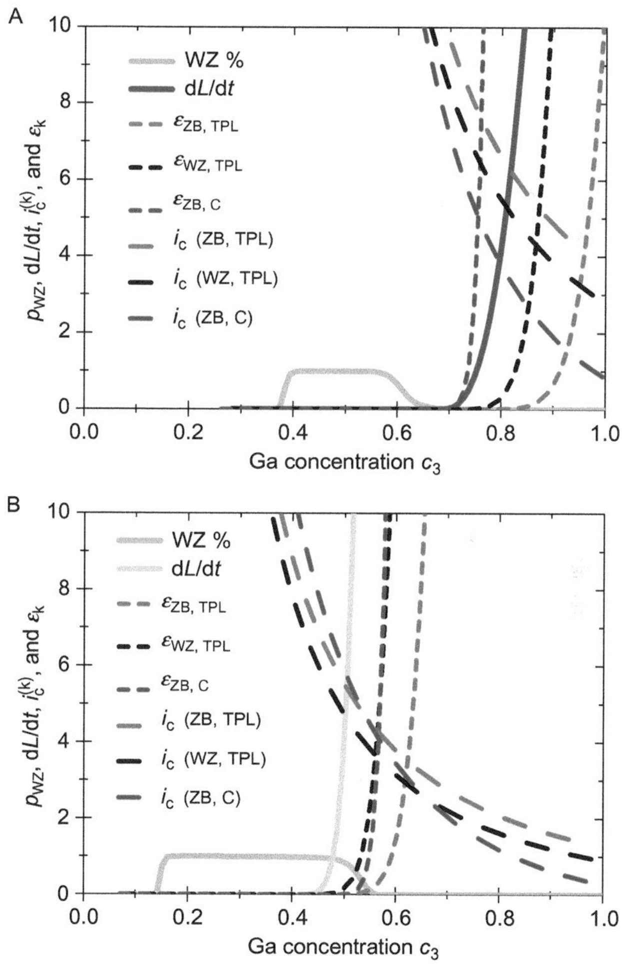


**Figure 30** Wurtzite percentage $p_{WZ}$ (light green (light gray in the print version) lines), nucleation-mediated elongation rate $dL/dt$ in nm/s (bold solid lines), polynucleation thresholds $\varepsilon_k$ (dotted lines), and critical sizes $i_c^{(k)}$ (dashed lines) versus Ga concentration for different growth scenarios $k=$(WZ,TPL), (ZB,TPL), and (ZB,C) of 30-nm radius Au-catalyzed GaAs nanowires at 550 °C for different $G$: $G=-5$ (A), $-1$ (B), 2 (C), and 4 (D). The key for the colors (different gray shades in the print version) is the same as in Fig. 29.

*(Continued)*

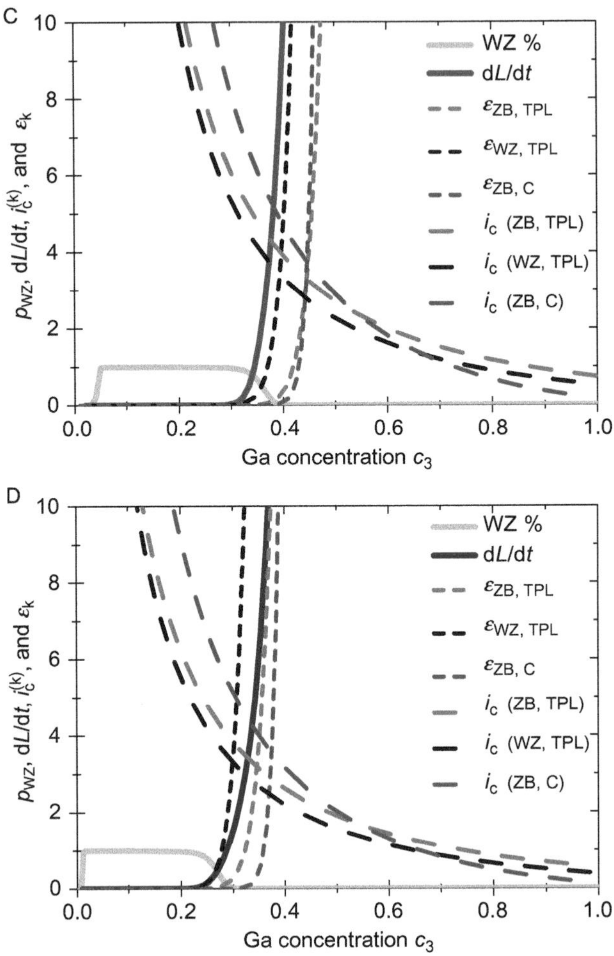


**Figure 30—Cont'd**

Let us now analyze the obtained dependences. Figure 30A shows the calculated curves at low As flows relating to $G = -4$. It is seen that the WZ region is quite narrow, extending only from ~0.37 to ~0.6 Ga concentration in the Au–Ga–As droplet. The typical nanowire elongation rate ~1 nm/s relates to a much larger Ga concentration >0.7. Such growth conditions yield pure and stable ZB structure. The VLS growth in this case is polynuclear due to a low surface energy of Ga-rich droplets. The corresponding value of the $\varepsilon_C$ is of the order of 10, meaning that about 10 2D islands will nucleate at the center of each NW monolayer and then merge together to form the continuous coverage of the top nanowire facet.

Typical critical size of (ZB,C) island equals 5 GaAs pairs. In fact, graphs in Fig. 29 show a possibility to completely suppress nucleation of WZ islands for whatever Ga concentration (for example, in nonstationary and transient growth steps) at very low As flows. Increasing the As flow to the value corresponding to $G=-1$ (Fig. 30B) changes the situation completely. With this As flow, the reasonable elongation rates of the order of 1–10 nm/s are within the WZ region, and the growth mode is polynuclear with a critical size of the order of 5. The curves in Fig. 30A and B explain the ZB-to-WZ phase transition observed in MBE and CBE with increasing the group V flux (Dheeraj et al., 2013; Fakhr et al., 2010; Husanu et al., 2014; Kelrich et al., 2015; Krogstrup et al., 2012).

Further increase of the As flow to $G=2$ narrows the WZ region of Ga concentrations to $c_3 < 0.35$. Such a narrowing takes place due to a combined effect of reduction of the TPL ring and polynucleation. Under these growth conditions, the preferred crystal structure becomes highly dependent on the elongation rate. At $dL/dt \sim 1\,\mathrm{nm/s}$, GaAs nanowires are predominantly WZ, while at $dL/dt \sim 10\,\mathrm{nm/s}$ they become almost pure ZB. This explains the benefits of rapid elongation rates for achieving the phase purity (Gil et al., 2014; Joyce et al., 2009; Ren et al., 2011). The VLS growth proceeds in mononuclear regime at $dL/dt = 1\,\mathrm{nm/s}$ and in polynuclear regime at $dL/dt = 10\,\mathrm{nm/s}$, with typical critical sizes lying between 3 and 4 pairs. Finally, when the increased As flow yields a high value of $G=4$, we observe pure ZB phase emerging in the polynuclear regime, with the critical size of 3 pairs. Thus, our results show that the As flow dependence of the preferred crystal structure is generally nonmonotonic and that the ZB phase is favored at high group V flows (due to polynucleation), the effect which is experimentally observed during MOCVD growth (Assali et al., 2013; Lehmann et al., 2012; Lehmann et al., 2013).

It is noteworthy that our model predicts the WZ structure appearing usually in the regimes with mononucleation, as in the Glas model (Glas et al., 2007). However, the reverse WZ-to-ZB phase transition at high As flows simply cannot be understood without polynucleation. The phase transitions in Au-catalyzed NWs as well as the mono-to-polynuclear growth transformation can be described within the macroscopic approach, although often reaching its limit with the critical sizes of only several III–V pairs. As seen from Figs. 29 and 30, the crystal phase purity in ZB phase usually relates to polynuclear growth (which was completely disregarded in previous studies: Dubrovskii and Grechenkov, 2015; Dubrovskii et al., 2008a; Gil et al., 2014; Glas et al., 2007; Johansson et al., 2009). Furthermore, in the regimes with high material inputs (MOCVD and HVPE) which yield high chemical potentials and rapid nanowire elongation rates, the VLS growth acquires purely atomistic character without any nucleation barrier. Then the crystal structure of the nanowires is expected to repeat the cubic stacking of the substrate from which the VLS growth starts. The situation for ternary nanowires might be more complex and requires a separate study.

As an example of structural control by the growth parameter tuning, Fig. 31 shows the TEM data on the crystal phase of InP nanowire growth by Au-assisted selective area CBE (Kelrich et al., 2015). The SEM images of these nanowires are presented in Fig. 12. It is seen that increasing the phosphine flow and decreasing temperature always favor the WZ structure, which is within the general trend for MBE and CBE as discussed above. Very importantly, the combination of high growth temperature (480 °C) and high phosphine flow (5 sccm and above) enables fabrication of

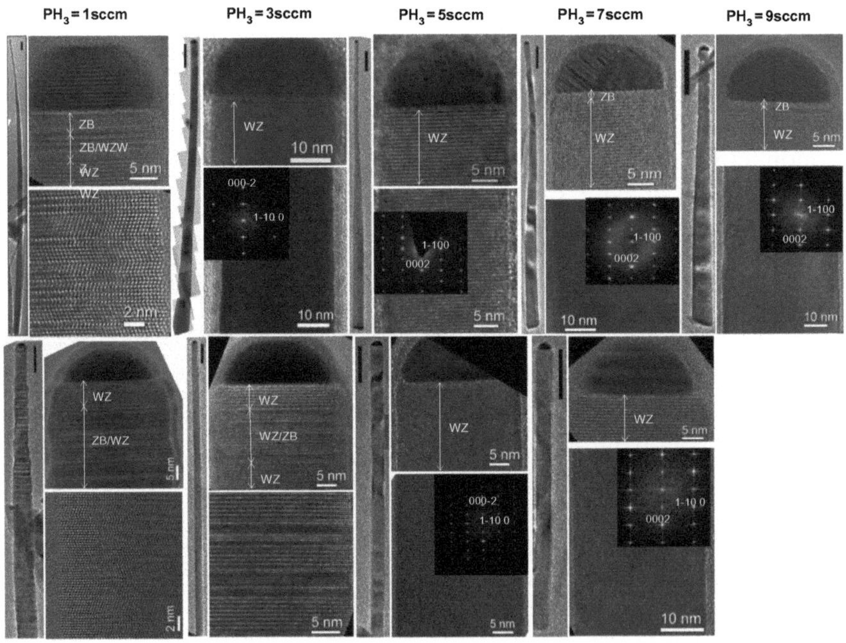


**Figure 31** Low magnification TEM and HRTEM images of InP nanowires grown at different phosphine fluxes and temperatures. The top row shows the nanowire crystal structure achieved at 420 °C, while the bottom row demonstrates nanowires grown at 480 °C. Each column represents different phosphine flow rate used. The crystal structures of the nanowires grown by residual indium from the catalyst during cooling under phosphine flow are indicated. The insets show respective SADP and FFT of the WZ structure. The scale bar in low magnification figures is 100 nm.

nontapered, pure WZ InP nanowires with only one stacking fault found per 2 μm length. The polytypism seen just beneath the droplets of some nanowires is related to nonstationary effects induced by the growth stop.

Thus, the refined structural model is capable of describing the transition from mononuclear to polynuclear regime and ultimately to atomistic growth. It shows that the WZ crystal structure usually emerge in mononuclear regime with more or less macroscopic nuclei consisting of several III–V pairs, while the ZB structure often forms in polynuclear or even atomistic growth regimes. The model also explains the nonmonotonic dependence of the crystal structure on the group V flow, with the ZB structure being preferred at very low and very high group V flows and the WZ structure forming in between. These results show that the VLS growth of Au-catalyzed GaAs NWs usually occurs near the WZ-to-ZB transition lines which can be crossed by modulating the As flow. With the known energetic parameters, the model allows for a precise description of the nucleation rates, critical sizes, polynucleation thresholds, elongation rates, and WZ percentages versus varying growth conditions. The model also enables us to determine the limits beyond which the VLS growth, crystal structure, and composition of III–V NWs are no longer limited by nucleation and follow the simple atomistic picture.

## REFERENCES

Akopian, N., Patriarche, G., Liu, L., Harmand, J.C., Zwiller, V., 2010. Crystal phase quantum dots. Nano Lett. 10, 1198.

Ameruddin, A., Caroff, P., Tan, H.H., Jagadish, C., Dubrovskii, V.G., 2015. Understanding the growth and composition evolution of gold-seeded ternary InGaAs nanowires. Nanoscale. http://dx.doi.org/10.1039/c5nr04129e.

André, Y., Lekhal, K., Hoggan, P., Avit, G., Cadiz, F., Rowe, A., Paget, D., Petit, E., Leroux, C., Trassoudaine, A., Ramdani, M.R., Monier, G., Colas, D., Ajib, R., Castelluci, D., Gil, E., 2014. Vapor liquid solid-hydride vapor phase epitaxy (VLS-HVPE) growth of ultra-long defect-free GaAs nanowires: ab initio simulations supporting center nucleation. J. Chem. Phys. 140, 194706.

Assali, S., Zardo, I., Plissard, S., Kriegner, D., Verheijen, M.A., Bauer, G., Meijerink, A., Belabbes, A., Bechstedt, F., Haverkort, J.E.M., Bakkers, E.P.A.M., 2013. Direct band gap wurtzite gallium phosphide nanowires. Nano Lett. 13, 1559.

Blakely, J.M., Jackson, K.A., 1962. Growth of crystal whiskers. J. Chem. Phys. 37, 428.

Bouravleuv, A.D., Sibirev, N.V., Statkute, G., Cirlin, G.E., Lipsanen, H., Dubrovskii, V.G., 2010. Influence of substrate temperature on the shape of GaAs nanowires grown by Au-assisted MOVPE. J. Cryst. Growth 312, 1676.

Burke, R.A., Weng, X., Kuo, M.-W., Song, Y.-W., Itsuno, A.M., Mayer, T.S., Durbin, S.M., Reeves, R.J., Redwing, J.M., 2010. Growth and characterization of unintentionally doped GaSb nanowires. J. Electron. Mater. 39, 355.

Chatillon, C., Hodaj, F., Pisch, A., 2009. Thermodynamics of GaAs nanowire MBE growth with gold droplets. J. Cryst. Growth 311, 3598.

Chou, Y.-C., Hillerich, K., Tersoff, J., Reuter, M.C., Dick, K.A., Ross, F.M., 2014. Atomic-scale variability and control of III-V nanowire growth kinetics. Science 343, 281.

Cirlin, G.E., Dubrovskii, V.G., Samsonenko, Y.B., Bouravleuv, A.D., Durose, K., Proskuryakov, Y.Y., Mendis, B., Bowen, L., Kaliteevski, M.A., Abram, R.A., Zeze, D., 2010. Self-catalyzed, pure zinc blende GaAs nanowires grown on Si(111) by molecular beam epitaxy. Phys. Rev. B 82, 035302.

Colombo, C., Spirkoska, D., Frimmer, M., Abstreiter, G., Fontcuberta i Morral, A., 2008. Ga-assisted catalyst-free growth mechanism of GaAs nanowires by molecular beam epitaxy. Phys. Rev. B 77, 155326.

Consonni, V., Dubrovskii, V.G., Trampert, A., Geelhaar, L., Riechert, H., 2012. Quantitative description for the growth rate of self-induced GaN nanowires. Phys. Rev. B 85, 155313.

Dalacu, D., Kam, A., Austing, D.G., Wu, X.H., Lapointe, J., Aers, G.C., Poole, P.J., 2009. Selective-area vapour–liquid–solid growth of InP nanowires. Nanotechnology 20, 395602.

Dayeh, S.A., Picraux, S.T., 2010. Direct observation of nanoscale size effects in Ge semiconductor nanowire growth. Nano Lett. 10, 4032.

Dheeraj, D.L., Patriarche, G., Zhou, H., Hoang, T.B., Moses, A.F., Grønsberg, S., van Helvoort, A.T.J., Fimland, B.-O., Weman, H., 2008. Growth and characterization of wurtzite GaAs nanowires with defect-free zinc blende GaAsSb inserts. Nano Lett. 8, 4459.

Dheeraj, D.L., Munshi, A.M., Scheffler, M., van Helvoort, A.T.J., Weman, H., Fimland, B.O., 2013. Controlling crystal phases in GaAs nanowires grown by Au-assisted molecular beam epitaxy. Nanotechnology 24, 015601.

Dick, K.A., Deppert, K., Martensson, T., Mandl, S., Samuelson, L., Seifert, W., 2005. Failure of the vapor-liquid-solid mechanism in Au-assisted MOVPE growth of InAs nanowires. Nano Lett. 5, 761.

Dick, K.A., Deppert, K., Samuelson, L., Wallenberg, L.R., Ross, F.M., 2008. Control of GaP and GaAs nanowire morphology through particle and substrate chemical modification. Nano Lett. 8, 4087.

Dick, K.A., Thelander, C., Samuelson, L., Caroff, P., 2010. Crystal phase engineering in single InAs nanowires. Nano Lett. 10, 3494.

Dubrovskii, V.G., 2009. Fluctuation-induced spreading of size distribution in condensation kinetics. J. Chem. Phys. 131, 164514.

Dubrovskii, V.G., 2013. Self-regulated pulsed nucleation in catalyzed nanowire growth. Phys. Rev. B. 87, 195426.

Dubrovskii, V.G., 2014a. Nucleation Theory and Growth of Nanostructures. Springer, Heidelberg–New York–Dordrecht–London.

Dubrovskii, V.G., 2014b. Influence of the group V element on the chemical potential and crystal structure of Au-catalyzed III-V nanowires. Appl. Phys. Lett. 104, 053110.

Dubrovskii, V.G., Grechenkov, J., 2015. Zeldovich nucleation rate, self-consistency renormalization, and crystal phase of Au-catalyzed GaAs nanowires. Cryst. Growth Des. 15, 340.

Dubrovskii, V.G., Hervieu, Yu.Yu., 2014. Diffusion-induced growth of nanowires: generalized boundary conditions and self-consistent kinetic equation. J. Cryst. Growth 401, 431.

Dubrovskii, V.G., Nazarenko, M.V., 2010. Nucleation theory beyond the deterministic limit. I. The nucleation stage. J. Chem. Phys. 132, 114507.

Dubrovskii, V.G., Sibirev, N.V., 2004. Growth rate of a crystal facet of arbitrary size and the growth kinetics of vertical nanowires. Phys. Rev. E 70, 031604.

Dubrovskii, V.G., Sibirev, N.V., 2007. General form of the dependences of nanowire growth rate on the nanowire radius. J. Cryst. Growth 304, 504.

Dubrovskii, V.G., Sibirev, N.V., 2014. Size distributions, scaling properties, and Bartelt-Evans singularities in irreversible growth with size-dependent capture coefficients. Phys. Rev. B 87, 195426.

Dubrovskii, V.G., Cirlin, G.E., Soshnikov, I.P., Tonkikh, A.A., Sibirev, N.V., Samsonenko, Y.B., Ustinov, V.M., 2005. Diffusion-induced growth of GaAs nanowhiskers: theory and experiment. Phys. Rev. B 71, 205325.

Dubrovskii, V.G., Sibirev, N.V., Cirlin, G.E., Harmand, J.C., Ustinov, V.M., 2006. Theoretical analysis of the vapor-liquid-solid mechanism of nanowire growth during MBE. Phys. Rev. E 73, 021603.

Dubrovskii, V.G., Sibirev, N.V., Suris, R.A., Cirlin, G.E., Harmand, J.C., Ustinov, V.M., 2007. Diffusion controlled growth of semiconductor nanowires: vapor pressure versus high vacuum deposition. Surf. Sci. 601, 4395.

Dubrovskii, V.G., Sibirev, N.V., Harmand, J.C., Glas, F., 2008a. Growth kinetics and crystal structure of semiconductor nanowires. Phys. Rev. B 78, 235301.

Dubrovskii, V.G., Sibirev, N.V., Cirlin, G.E., Tchernycheva, M., Harmand, J.C., Ustinov, V.M., 2008b. Shape modification of III-V nanowires: the role of nucleation on sidewalls. Phys. Rev. E 77, 031606.

Dubrovskii, V.G., Sibirev, N.V., Cirlin, G.E., Soshnikov, I.P., Chen, W.H., Larde, R., Cadel, E., Pareige, P., Xu, T., Grandidier, B., Nys, J.-P., Stievenard, D., Moewe, M., Chuang, L.C., Chang-Hasnain, C., 2009a. Gibbs-Thomson and diffusion-induced contributions to the growth rate of Si, InP and GaAs nanowires. Phys. Rev. B 79, 205316.

Dubrovskii, V.G., Sibirev, N.V., Cirlin, G.E., Bouravleuv, A.D., Samsonenko, Y.B., Dheeraj, D.L., Zhou, H.L., Sartel, C., Harmand, J.C., Patriarche, G., Glas, F., 2009b. Role of non-linear effects in nanowire growth and crystal phase. Phys. Rev. B 80, 205305.

Dubrovskii, V.G., Cirlin, G.E., Sibirev, N.V., Jabeen, F., Harmand, J.C., Werner, P., 2011. New mode of vapor-liquid-solid nanowire growth. Nano Lett. 11, 1247.

Dubrovskii, V.G., Xu, T., Lambert, Y., Nys, J.-P., Grandidier, B., Stievenard, D., Chen, W., Pareige, P., 2012a. Narrowing the length distribution of Ge nanowires. Phys. Rev. Lett. 108, 105501.

Dubrovskii, V.G., Bolshakov, A.D., Williams, B.L., Durose, K., 2012b. Growth modeling of CdTe nanowires. Nanotechnology 23, 485607.

Dubrovskii, V.G., Timofeeva, M.A., Kelrich, A., Ritter, D., 2015a. Growth and morphological modeling of InP nanowires obtained by Au-catalyzed selective area MOMBE. J. Cryst. Growth 413, 25.

Dubrovskii, V.G., Xu, T., Díaz Álvarez, A., Plissard, S., Caroff, P., Glas, F., Grandidier, B., 2015b. Self-equilibration of the diameter of Ga-catalyzed GaAs nanowires. Nano Lett. 15, 5580.

Einax, M., Dieterich, W., Maass, P., 2013. Cluster growth on surfaces: densities, size distributions, and morphologies. Rev. Mod. Phys. 85, 921.

Fakhr, A., Haddara, Y.M., LaPierre, R.R., 2010. Dependence of InGaP nanowire morphology and structure on molecular beam epitaxy growth conditions. Nanotechnology 21, 165601.

Fontcuberta i Morral, A., Colombo, C., Abstreiter, G., Arbiol, J., Morante, J.R., 2008. Nucleation mechanism of gallium-assisted molecular beam epitaxy growth of gallium arsenide nanowires. Appl. Phys. Lett. 92, 063112.

Fröberg, L.E., Seifert, W., Johansson, J., 2007. Diameter-dependent growth rate of InAs nanowires. Phys. Rev. B 76, 153401.

Gamalski, A.D., Ducati, C., Hofmann, S., 2011. Cyclic supersaturation and triple phase boundary dynamics in germanium nanowire growth. J. Phys. Chem. C 115, 4413.

Gil, E., Dubrovskii, V.G., Avit, G., André, Y., Leroux, C., Lekhal, K., Grecenkov, J., Trassoudaine, A., Castelluci, D., Monier, G., Ramdani, R.M., Robert-Goumet, C.,

Bideux, L., Harmand, J.C., Glas, F., 2014. Record pure zincblende phase in GaAs nanowires down to 5 nm in radius. Nano Lett. 14, 3938.
Givargizov, E.I., 1987. Highly Anisotropic Crystals. Springer, Berlin.
Glas, F., 2010a. Chemical potentials for Au-assisted vapor-liquid-solid growth of III-V nanowires. J. Appl. Phys. 108, 073506.
Glas, F., 2010b. Vapor fluxes on the apical droplet during nanowire growth by molecular beam epitaxy. Phys. Status Solidi B 247, 254.
Glas, F., 2014. Statistics of sub-Poissonian nucleation in a nanophase. Phys. Rev. B 90, 125406.
Glas, F., Harmand, J.C., Patriarche, G., 2007. Why does wurtzite form in nanowires of III-V zinc-blende semiconductors? Phys. Rev. Lett. 99, 146101.
Glas, F., Harmand, J.C., Patriarche, G., 2010. Nucleation antibunching in catalyst-assisted nanowire growth. Phys. Rev. Lett. 104, 135501.
Glas, F., Ramdani, M.R., Patriarche, G., Harmand, J.C., 2013. Predictive modeling of self-catalyzed III-V nanowire growth. Phys. Rev. B 88, 195304.
Guo, Y.-N., Xu, H.-Y., Auchterlonie, G.J., Burgess, T., Joyce, H.J., Gao, Q., Tan, H.H., Jagadish, C., Shu, H.-B., Chen, X.-S., Lu, W., Kim, Y., Zou, J., 2013. Phase separation induced by Au catalysts in ternary InGaAs nanowires. Nano Lett. 13, 643.
Harmand, J.C., Patriarche, G., Péré-Laperne, N., Mérat-Combes, M.-N., Travers, L., Glas, F., 2005. Analysis of vapor-liquid-solid mechanism in Au-assisted GaAs nanowire growth. Appl. Phys. Lett. 87, 203101.
Harmand, J.C., Glas, F., Patriarche, G., 2010. Growth kinetics of a single $InP_{1-x}As_x$ nanowire. Phys. Rev. B 81, 235436.
Husanu, E., Ercolani, D., Gemmi, M., Sorba, L., 2014. Growth of defect-free GaP nanowires. Nanotechnology 25, 205601.
Jabeen, F., Grillo, V., Rubini, S., Martelli, F., 2008. Self-catalyzed growth of GaAs nanowires on cleaved Si by molecular beam epitaxy. Nanotechnology 19, 275711.
Jasper, J.J., 1972. The surface tension of pure liquid compounds. J. Phys. Chem. Ref. Data 1 (4), 841.
Johansson, J., Svensson, C.P.T., Mårtensson, T., Samuelson, L., Seifert, W., 2005. Mass transport model for semiconductor nanowire growth. J. Phys. Chem. B 109, 13567.
Johansson, J., Karlsson, L.S., Dick, K.A., Bolinsson, J., Wacaser, B.A., Deppert, K., Samuelson, L., 2009. Effects of supersaturation on the crystal structure of gold seeded III–V nanowires. Cryst. Growth Des. 9, 766.
Joyce, H.J., Gao, Q., Tan, H.H., Jagadish, C., Kim, Y., Fickenscher, M.A., Perera, S., Hoang, T.B., Smith, L.M., Jackson, H.E., Yarrison-Rice, J.M., Zhang, X., Zou, J., 2009. Unexpected benefits of rapid growth rate for III–V nanowires. Nano Lett. 9, 695.
Kashchiev, D., 2000. Nucleation: Basic Theory with Applications. Butterworth-Heinemann, Oxford.
Kashchiev, D., 2006. Dependence of the growth rate of nanowires on the nanowire diameter. Cryst. Growth Des. 6, 1154.
Kashchiev, D., 2008. Toward a better description of the nucleation rate of crystals and crystalline monolayers. J. Chem. Phys. 129, 164701.
Kelrich, A., Calahorra, Y., Greenberg, Y., Gavrilov, A., Cohen, S., Ritter, D., 2013. Shadowing and mask opening effects during selective-area vapor–liquid–solid growth of InP nanowires by metalorganic molecular beam epitaxy. Nanotechnology 24, 475302.
Kelrich, A., Dubrovskii, V.G., Calahorra, Y., Cohen, S., Ritter, D., 2015. Control of morphology and crystal purity of InP nanowires by variation of phosphine flux during selective area MOMBE. Nanotechnology 26, 085303.
Kim, Y.H., Woo Park, D.W., Lee, S.J., 2012. Gallium-droplet behaviors of self-catalyzed GaAs nanowires: a transmission electron microscopy study. Appl. Phys. Lett. 100, 033117.

Kodambaka, S., Tersoff, J., Reuter, M.C., Ross, F.M., 2006. Diameter-independent kinetics in the vapor-liquid-solid growth of Si nanowires. Phys. Rev. Lett. 96, 096105.
Kodambaka, S., Ngo, C., Palisaitis, J., Mayrhofer, P.H., Hultman, L., Persson, P.O.A., 2013. Kinetics of Ga droplet decay on thin carbon films. Appl. Phys. Lett. 102, 161601.
Krogstrup, P., Madsen, M.H., Wen, H., Miwa, K., Nygård, J., Masamitu, T., Feidenhans, R.K., 2012. In-situ x-ray characterization of wurtzite formation in GaAs nanowires. Appl. Phys. Lett. 100, 093103.
Krogstrup, P., Jørgensen, H.I., Johnson, E., Madsen, M.H., Sørensen, C.B., Fontcuberta i Morral, A., Aagesen, M., Nygård, J., Glas, F., 2013. Advances in the theory of III–V nanowire growth dynamics. J. Phys. D Appl. Phys. 46, 313001.
Kukushkin, S.A., Osipov, A.V., 1996. New phase formation on solid surfaces and thin film condensation. Prog. Surf. Sci. 51, 1.
Landau, L.D., Lifshitz, E.M., 1980. Statistical physics. In: Course of Theoretical Physics, vol. 5. Pergamon, Oxford, New York, Toronto – Sydney – Paris – Frankfurt.
Lehmann, S., Jacobsson, D., Deppert, K., Dick, K., 2012. High crystal quality wurtzite-zincblende heterostructures in metal-organic vapor phase epitaxy-grown GaAs nanowires. Nano Res. 5, 470.
Lehmann, S., Wallentin, J., Jacobsson, D., Deppert, K., Dick, K.A., 2013. A general approach for sharp crystal phase switching in InAs, GaAs, InP, and GaP nanowires using only group V flow. Nano Lett. 13, 4099.
Lu, Y.Y., Cui, H., Yang, G.W., Wang, C.X., 2012. Diameter-dependent or independent: toward a mechanistic understanding of the vapor–liquid–solid Si nanowire growth rate. Nano Lett. 12, 4032.
Markov, I., 2003. Crystal Growth for Beginners, second ed. World Scientific, New Jersey–London–Singapore–Hong Kong.
Massalski, T.M., 1986. In: Binary Alloy Phase Diagrams, vol. 1. American Society for Metals, Metals Park, OH.
Matteini, F., Dubrovskii, V.G., Rüffer, D., Tütüncüoğlu, G., Fontana, Y., Fontcuberta i Morral, A., 2015. Tailoring the diameter and density of self-catalyzed GaAs nanowires on silicon. Nanotechnology 26, 105603.
Moewe, M., Chuang, L.C., Dubrovskii, V.G., Chang-Hasnain, C., 2008. Growth mechanisms and crystallographic structure of InP nanowires on lattice-mismatched substrates. J. Appl. Phys. 104, 044313.
Nebol'sin, V.A., Shchetinin, A.A., 2003. Role of surface energy in the vapor–liquid–solid growth of silicon. Inorg. Mater. 39, 899.
Noborisaka, J., Motohisa, J., Fukui, T., 2005. Catalyst-free growth of GaAs nanowires by selective-area metalorganic vapor-phase epitaxy. Appl. Phys. Lett. 86, 213102.
Patriarche, G., Glas, F., Tchernycheva, M., Sartel, C., Largeau, L., Harmand, J.-C., Cirlin, G.E., 2008. Wurtzite to zinc blende phase transition in GaAs nanowires induced by epitaxial burying. Nano Lett. 8, 1638.
Persson, A.I., Larsson, M.W., Stenstrom, S., Ohlsson, B.J., Samuelson, L., Wallenberg, L.R., 2004. Solid-phase diffusion mechanism for GaAs nanowire growth. Nat. Mater. 3, 678.
Plante, M.C., LaPierre, R.R., 2006. Growth mechanisms of GaAs nanowires by gas source molecular beam epitaxy. J. Cryst. Growth 286, 394.
Plante, M.C., LaPierre, R.R., 2009. Analytical description of the metal-assisted growth of III–V nanowires: axial and radial growths. J. Appl. Phys. 105, 114304.
Plissard, S., Larrieu, G., Wallart, X., Caroff, P., 2011. High yield of self-catalyzed GaAs nanowire arrays grown on silicon via gallium droplet positioning. Nanotechnology 22, 275602.
Poole, P.J., Dalacu, D., Wu, X., Lapointe, J., Mnaymneh, K., 2012. Interplay between crystal phase purity and radial growth in InP nanowires. Nanotechnology 23, 385205.

Priante, G., Ambrosini, S., Dubrovskii, V.G., Franciosi, A., Rubini, S., 2013. Stopping and resuming at will the growth of GaAs nanowires. Cryst. Growth Des. 13, 3976.

Ramdani, M.R., Harmand, J.C., Glas, F., Patriarche, G., Travers, L., 2013. Arsenic pathways in self-catalyzed growth of GaAs nanowires. Cryst. Growth Des. 13, 91.

Ren, X., Huang, H., Dubrovskii, V.G., Sibirev, N.V., Nazarenko, M.V., Bolshakov, A.D., Ye, X., Wang, Q., Huang, Y., Zhang, X., Guo, J., Liu, X., 2011. Experimental and theoretical investigations on the phase purity of GaAs zincblende nanowires. Semicond. Sci. Technol. 26, 014034.

Ross, F.M., Tersoff, J., Reuter, M.C., 2005. Sawtooth faceting in silicon nanowires. Phys. Rev. Lett. 95, 146104.

Ruth, V., Hirth, J.R., 1964. Kinetics of diffusion-controlled whisker growth. J. Chem. Phys. 41, 31.

Sabelfeld, K.K., Kaganer, V.M., Limbach, F., Dogan, P., Brandt, O., Geelhaar, L., Riechert, H., 2013. Height self-equilibration during the growth of dense nanowire ensembles: order emerging from disorder. Appl. Phys. Lett. 103, 133105.

Schubert, L., Werner, P., Zakharov, N.D., Gerth, G., Kolb, F.M., Long, L., Gösele, U., Tan, T.Y., 2004. Silicon nanowhiskers grown on <111> Si substrates by molecular-beam epitaxy. Appl. Phys. Lett. 84, 4968.

Seifert, W., Borgstrom, M., Deppert, K., Dick, K.A., Johansson, J., Larsson, M.W., Martensson, T., Skold, N., Svensson, C.P.T., Wacaser, B.A., Wallenberg, L.R., Samuelson, L., 2004. Growth of one-dimensional nanostructures in MOVPE. J. Cryst. Growth 272, 211.

Shawarz, R.I., 1980. Diffusion coefficient of arsenic in gallium-rich melts at 720°C. Thin Solid Films 66, L3.

Sibirev, N.V., Nazarenko, M.V., Zeze, D.A., Dubrovskii, V.G., 2014. Modeling the nucleation statistics in vapor–liquid–solid nanowires. J. Cryst. Growth 401, 51.

Spirkoska, D., Arbiol, J., Gustafsson, A., Conesa-Boj, S., Glas, F., Zardo, I., Heigoldt, M., Gass, M.H., Bleloch, A.L., Estrade, S., Kaniber, M., Rossler, J., Peiro, F., Morante, J.R., Abstreiter, G., Samuelson, L., Fontcuberta i Morral, A., 2009. Structural and optical properties of high quality zinc-blende/wurtzite GaAs nanowire heterostructures. Phys. Rev. B 80, 245325.

Stringfellow, G.B., 1972. Calculation of ternary phase diagrams of III-V systems. J. Phys. Chem. Solids 33, 665.

Stringfellow, G.B., 2001. Fundamental aspects of organometallic vapor phase epitaxy: review. Mater. Sci. Eng. B 87, 97.

Tchernycheva, M., Travers, L., Patriarche, G., Harmand, J.C., Cirlin, G.E., Dubrovskii, V.G., 2007. Au-assisted molecular beam epitaxy of InAs nanowires: growth and theoretical analysis. J. Appl. Phys. 102, 094313.

Wagner, R.S., Ellis, W.C., 1964. Vapor-liquid-solid mechanism of singe crystal growth. Appl. Phys. Lett. 4, 89.

Wagner, R.S., Ellis, W.C., 1965. The vapor-liquid-solid mechanism of crystal growth and its application to silicon. Trans. Metall. Soc. AIME 233, 1053.

Wen, C.Y., Tersoff, J., Hillerich, K., Reuter, M.C., Park, J.H., Kodambaka, S., Stach, E.A., Ross, F.M., 2011. Periodically changing morphology of the growth interface in Si, Ge, and GaP nanowires. Phys. Rev. Lett. 107, 025503.

Yee, R.J., Gibson, S.J., Dubrovskii, V.G., LaPierre, R.R., 2012. Effects of Be doping on InP nanowire growth mechanisms. Appl. Phys. Lett. 101, 263106.

Zeldovich, J., 1942. Theory of the formation of a new phase. Cavitation. Sov. Phys. JETP (Eng. Transl.) 12, 525.

CHAPTER TWO

# Strain in Nanowires and Nanowire Heterostructures

**Frank Glas**[1]
CNRS—Laboratoire de Photonique et de Nanostructures, Marcoussis, France
[1]Corresponding author: e-mail address: frank.glas@lpn.cnrs.fr

## Contents



## 1. INTRODUCTION

### 1.1 Scope

If we call stress-free (SF) the elastic state of a homogeneous single crystal free of any external force (i.e., the intrinsic state of the bulk crystal), the main reason for strain to appear in a NW is that it associates two or more materials that present different SF states. Frequently, but not necessarily, the materials

*Semiconductors and Semimetals*, Volume 93
ISSN 0080-8784
http://dx.doi.org/10.1016/bs.semsem.2015.09.004

are crystals having the same crystalline structure but different lattice parameters (they present a *lattice mismatch*), but there also exist hybrid NWs associating crystals with different structures (Conesa-Boj et al., 2013; Hocevar et al., 2012) or crystalline and amorphous materials. Usually, the heterogeneity is introduced deliberately. One of the main current aims of NW growth is indeed to fabricate tailored heterostructures. Two main types of heterostructures can be formed in NWs, namely axial and radial heterostructures. If, during growth, one switches from material A to material B, while maintaining growth along the NW axis, one obtains an axial heterostructure (Fig. 1A). If, on the other hand, one initiates radial growth while at the same time switching from A to B, one forms a radial (or core–shell) heterostructure, with B surrounding A (Fig. 1B and C). In these two types of heterostructure, the NW itself is composed of different materials. A third type of heterostructure occurs even for a homogeneous B NW when it is grown on a substrate of a different material A, a fairly frequent situation. The heterointerface is then located at the base of the NW (Fig. 1D). Any combination of these three basic types may be considered.

## 1.2 Heterostructures, Mismatch, and Accommodation

Quite generally, an A/B interface is said to be *coherent* if the lattice planes not parallel to the interface are equally spaced in A and B and continuous across

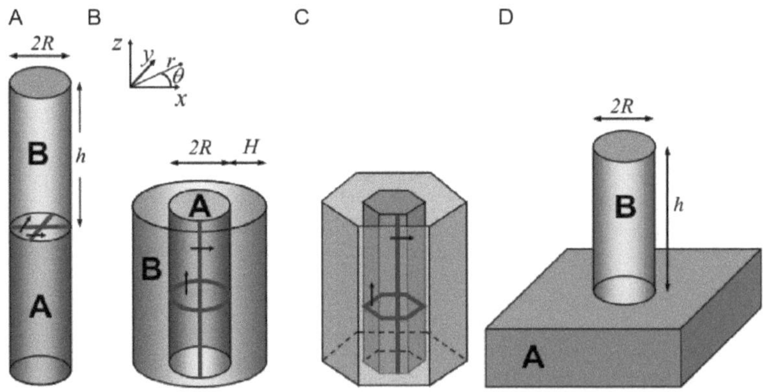


**Figure 1** Schematics of the basic types of heterostructure in a NW: (A) axial and (B) core–shell (or radial). The thick blue (dark gray in the print version) lines show the interfacial dislocations considered in simple models of plastic relaxation [(A) edge segments and (B) basal loop and axial dislocation], their Burgers vectors being indicated by arrows. (C) A more realistic core–shell NW with hexagonal cross section. (D) NW on substrate. Various coordinates systems are indicated: orthonormal *xyz* axes with *z* along NW axis, cylindrical coordinates for NWs with circular section.

the interface. A more general definition, valid even for noncrystalline materials, is the following. Imagine that, starting, e.g., from their SF states, one applies forces to A and B such that they can be physically assembled, with atomic bonds forming across the interface. When these forces are released, the system will *relax*: by this, we mean that it will evolve toward a state of lower energy. The relaxation is coherent if the extra deformation that generally appears at this stage in both A and B is such that the displacements along the interface are continuous across the interface. Obviously, a lattice mismatch between crystals A and B makes it impossible for A and B to retain their intrinsic SF state if the interface is coherent. Hence, the necessity of an *accommodation* of the lattice mismatch, which can take two extreme forms.

Recall the common case of a thin layer deposited on a bulk semi-infinite substrate, later referred to as "the 2D case." Either the coherence at the interface is preserved thanks to a deformation of one or both materials; the accommodation is then purely *elastic*. One can even safely consider (ignoring possible curvature effects) that, due to the large difference of thickness between layer and substrate, only the former is strained while the latter retains its bulk SF lattice constants; the deposit then adopts the lattice parameter(s) of the substrate in the interface plane and extends or contracts in the normal direction. This strain affects uniformly the whole layer. Conversely, accommodation may be realized *plastically*, via the formation of a network of misfit dislocations (fully or partly of edge character) at the interface, which is then *incoherent*. The density of dislocations of a given type that accommodates a given mismatch is proportional to the relative mismatch (Hirth and Lothe, 1982). If the mismatch is fully accommodated by dislocations, both crystals are strained in the vicinity of the dislocation cores but rapidly tend to their respective SF state away from the latter. If the density of dislocations is insufficient, the interface is only partially coherent and the entire thin layer remains elastically strained, albeit less than in the fully coherent case.

A major question regarding strain accommodation in heterostructures is to find out which factors determine the mode of relaxation (elastic or plastic). In the 2D case, it is usually observed that dislocations do not form until the growing layer reaches some critical thickness $h_c$. Such a critical thickness exists because, as seen above, the energy in the coherent state is uniformly distributed in the layer (so that the energy per unit area scales with layer thickness $h$), whereas the energy in the fully plastically relaxed state is concentrated in the interface region (the energy per unit area depends only weakly on $h$ via the strain field of the dislocations). In many applications of heterostructures, in particular those based on the electrical and optical properties of semiconductor materials, dislocations have a deleterious effect, so that many methods have been devised to eliminate them, or at least to reduce their density. One obvious strategy is to keep the deposit thickness below $h_c$. This is however not always possible since $h_c$ decreases rapidly when the misfit increases (Fitzgerald, 1991).

Another approach is to play with the dimensionality and dimensions of the deposit (and sometimes of the substrate) to prevent or hinder dislocation formation. Indeed, when the constraint of infinite lateral extension (2D case) is lifted, the deposit may recover its SF state even if the interface remains coherent. The lattice planes may then deform continuously from the spacing of the substrate toward the intrinsic spacing of the deposit over some distance from the interface. This is realized in uncapped quantum dots. Formation of quantum dots usually starts by growing coherent nanoislands of a highly mismatched semiconductor onto a bulk substrate. It is observed that, at given mismatch, the A/B interface remains coherent up to much larger nominal deposit thicknesses when B forms as quantum dots rather than as a 2D layer (Glas et al., 1987). The reason is that the elastic energy stored in a given volume of B may be much less when B is an island than when it is a 2D layer. This in turn stems from the fact that elastic lattice relaxation is more efficient in such islands because these have additional lateral free surfaces (where no forces apply) as compared with a 2D layer. A similar but weaker effect occurs for quantum wires parallel to the substrate.

## 1.3 Nanowire Specificities

These ideas can be transferred to axial heterostructures in NWs, where there is a single interface plane (Fig. 1A). The very limited size of this interface profoundly alters the balance between elastic and plastic relaxation with respect to the 2D case. Now, the whole structure (deposit and effective substrate, i.e., NW stem) can relax at the sidewalls of the NW. The net effect is that plastic relaxation is delayed with respect to the 2D case in terms of layer thickness, and possibly even totally suppressed. The situation is more complex for core–shell heterostructures (Fig. 1B and C). In a NW with hexagonal cross section (Fig. 1C), the interface is still locally planar, but these planes have a limited transverse extension, whereas they are usually very long in the axial direction. Moreover, adjacent facets have different orientations and this affects elastic relaxation in the corner regions. The same ideas apply to NWs with circular cross sections (often considered because they are simpler to model although they do not usually occur, except maybe at very small NW radii). We can thus already guess that core–shell heterostructures present advantages with respect to 2D structures but probably not as marked as in the case of axial heterostructures. This will be discussed in detail.

A major interest of NWs is thus that their geometry allows one to form coherent heterostructures using thicker layers or higher mismatches than in other geometries. Although this easier accommodation of stress in NWs compared with planar structures is often cited as an important advantage, the literature devoted only or principally to stress relaxation in NWs is rather scarce. Most of it deals with calculations and only recently appeared the first serious studies aiming either at measuring elastic strains or at identifying the actual mechanisms of plastic relaxation.

Another specificity of semiconductor NWs relates to their crystalline structure. Whereas elemental semiconductors and III–V compounds (with the exception of the group-III nitrides) adopt a face-centered-cubic (fcc) structure in bulk form and even in other types of nanostructures, NWs of these materials may be found with a hexagonal structure. This is especially true of the III–V compounds, which frequently adopt the hexagonal wurtzite (WZ) structure as well as the standard cubic zinc blende (ZB) structure (Glas et al., 2007; Hiruma et al. 1996). The two structures may even coexist in the same NW. So far, the vast majority of NWs has been grown along a [111] axis (cubic structure) or along the $\langle 0001 \rangle$ (or **c**) axis (hexagonal structure). An important difference between the two is that only the hexagonal structure displays transverse elastic isotropy around its growth axis (Sadd, 2005). These facts have to be borne in mind when strain is discussed.

The present review focuses on strain *per se* and not on its effects on the physical properties of NWs. For this, the reader is referred to specialized publications, such as (among many others) those of Niquet (2007) and Niquet et al. (2012) or the two volumes on wide band gap semiconductor NWs edited by Consonni and Feuillet (2014). The chapter is organized as follows. We first present in Section 2 a brief introduction to the calculation and measurement of elastic and plastic strain in NWs. Sections 3–5 are devoted, respectively, to axial heterostructures, NWs on a misfitting substrate (in epitaxial growth, the corresponding interface is of course formed before any axial heterostructure), and core–shell heterostructures. We first review the calculations of coherent elastic relaxation and the main characters of the strain distribution, then the predictions about plastic relaxation, and confront these results to the rather scarce experiments that exist. Such interfaces between different materials have been studied for decades in planar semiconductor structures and in other types of nanostructures. In addition, NWs present other instances where strain may occur. We will examine some of these in Section 6, in particular, the joining of two regions where the *same* material adopts the two different structures mentioned above. There, we will also consider if the sidewalls surrounding the NW might be a source of strain.

## 2. METHODS OF CALCULATION AND MEASUREMENT OF STRAIN IN NANOWIRES

### 2.1 Calculation of Elastic Strain

Elastic relaxation in NWs is usually calculated in the framework of linear elasticity. Because linear elasticity does not have any built-in length scale, some calculations performed long ago for macroscopic systems apply to NWs, as long as relaxation remains coherent. If however line defects appear, the absolute dimensions matter. As regards plastic relaxation, nanostructures, in particular NWs, may indeed behave very differently from macroscopic structures, as demonstrated in Sections 3–5.

The calculations may be analytical or numerical and assume either isotropic or anisotropic elasticity. Analytical approaches usually involve more restrictive approximations (in particular, as regards NW geometry) than numerical simulations. One first derives differential equations verified by some quantities, e.g., the displacements or a stress function (Sadd, 2005), given the symmetries of the problem. The general solutions of these equations depend on a few parameters in each domain (core and axial deposit, or core and shell), which are found by applying the boundary conditions (interfacial coherency, free external sidewalls, free top surface for a NW of finite length, etc.).

Numerical methods give access to more realistic NWs. The typical hexagonal cross section (Fig. 1C) and the difference of elastic constants between the different parts of the heterostructure may easily be taken into account. The calculations must however be performed for particular materials and the transferability of the results from one system to another is not guaranteed. The finite elements (FE) method is frequently used in elasticity; it consists in finding an approximate solution to the equations of elasticity after discretization of the domain of interest into elements of various sizes and shapes. Other methods describe the system at the atomic level. The valence force field (VFF) method, which has been used for decades for semiconductors (Glas et al., 1990; Keating, 1966; Martin, 1970), minimizes the total energy of the system, taken as a sum of terms depending on the distortions of bond lengths and bond angles with respect to an ideal configuration (the SF state for simple and homogeneous materials). At variance with VFF, which can handle millions of atoms, the density functional method (DFT) is restricted to very small systems. It has however been applied to strain calculation in narrow core–shell NWs (Musin and Wang, 2005).

Obviously, when an axially symmetric NW is modeled (Fig. 1 A, B, D), one tends to use cylindrical coordinates and to compute radial ($\epsilon_{rr}$) and hoop ($\epsilon_{\theta\theta}$) strains, whereas for a NW with faceted sidewalls (Fig. 1C), one uses either cartesian coordinates or axes adapted to the crystal symmetry.

## 2.2 Experimental Assessment of Elastic Strain and Plastic Relaxation

Most experimental methods used to assess elastic relaxation in NWs measure strains rather than stresses. Since NWs are often fabricated as ensembles displaying variable geometries (and in particular a distribution of radii), the methods able to address individual NWs are particularly interesting.

Foremost among these are the various techniques of transmission electron microscopy (TEM). Electron diffraction provides easy access to the difference of lattice spacing in multiple directions (Popovitz-Biro et al., 2011). The spacings are however averaged over some volume of the NW (which may be as small as 1 nm in the transverse direction; Béché et al., 2011) and, because the spots are broadened by various electron scattering processes, the angular resolution and hence the parameter measurement accuracy is much worse than in X-ray diffraction (XRD). Standard high-resolution TEM (HREM) offers resolution at the level of the atomic column, but the information results from a complex dynamical scattering process (the method is based on interferences between a transmitted and several diffracted beams) and the information is averaged in the thickness of the sample along the electron beam direction. In the recent years, the high-angle annular dark field (HAADF) technique has gained increasing popularity. With the advent of aberration-corrected scanning TEM instruments, it routinely achieves atomic column resolution and the interpretation of the images is somewhat simpler than in standard HREM. In particular, the local intensity in the image of an alloy can be related to its composition. This technique has been applied widely to NWs at various levels of resolution (Arbiol et al., 2012; Hocevar et al., 2012; Ramdani et al., 2013; Rigutti et al., 2011). Energy dispersive X-ray spectroscopy and cathodoluminescence (Tourbot et al., 2012) can also be performed at high resolution in a TEM and provide useful information about local composition or local bandgap. In the case of alloys, compositional information is essential to evaluate critical dimensions (see below), since the SF misfit relates directly to alloy composition. Standard and HAADF high-resolution imaging are also choice techniques to identify and locate extended defects such as dislocations, stacking faults, or twin planes (Priante et al., 2014). They are however inaccurate for the measurement of atomic displacements and hence of strain. The geometrical phase analysis method has been developed to obviate these shortcomings (Hÿtch et al., 1998). It is now widely used to measure strain with near atomic resolution. The method is based on filtering a HREM image (be it standard or HAADF) around a given lattice spatial frequency to obtain information about the displacement field of the corresponding planes, which modifies locally the phase of the electron wave function. The method has been applied, in particular, to NWs with heterostructures in the group-III nitrides (Arbiol et al., 2012; Bougerol et al., 2010; Hestroffer et al., 2010; Kehagias et al., 2013; Tourbot et al., 2011), III-V (Hocevar et al., 2013) and II–VI (Bellet-Amalric et al., 2010) compounds and also in mixed III–V/Si heterostructures (Conesa-Boj et al., 2014). If one is not interested in high spatial resolution, simple Fourier filtering of extended areas of HREM images may reveal the identity (Nazarenko et al., 2013) or on the contrary the difference (Lauhon et al., 2002) of lattice parameters between various parts of a NW, e.g., core and shell. Note finally that, most often, the NWs are scraped off their substrate and dispersed on a thin membrane, so that the NW axis is normal to the electron beam. Therefore, the interpretation may be trickier for core–shell heterostructures (in parts of the NW, the electron beam traverses both shell and core) than for axial heterostructures. Ambiguities may be lifted by complementing such images of a core–shell structure by others obtained by sectioning the NW transversely (Popovitz-Biro et al., 2011).

XRD also provides valuable information on strained heterostructures. At variance with TEM, the NWs are usually kept on their substrate. Standard (high-angle) XRD can then be used to measure the lattice spacings along the growth axis, while grazing incidence XRD gives access to the in-plane lattice spacings (Eymery et al., 2007; Keplinger et al., 2009). Coherent XRD is a rapidly progressing technique. High brightness synchrotron sources now permit to illuminate coherently single NWs with diameters below 100 nm. Scattering from this single object is recorded around a Bragg peak and the three-dimensional reciprocal space data can be inverted to recover the shape of the scattering object (Favre-Nicolin et al., 2009) and, by comparison with calculations, the strain field in inhomogeneous NWs (Favre-Nicolin et al., 2010; Haag et al., 2013).

Finally, various methods of optical spectroscopy may be used to probe strain in single NWs dispersed on a membrane. Microphotoluminescence (μPL) allowed Thillosen et al. (2006) to check that their GaN NWs were in their SF state, by measuring the donor-bound exciton peak at the position expected from fully relaxed material. There is growing interest in correlating optical and structural measurements (the latter carried out by TEM) successively on the same NW. For instance, Rigutti et al. (2011) could correlate the shift of the $X_c$ exciton peak in the μPL spectra of individual GaN NWs with AlN shells of increasing thicknesses to the strain calculated via VFF from precise HRTEM measurements of core and shell dimensions on the same NWs (see Section 5.1). Cathodoluminescence can be used to the same effect, with the added advantage of performing optical and structural studies in the same TEM instrument (Lim et al., 2009; Tourbot et al., 2012; Zagonel et al., 2011).

Micro-Raman spectroscopy is also capable of probing strain in individual NWs. In this case, one measures the shifts of specific phonon-related peaks induced by a strain field that alters the phonon frequencies. However, the quantitative analysis is not straightforward since the strain in a NW usually varies on a length scale well below the experimental resolution of the method and since the Raman shifts depend on several components of the strain tensor (Singh et al., 2011). Therefore, strain cannot be extracted directly; instead, one has to compare simulations to experiments. Raman spectroscopy has been applied to NW heterostructures, in particular in group-III nitrides (ensembles of NWs; Bougerol et al., 2010; Cros et al., 2013; Hestroffer et al., 2010) or single NWs (Laneuville et al., 2011) and in individual NWs of elementary (Dillen et al., 2012) or compound (Alarcón-Lladó et al., 2013; Montazeri et al., 2010; Zardo et al., 2009) semiconductors.

# 3. AXIAL HETEROSTRUCTURES

## 3.1 Calculation of Elastic Relaxation in Axial Heterostructures

Let us consider a NW of radius $R$, constituted of a foot of material A topped by a B layer of thickness $h$ (Fig. 1A). A foot much taller than the NW diameter can be considered as a NW substrate of effectively semi-infinite extension and we may then forget about the actual bulk substrate which may support it (for this, see Section 4). To investigate theoretically the mechanisms of strain relaxation during growth, one keeps the top surface of the B layer free. If, on the other hand, one is interested in the effect of strain on the electronic properties of NWs, one usually considers a double A/B/A heterostructure (Geng et al., 2012), possibly wrapped by a shell. The elastic state of axial heterostructures has been calculated by several authors. The differences relate to the shape of the section of the NW, the elastic properties of the materials, and the method of solution of the equations of elasticity (Section 2.1). The boundary conditions consist in coherency at the A/B interface and cancellation of the forces on all free surfaces.

Let us first examine the calculations for a B layer *without* a free top surface. In addition to coherency, it then suffices to cancel radial and tangential forces on the sidewalls. The simplest case is that of a single heterostructure made of two semi-infinite materials A and B. Such NWs were treated by Ertekin et al. (2002, 2005) by way of a variational method: the authors posit analytical forms for the elastic strains, specified by a few parameters, compute the total elastic energy, and find the parameters that minimize it. The results should be treated cautiously since the trial strain fields are not solutions of the equations of elasticity and do not even produce free lateral surfaces.

Ertekin et al. overlooked that a more general problem had been treated rigorously by Barton long before (Barton, 1941; Timoshenko and Goodier, 1951). In this pre-nano-era, Barton actually considered a cylindrical bar subjected to a uniform pressure along part $h$ of its infinite length. However, in the framework of continuum linear elasticity (where the system state depends only on the ratios of its dimensions), Barton's solution remains entirely valid for an infinite NW A including a misfitting B section (from which one can recover the solution for a semi-infinite B layer by taking the $h \to \infty$ limit). Barton uses a Fourier synthesis method common in linear elasticity (Sadd, 2005). He first finds an analytical solution for an elementary sinusoidal axial modulation of SF parameter (assuming an elastically isotropic continuum with uniform elastic constants). Any distribution of SF parameter, and in particular any distribution of axial inclusions, can then be solved by superimposing these elementary solutions, properly weighted. The same problem was treated by Kästner and Gösele (2004), who also overlooked Barton's work.

Barton's solution has been adapted to NWs of WZ structure with circular section and identical elastic constants. Zhong and Sun (2002) considered a single axial inclusion, assuming transverse elastic anisotropy. Kaganer and Belov (2012) generalized this by first deriving the elementary solution for a single sinusoidal component and then applying Fourier synthesis to the case of an arbitrary distribution of SF lattice parameter. For the particular case of an axial superlattice (SL), Wölz et al. (2013) proposed a "2D approximation" whereby, neglecting relaxation at the free sidewalls, the in-plane lattice parameter is uniform in the whole structure. The in-plane and axial strains (with respect to the SF states of A and B) are then uniform in the alternating wells and barriers that constitute the SL and proportional to the thicknesses of barrier and well, respectively. This is nothing else than the solution for a laterally infinite SL without substrate. The authors show that the average axial strains so calculated are a good approximation if the total height of the SL is of the order of the NW diameter but deviate from the exact calculation for smaller heights, which suggests a means of engineering strain by playing on the number of SL periods and ratio of well to barrier heights, as done in planar structures. Finally, Boxberg et al. (2012) compared FE calculations of the strain fields induced by periodic distributions of axial inclusions in ZB and WZ crystals and found that the main differences between these structures reside in the shear strains in the basal planes (those which are normal to the axis).

As regards atomistic approaches, Swadener and Picraux (2009) used an interatomic potential to find the equilibrium state of Si/Ge axial heterostructures. The VFF method has also been widely used to compute strains for axial heterostructures in infinitely long NWs. Niquet and coworkers first considered ZB GaAs/InAs and InAs/InP SLs (Niquet, 2006; Niquet and Camacho Mojica, 2008) and then GaN nanodisks in an AlN NW (Camacho Mojica and Niquet, 2010; Jalabert et al., 2012; Landré et al., 2010).

The whole NW with its axial insertion may be wrapped in a shell. This is frequently done in functional NWs, for instance in light emitting devices; then, the shell may act as a barrier preventing the charge carriers generated in the insertion to reach the sidewall surface where they might recombine nonradiatively. Such mixed heterostructures were treated by several authors (Furtmayr et al., 2011; Niquet, 2007; Rajadell et al., 2012). The shell is usually assumed to be of the same material as the semi-infinite NW foot. As expected, the shell impedes elastic relaxation of the inclusion. The elastic state of the system is highly sensitive to the aspect ratio of the inclusion and to the shell thickness (Rajadell et al., 2012).

The model of an infinite NW is not relevant for growth: one is then interested in the evolution with increasing thickness of layers that retain a free top surface. The case of a layer B of finite height with a free surface

(Fig. 1A) was treated by Glas (2006), assuming an isotropically elastic continuum. For lack of an analytical solution, the strains and the elastic energy were determined numerically in two ways, either by FE calculations (Fig. 2A and B) or by using Barton's solution. In the latter case, to find the solution for the A/B heterostructure with a free top surface, an infinitely long NW is considered, formed of a semi-infinite A foot, the B layer of interest, and a semi-infinite A cap in which a small number of additional misfitting layers of finite thicknesses are inserted. Superimposing Barton's solution for each layer of this composite NW (as allowed in linear elasticity) ensures that the lateral surfaces remain free. The number, positions, lengths, and misfits of the additional inclusions are optimized to cancel as best as possible the forces on the top facet of layer B (which is now an interface internal to the system). The two methods give very similar results, except for the thinnest layers (Glas, 2006). The second one has the advantage of providing analytical expressions for the elastic fields so approximated. FE calculations for heterostructures involving specific materials have also been performed (Hanke et al., 2007; Ye and Yu, 2014), some of them including the effect of a rigid

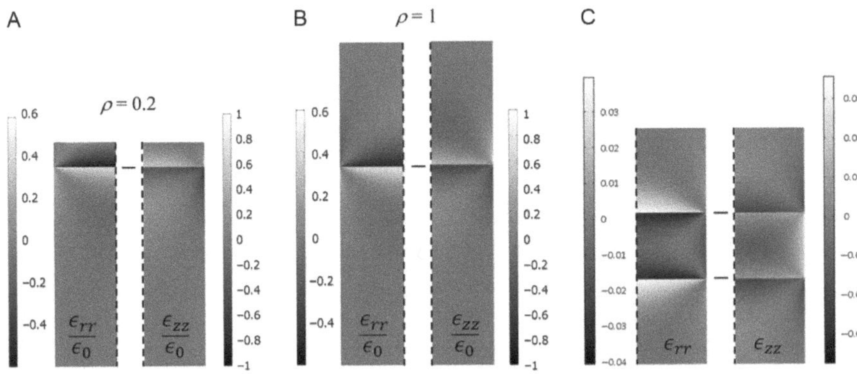


**Figure 2** Finite element calculations for axial heterostructures in elastically isotropic materials. (A,B) Layer with a free top surface of aspect ratio $\rho = 0.2$ (A) and $\rho = 1$ (B). Radial ($\epsilon_{rr}$) and axial ($\epsilon_{zz}$) extensional strains with respect to the SF state of each material, normalized to the intrinsic misfit $\epsilon_0$ of the layer relative to the semi-infinite stem of the NW. The map is drawn in a plane containing the NW axis (dashed lines) and only half-NWs are shown, the strains being symmetric about this axis. Poisson ratio of both materials is $\nu = 0.33$. (C) Double heterostructure: axial insertion of aspect ratio $\rho = 0.5$ having a misfit of +7.2% relative to the rest of the infinite NW in which it is inserted (approximating a CdSe insertion in ZnSe). Radial and axial strains (not normalized) with respect to the SF state of each material. Only the right half of the NW is shown (dashed line: NW axis). Poisson ratio of both materials is $\nu = 0.28$. In (A–C), the positions of the interfaces are marked by short segments.

substrate relatively close to the heterostructure (Kremer et al., 2014; Ye et al., 2009a,b).

These calculations substantiate the main features expected for an uncapped axial NW heterostructure. Namely, it is only when $h \ll R$ that the system state is close to that of a thin B layer deposited on a laterally infinite bulk A substrate (2D case): except at the very edge of the NW, the layer adapts its parameter to that of the NW foot (effective substrate) in the interface and extends or contracts (depending on the sign of the misfit) along the NW axis, while the substrate remains in its SF state. However, when $h$ increases, even modestly (Fig. 2A), the strains distribute between NW substrate and layer, which relax in all directions thanks to the free top and lateral surfaces. This is actually very effective well before the layer reaches an aspect ratio of 1, as shown in Fig. 2A. Any additional B material is deposited on a nearly strain-free surface and thus tends to adopt its own SF state (Fig. 2B), so that the elastic energy does not increase anymore with $h$. This saturation of the elastic energy, illustrated in Fig. 3 (full symbols), is the very reason for the existence of a critical radius for plastic relaxation, as will be seen in Section 3.2.1. The fact that the strain only penetrates the deposit over a distance of the order of the linear size

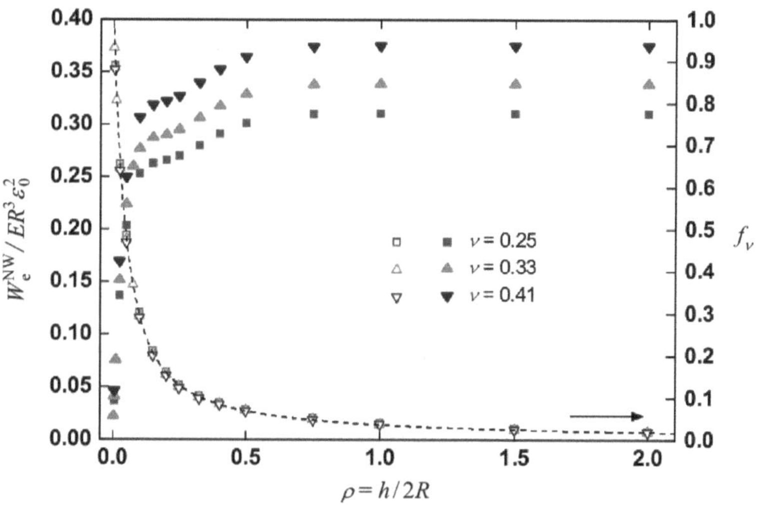


**Figure 3** Variation with the aspect ratio of the layer of the total energy $W_e^{NW}$ stored in the coherent NW with axial heterostructure (full symbols, left scale) and of the ratio $f_\nu$ of this energy to the energy stored in the same volume cut in a 2D system with identical layer thickness (empty symbols, right scale), for three values of the Poisson ratio $\nu$. The full line is a fit of the ratio using an approximation of function $f_\nu$ proposed by Glas and Daudin (2012), for $\nu = 0.33$.

of the source of disturbance (the interface) may be understood as an instance of Saint Venant's principle (Sadd, 2005).

To quantify this effect, we define function $f(\rho) = W_e^{NW}/W_e^{2D}$, where $W_e^{NW}$ is the total elastic energy stored in the whole NW with a coherent interface (state Se) and $W_e^{2D}$ the energy stored in the same volume cut in a planar heterostructure with a B epilayer of identical thickness $h$ coherently deposited on a bulk A substrate. Function $f$ must satisfy two constraints (Glas, 2006). First, $f(\rho) \to 1$ when $h \to 0$ (2D limit). At the opposite, when $h \gg R$, since the system is only strained over a height of order $R$, then $W_e^{NW} \sim R^3$, whereas $W_e^{2D}$ scales with the volume $\pi R^2 h$ of the uniformly strained 2D layer (see Section 1.2), hence $f(\rho) \sim \rho^{-1}$ at large $\rho$. If A and B are elastically isotropic continua with the same Young's modulus $E$ and Poisson ratio $\nu$, $W_e^{2D} = \pi R^2 h E \epsilon_0^2/(1-\nu)$ and function $f$ is independent of $E$ (we then note it $f_\nu$). Figure 3 also illustrates the variations of $f_\nu$ and shows that it depends only weakly on $\nu$. We proposed two simple approximations of $f_\nu$ that involve, respectively, one and three $\nu$-dependent parameters (Glas, 2006; Glas and Daudin, 2012); the second one reproduces very well the nontrivial variations of $W_e^{NW}$ around $\rho = 0.3$; Fig. 3). Note that the elastic energy is considerably reduced with respect to the 2D case even for modest aspect ratios, for instance to about a third of its 2D value for $\rho \approx 0.08$.

Turning briefly to the axial insertion embedded in a long NW (double axial heterostructure), we see in Fig. 2C that, as could be expected from the previous discussion, provided the insertion is not too thin, the strain fields generated at the top and bottom interfaces (due to the coherency constraint) hardly interact, so that the central part of the insertion is close to its SF state. Since the regions near the interfaces and sidewalls are on the contrary highly strained, the strain may be very inhomogeneous in insertions of aspect ratios around 1 (Fig. 2C).

## 3.2 Critical Dimensions for the Plastic Relaxation of Axial Heterostructures

### *3.2.1 Theory*

As regards plastic relaxation, axial heterostructures present a feature well known for epitaxial 2D layers, namely the existence of a critical layer thickness above which the introduction of dislocations at the interface between NW stem and layer becomes energetically favorable. However, this critical thickness depends not only on misfit but also on NW radius. Moreover, for small enough radii, the critical thickness becomes infinite. In other words, for a given misfit, there exists a critical NW radius $R_c$ such that arbitrarily thick layers should grow without dislocations on top of any NW of radius $R \leq R_c$.

The very existence of a critical radius can be asserted without any calculation (Glas, 2006). We saw in Section 3.1 that, provided aspect ratio $\rho$ is large enough, the elastic energy $W_e^{NW}$ in state Se (with pure elastic relaxation) scales with $R^3$, independently of the actual layer height. So then does its limit for $\rho \to \infty$ (Fig. 3). On the other hand, the energy of state Sd (with one or several dislocations) is for sure larger than the energy of the dislocation cores, which (neglecting possible atomic rearrangements at the intersection with the sidewalls) is proportional to the dislocation length (Hirth and Lothe, 1982), and hence to $R$, and independent of $h$, of the precise location of the defects in the interface and even of their number. From these different power-law dependences in $R$, it follows that, if the layer is thick enough, the dislocation energy always dominates, provided $R$ is small enough. In other words, there exists a critical radius $R_c$ below which plastic relaxation is unfavorable and coherent layers of arbitrary thickness should be obtainable.

We now review various quantitative estimates of the critical dimensions, limiting ourselves to axial deposits with a free surface (Figs. 1A and 2A and B) and starting with the critical thickness. For a NW of radius $R$, we may adopt the same "equilibrium" criterion as in the 2D case, namely that the critical layer thickness $h_c(R)$ is that at which the energy of the fully coherent state Se (Section 3.1) becomes larger than that of state Sd, with interfacial misfit dislocations (having an in-plane edge component). Calculating the latter energy imposes to specify these dislocations. One often assumes dislocations of maximal length, i.e., passing through the center of the interfacial disk (as in Fig. 1A). Kästner and Gösele (2004) noted that this is somehow justified by the fact that, since the misfit-induced forces acting on the dislocation are radial, this is an equilibrium position for an interfacial dislocation. This scheme is clearly not adapted to NWs with hexagonal cross sections and was even shown recently to be sometimes inadequate for a circular section (see below). The critical thickness is defined by the introduction of the first dislocation. Some studies have indeed considered a single interfacial defect (Ye et al., 2009b). However, since it is necessary to accommodate the misfit along two independent directions in the planar interface, most authors assume a pair of orthogonal dislocations (Fig. 1A) (Ertekin et al., 2002, 2005; Glas, 2006; Ye et al., 2009a).

Whereas the purely elastic energy of state Se may be safely calculated using linear elasticity, as reviewed in Section 3.1, evaluating the energy of state Sd is more tricky. One must then account for the modification of the strain and stress fields of state Se by the dislocation. In linear elasticity, these modified fields are the sums of the Se fields and of those induced by the dislocation, which are independent of the former but not of the geometry of the system (in particular, they must also keep the sidewalls free of forces). The elastic energy is obtained by integrating the products of the total strain and stress components (Sadd, 2005). It is thus comprised of three terms, namely the elastic energy in state Se, the dislocation self-energy, and the interaction energy, which involves the strain–stress cross-products of the Se and dislocation fields.

For an analytical estimate of the critical dimensions, one may proceed as sometimes done in 2D structures, namely approximate the total energy by the sum of the dislocation self-energy and of a reduced elastic energy allowing for partial misfit accommodation by the dislocation (Glas, 2013). However, even in such a simple scheme, the calculation poses some problems. First, the strain field of a dislocation segment normal to the axis of a cylinder has not been calculated analytically so far. Second, the reduced misfit is not easy to define in NWs, because the interface is finite (in 2D systems, it can be defined for a periodic array of dislocations). Despite these difficulties, Glas (2006) used this method to calculate the critical dimensions. The reduced misfit is set to $\epsilon_0 - \epsilon_a$, with $\epsilon_0$ the SF misfit and $\epsilon_a$ that part of the misfit which is accommodated by the dislocations. More precisely, $\epsilon_a$ is taken as the misfit that would be accommodated in the 2D case by a dislocation array with the same length of dislocation per unit area. Building on an idea proposed by Ovid'ko (2002) for NWs lying on a substrate, the dislocation self-energy is evaluated by using the 2D formula (Fitzgerald, 1991) for a layer thickness equal to the distance $\widetilde{h}$ between the dislocation and the closest free surface. Since the distance to the NW sidewall varies along the dislocation, Glas proposed to use its average $\bar{d} = 2R/\pi$. When $h$ is small, $\widetilde{h} = h$, whereas $\widetilde{h} = \bar{d}$ at large $h$. Using this approach and the first expression of $f_\nu$ mentioned above, Glas (2006) derived an implicit equation from which the radius-dependent critical thickness $h_c(R)$ may be calculated for given material parameters.

The variations of this critical thickness with NW radius are shown by full lines in Fig. 4, assuming plastic relaxation by 60° dislocations of the fcc structure with Burgers vector modulus $b = 0.4$nm, a value pertaining to ZB GaAs and a good approximation for many cubic semiconductors. For each misfit, the $h_c(R)$ curve separates, in the $(R, h)$ plane of layer dimensions, the

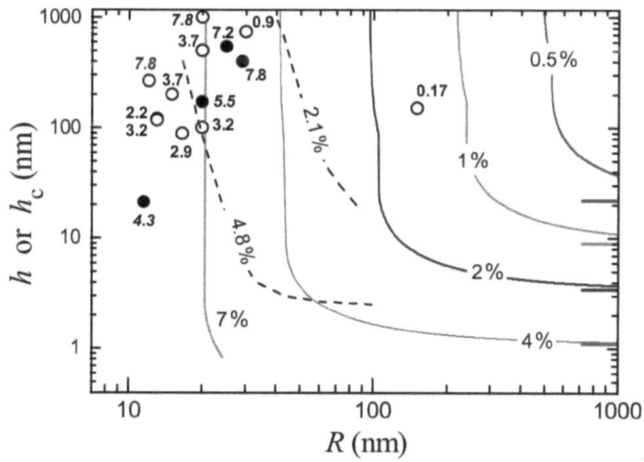


**Figure 4** Full lines: variations of the critical thickness $h_c$ of a misfitting layer growing on top of a NW as a function of the NW radius, for various values of the misfit (indicated for each curve), calculated for elastically isotropic materials with $\nu = 1/3$ (Glas, 2006). In the plastically relaxed state, misfit is accommodated by 60° dislocations ($b = 0.4$nm). Each segment at the right gives the asymptote of the curve immediately above, i.e., the 2D critical thickness. Dashed lines: Critical thickness calculated by Ye et al. (2009b) assuming pure edge dislocations with $b = 0.3$nm for the two misfits indicated. The symbols give the radii and thicknesses of layers grown without (circles) or with (disks) dislocations, for various values of the misfit (indicated in percent near each point), by the following experimentators: 0.17% (Wu et al., 2002), 0.9% (Svensson et al., 2005), 2.2% and 2.9% (Glas, 2006), 3.2% (Björk et al., 2002b; Larsson et al., 2007), 3.7% (Verheijen et al., 2006, Zhang et al., 2010), 4.3% (Tourbot et al., 2011), 5.5% (de la Mata et al., 2014), 7.2% (Hiruma et al., 1996), 7.8% (de la Mata et al., 2014; Guo et al., 2006; Jeppsson et al., 2008). The italics signal the few results not in agreement with the calculation of Glas (2006).

coherency domain (below, left) from the domain of plastic relaxation (above, right). The critical thickness increases rapidly as the NW radius decreases. Moreover, for each misfit, $h_c(R) \to \infty$ when $R$ tends to some finite value. This value is the critical radius $R_c(\epsilon_0)$, to be further discussed below.

Ye and coworkers (2009a,b, 2014) adopted another approach, based on a single-step numerical calculation of the elastic energy of state Sd, to which it suffices to add the dislocation core energy. This yields a better estimate of the Sd energy, at the expense of generality and analyticity of the calculations. Specifically, the energy of state Sd was first obtained by FE calculations, introducing extra half-planes corresponding to a single edge dislocation (Ye et al., 2009a) or a pair of crossed edge dislocations (Ye et al., 2009b) and, more recently, using the Peach-Koehler formula (Hirth and Lothe, 1982) which gives at first order the energy as the work performed during the glide of the dislocation from the surface in the preexisting Se strain field, itself calculated by FE (Ye and Yu, 2014). Using this last approach, the authors could treat the 60° dislocations of the fcc structure and demonstrate that their equilibrium position is displaced from the center of the interfacial disk (as found before by Spencer and Tersoff, 2000 for quantum dots, using the same approach). The critical thicknesses determined in this way (dashed lines in Fig. 4) are comparable to those calculated by Glas (2006) but vary less abruptly with NW radius.

Dealing with infinite A/B NWs, Ertekin et al. (2002, 2005) could not calculate any critical thickness, but they were the first to determine a critical radius. The calculation is again based on the comparison of the energies of the NW with and without a single pair of misfit dislocations. In the former case, the authors evaluated the dislocation self-energy by replacing the layer thickness of the 2D formula (Fitzgerald, 1991) by the NW radius $R$ and computed the elastic energy for a residual misfit $\epsilon_a = b/(2R)$. Conversely, in models allowing for a variable finite layer height, the critical radius is simply obtained as the radius at which the critical thickness becomes infinite. In this way, Ye and coworkers (2009a,b, 2014) could determine the critical radii for a particular set of heterostructures. On the other hand, Glas (2006) derived a second implicit equation which gives the critical radius for given misfit and material parameters. The respective merits of these two approaches are the same as for the critical thickness.

Let us now examine the quantitative estimates of the critical dimensions, which are summarized in Figs. 4 and 5. Those of Ye et al. (2009a,b) tend to be lower than those of Glas (2006), presumably because the energy of state Sd is lower. This comparison is especially meaningful when the same Burgers vectors are considered (Ye and Yu, 2014). Whatever the authors, the critical radii remain of several tens of nm (which is easily achievable in most systems) for misfits of several %. This confirms that, because of elastic relaxation at the lateral free surfaces, the NW geometry is indeed very favorable to the formation of defect-free axial heterostructures. As will be seen in Section 5, the situation is less favorable for core–shell heterostructures.

### 3.2.2 Experiments

As yet, there is no systematic experimental determination of critical thicknesses in NWs with axial heterostructures, although the state (coherent or not) of the interface is frequently mentioned. In most strained axial heterostructures fabricated, so far, the layers have been prudently kept below their 2D critical thicknesses (Lauhon et al., 2004; Poole et al., 2003). However,

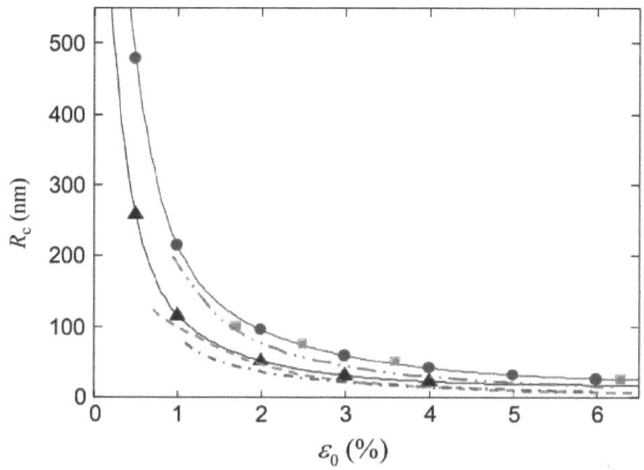


**Figure 5** Symbols and full lines: critical NW radius $R_c$ under which arbitrarily thick misfitting layers should grow coherently, as a function of misfit $\epsilon_0$, as calculated by Glas (2006). For $R \geq R_c$, plastic relaxation is assumed to occur via edge dislocations (blue (dark gray in the print version) triangles and line) or 60° dislocations (red (gray in the print version) disks and line). $b = 0.4$ and 0.33 nm. Dashed and dash-dotted lines: calculations of Ye et al. (2009b) and Ertekin et al. (2005) for pure edge dislocations with $b = 0.3$ nm. Double-dot-dash lines: calculations of Ye and Yu (2014) for a 60° dislocation with $b = 0.4$ nm. Green (light gray in the print version) squares: critical radii measured by Chuang et al. (2007) for NWs on misfitting substrates.

there are reports of layers grown beyond this limit (Fig. 4), most of which claim the absence of misfit dislocations (Björk et al., 2002a; Clark et al., 2008; Glas, 2006; Guo et al., 2006; Jeppsson et al., 2008; Larsson et al., 2007; Svensson et al., 2005; Verheijen et al., 2006; Wu et al., 2002; Zhang et al., 2010). On the other hand, there is an early report by Hiruma et al. (1996) of misfit dislocations in a cubic GaAs/InAs axial heterostructure. More recently, Tourbot et al. (2011) observed by TEM a single edge dislocation lying end on in a $Ga_{0.57}In_{0.43}N$ interface, without capping layer. Interestingly, the dislocation was located midway between the sidewalls, as considered in the simple models (Fig. 1A). In most cases, the calculations of Glas (2006) and Ye et al. (2009b) appear to predict correctly the state (coherent or not) experimentally observed (symbols in Fig. 4).

There seems to be no experimental determination of the critical radii of axial heterostructures in NWs. Note that the absence of dislocation for radius $R$ and deposit thickness $h$ only means that $h_c(R) > h$, whereas the presence of dislocations implies that $R_c < R$ and $h_c(R) < h$. Determining the critical NW radius may seem difficult, since it should involve the growth of very tall deposits. In practice, we might content ourselves with the growth of thicknesses of the order of the NW diameter, since, as discussed above, the upper portion of thicker layers is effectively free of strain and stress (Fig. 2B).

## 4. NANOWIRES ON A MISFITTING SUBSTRATE

NWs offer great flexibility for growing not only A/B axial heterostructures in the body of the NW (Section 3) but also a NW of material B on a substrate of another material A (Fig. 1D). Along with chemical synthesis, epitaxy is actually the main route for elaborating semiconductor NWs. A and B may be both elementary or compound semiconductors but the growth of compound NWs on Si substrates, which opens an alternative route to the integration of optical and microelectronic functions, has developed rapidly, in particular with the advent of self-catalyzed III–V NW growth (see e.g., Ramdani et al., 2013 and references therein).

The case of a cylindrical NW B grown directly on a misfitting bulk substrate A (Fig. 1D) is closely related to that of an axial A/B heterostructure in a NW (Section 3 and Fig. 1A). We may however expect that lateral relaxation will not be as efficient because the substrate is devoid of lateral free surfaces and thus has less latitude to relax. This is true only to a limited extent. FE calculations for a coherent interface in such a structure show that elastic strain relaxation is also very efficient in this case. Assuming elastically isotropic materials with identical elastic constants, this is now measured by ratio $f_\nu'(\rho) = W_e^{'NW} / W_e^{2D}$, where $W_e^{'NW}$ is the elastic energy stored in the whole system and $W_e^{2D}$ the energy of a section $\pi R^2$ of the 2D system for the same deposit height $h$, as in Section 3.1. When $h \rightarrow 0$, the deposit tends to its 2D elastic state in both cases, hence $f_\nu' \rightarrow 1$ and $f_\nu'/f_\nu \rightarrow 1$. As $h$ increases, $f_\nu'/f_\nu$ rapidly tends to an asymptotic value typically close to 1.5. Using the same energy estimate as in the axial case for the configuration with a pair of interfacial dislocations (Glas, 2006) yields critical radii which are about 2/3 of the critical radii for the axial heterostructure. The misfit does not affect much the NW properties, since relaxation is again nearly complete above a distance of order $R$ from the substrate. Because of this enhanced elastic relaxation, the critical thickness for plastic relaxation is also larger than in the 2D case. One also finds that, below a certain misfit-dependent critical radius, the NW should grow without plastic relaxation. However, the critical dimensions remain lower than for the genuine axial heterostructure (Section 3.2). Several recent calculations confirm this trend (Sburlan et al., 2012; Ye and Yu, 2014; Zhang et al., 2011).

At the beginning of growth, a misfitting NW on a substrate resembles a quantum island grown on a substrate in the Volmer–Weber or Stranski–Krastanow modes. The main difference is that the lateral extension of the NW is limited to a small diameter, fixed by the catalyst nanoparticle in the case of vapor–liquid–solid (VLS) growth, by some self-limiting growth mechanism, as in the case of group-III nitride NWs (Consonni, 2013), or by the openings of a mask. On the contrary, quantum islands are not limited laterally (except by the surrounding strain field that impedes the surface diffusion and attachment of species). Their width thus tends to increase with their height, which inevitably leads to plastic relaxation beyond a certain volume. This is an important difference between quantum dots and NWs.

There are only a few experiments that purport to have determined critical radii in this case. Unfortunately, this seems to have been done not by actually observing the structure of the interface, but by recording a drastic change of growth mode for some critical NW radius, possibly accompanied by a dramatic decrease of the photoluminescence efficiency of the structure. It is actually often difficult to observe defects lying in the substrate/NW interface since this part of the structure may be buried into a parasitic layer growing between the NWs. Chuang et al. (2007) and Cirlin et al. (2009) grew various compound semiconductors on mismatched Si or GaAs substrates. The results of these two groups for the same A/B couples are very close. Their experimental critical radii decrease with increasing mismatch and are approximately twice as large as those calculated by Ertekin et al. (2005), but surprisingly close to the values of Glas (2006) for axial heterostructures and 60° dislocations (Fig. 5). However, the values which can be calculated for the same defects following the method outlined above for the actual case of a NW on a mismatched substrate are about 40% smaller, whereas the values recently calculated by Ye and Yu (2014) appear to be closer.

# 5. CORE–SHELL HETEROSTRUCTURES

## 5.1 Elastic Relaxation in Core–Shell Heterostructures: Theoretical Considerations

When modeling a core–shell heterostructure, we may treat the interface as a set of planar sidewalls (Fig. 1C) or else as a single-curved surface (Fig. 1B). In any case, the core–shell misfit has an axial component and an "in-plane tangential component" (the latter normal to the axis and tangent to the sidewall), both of which must be accommodated. In elastically isotropic or cubic

NWs, these misfits are equal (in case of partial coherence, the level of elastic accommodation and the mode of plastic relaxation may however be different). In NWs with hexagonal structure grown along the **c** axis, these components are *a priori* different and may even have different signs (Perillat-Merceroz et al., 2012). In the calculations, the NW is usually taken as infinite along its axis $z$. This remains a good approximation as long as the length of the NW is large compared to its diameter, provided one is not interested in the strains near the extremities.

Analytical calculations of strain relaxation in core–shell heterostructures, invariably limited to NWs with circular section, have been performed by many authors under various hypotheses. Considering isotropically elastic materials, Ovid'ko and coworkers (Gutkin et al., 2000; Ovid'ko and Sheinerman, 2004) and Nazarenko et al. (2013) assume identical elastic constants for core and shell; this was generalized to different shear moduli and identical (Aifantis et al., 2007; Menéndez et al., 2011) or different (Kloeffel et al., 2014) Poisson ratios. Various cases of elastic anisotropy were also considered: transverse elastic isotropy about the system axis (which, as recalled in Section 1.3, includes hexagonal crystals) (Warwick and Clyne, 1991), hexagonal crystals with the NW axis along **c** (Haapamaki et al., 2012; Raychaudhuri and Yu 2006a,b; Salehzadeh et al., 2013), cubic crystals with the NW axis along [111] (Raychaudhuri and Yu, 2006a), or [001] (Trammell et al., 2008). Some of these studies do not provide a rigorous solution of the elasticity problem and make undue assumptions (for instance, Raychaudhuri and Yu (2006a,b) take all strain components uniform in core and shell). Ferrand and Cibert (2014) recently gave a very comprehensive calculation that generalizes the previous treatments and corrects some of them. They also treat the case of multiple shells and discuss the effect of strain on the electronic properties of the composite NWs.

Numerical calculations present the advantage of handling easily elastic and geometric anisotropies, and in particular to account for the hexagonal cross section adopted by most semiconductor NWs (Fig. 1C). Søndergaard et al. (2009) and Hocevar et al. (2013) applied the FE method to NWs of cubic structure. Grönqvist et al. (2009) performed joint FE and VFF calculations for ZB crystals with hexagonal cross section and for circular cylinders, and Boxberg et al. (2012) gave a comprehensive treatment of ZB and WZ NWs using the FE method. Niquet and coworkers developed a VFF model for WZ and applied it to core–shell GaN/AlN NWs with hexagonal cross sections (Camacho Mojica and Niquet, 2010; Camacho and Niquet, 2010; Hestroffer et al., 2010). Kavanagh et al. (2012) applied molecular dynamics to InAs/GaAs core–shell structures, without however giving details of the calculations.

The broad features of elastic relaxation in core–shell heterostructures depend little on crystal structure and sidewall geometry. For infinitely long NWs, as a consequence of the translational invariance along the axis, the elastic strains are independent of $z$ and $\epsilon_{zz}$ is uniform in core and shell. Moreover, the difference $\epsilon^{s}_{zz} - \epsilon^{c}_{zz}$ between axial extensional strains in shell (s) and core (c) with respect to the SF materials must equal the opposite of the SF axial shell/core relative lattice mismatch (Glas, 2013). As a first approximation, one may assume that $\epsilon_{zz}$ is partitioned between core and shell in inverse proportion to the areas of their sections, possibly corrected by the inverse ratio of the axial elastic constants $C_{zz}$ (Hestroffer et al., 2010). In addition, for isotropic elasticity and for hexagonal and cubic NWs with, respectively, **c** and [111] axes, all strain components are uniform in the core and the in-plane radial extensional strain $\epsilon_{rr}$ is uniform in the shell (assuming a circular section; Fig. 1B). In the less symmetric case of cubic NWs along [001], the strains are nonuniform in the core and the shell is warped (Boxberg et al, 2012; Ferrand and Cibert, 2014).

We illustrate the strain field and its dependence on the relative thicknesses of core and shell in the case of an InAs core of radius $R$ wrapped by an InP shell of thickness $H$, both of WZ structure, the axis of the NW being along **c** and its section hexagonal. The shell lattice parameters are smaller than those of the core by about 3.25% (assuming equiaxed misfits equal to the ZB value). If $H << R$, the situation resembles the 2D case, with most of the core, playing the role of a thick substrate, close to its SF state and the strain mainly localized in the shell. However, even for a modest shell thickness (Fig. 6, $H/R = 0.3$), core and shell start sharing the axial strain $\epsilon_{zz}$ (here in a ratio of about ½, close to the approximation mentioned above). The shell thus dilates and the core contracts along the NW axis while in the other (short) directions, the shell is able to relax and therefore accommodates most of the mismatch, as in the 2D case: along each planar portion of sidewall, it is tangentially dilated and, due to the Poisson effect, radially contracted (Fig. 6, note, e.g., that $\epsilon^{s}_{xx} > 0$, $\epsilon^{s}_{yy} < 0$ for the $xz$-oriented sidewalls). Although in the shell $\epsilon_{xx}$ and $\epsilon_{yy}$ are of the same order as the misfit, their half sum (the in-plane strain $\epsilon_{\perp}$) is much lower. As expected, when $H$ increases (Fig. 7, $H/R = 2.1$), the axial strain localizes mainly in the core. On the contrary, the partition of the transverse strains between core and shell does not change much in the vicinity of the interface, with the core slightly contracted and the shell highly dilated in the tangential direction. The strains

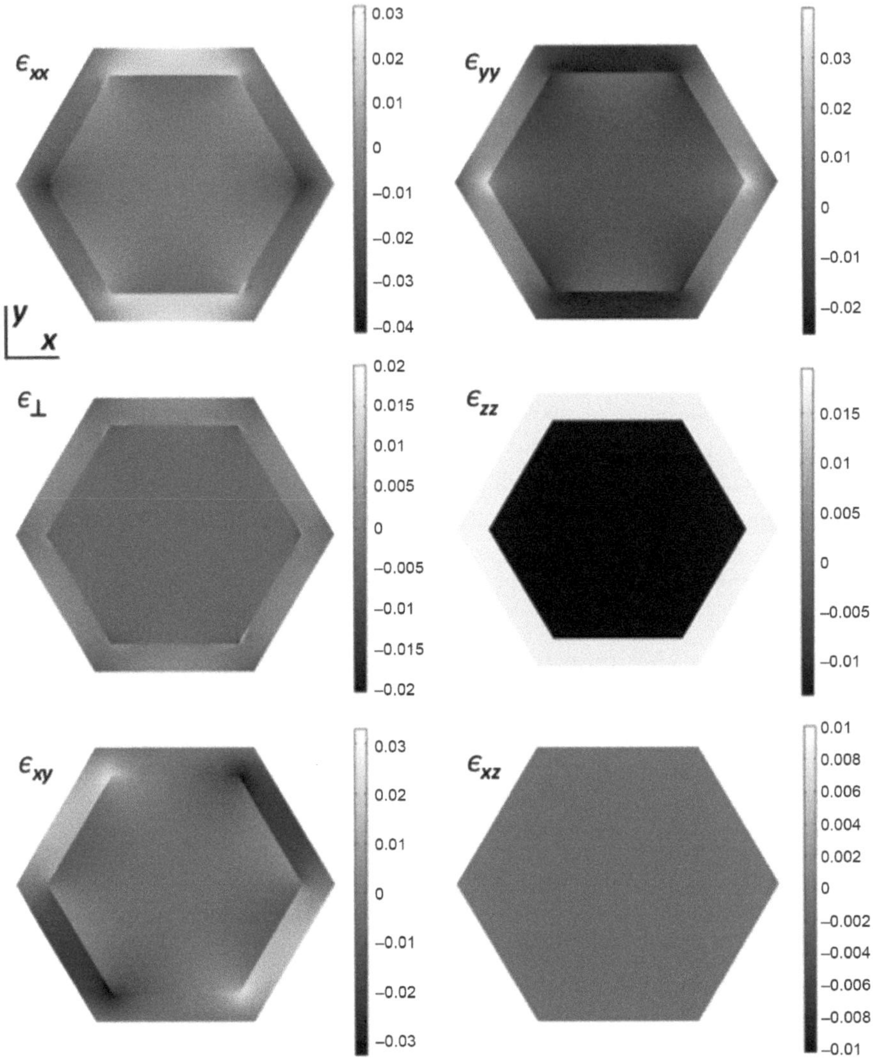


**Figure 6** Finite element calculations for an InAs core/InP shell NW with $H/R = 0.3$. The intrinsic lattice misfit is 3.25%. Maps of various components of the strain field in the *xy* plane (axes *x* and *y* as indicated) normal to the **c** (*z*) axis. $\epsilon_{\perp} = (\epsilon_{xx} + \epsilon_{yy})/2$.

normal to the sidewalls have the opposite sign in the shell, which is still largely free to contract in this direction. The most important point is that, at variance with the uniform $\epsilon_{zz}$ components, the in-plane strains decrease rapidly in the shell away from the interface over a distance of the order of $R$ (Fig. 7, left). Here again, the outer free surface of the NW and its particular

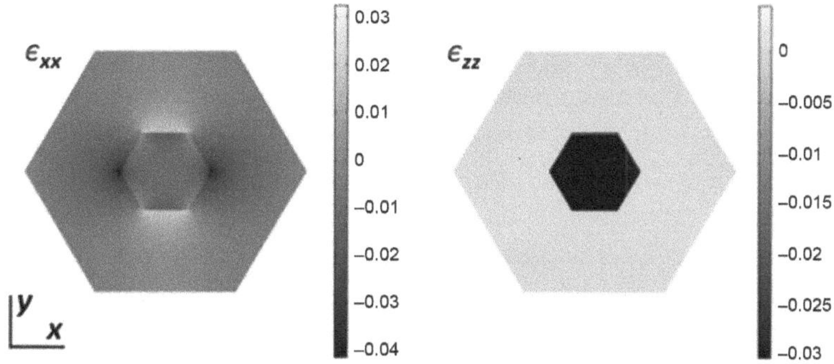


**Figure 7** Same as Fig. 6 for $H/R = 2.1$. Only two strain components are shown.

geometry insure efficient stress relaxation. As a consequence, the elastic energy stored in the NW per unit axial length increases with $H$ but rapidly saturates. The core has then reached its maximum deformation and, as in the case of the axial heterostructure, any extra material is added at the periphery of a shell which is virtually in its SF state, and therefore also adopts this state.

The shear strain in the transverse plane, $\epsilon_{xy}$, is of the same order as the extensional strains $\epsilon_{ii}$ (Fig. 6). This is yet another manifestation of the increased latitude for relaxation granted to the NW by its narrow lateral size. On the other hand, the shear strains $\epsilon_{iz}$ $(i = x, y)$ are zero. Boxberg et al. (2012) remarked that this is a major difference between the WZ and ZB structures. For the latter, the shear strains $\epsilon_{iz}$ are actually of the same order as the other strain components in the shell (Grönqvist et al., 2009), while they remain small in the core.

In the system illustrated, because of the transverse elastic isotropy of hexagonal crystals, the orientation of the NW sidewalls is irrelevant as long as continuum elasticity is used, and the strain field is sixfold symmetric. In atomistic VFF calculations, where the termination of the sidewall facets matters, a slight reduction of symmetry to threefold may be observed (Hestroffer et al., 2010).

## 5.2 Plastic Relaxation and Critical Dimensions in Core–Shell Heterostructures

### *5.2.1 Theoretical Considerations*

Predictions about the plastic relaxation of core–shell NW heterostructures have been available for at least 15 years but it is only recently that experiments started to appear, and they remain scarce. We first review the theoretical background. The heterostructure is defined by its dimensions $(R, H)$

(Fig. 1B and C) and by the core–shell misfit(s). As in the case of axial heterostructures (Section 3.2), most studies aim at determining possible critical values of these parameters, at which the system should undergo a transition from elastic to plastic. Following the "equilibrium" approach (Section 3.2.1), this is usually done by comparing the energies of states Se (pure elastic relaxation; see Section 5.1) and Sd, with one or several extended defects, most often interfacial dislocations. The onset of plastic relaxation is again marked by the introduction of the first dislocation, which now may *a priori* relieve partly either the axial misfit, the in-plane tangential misfit (Section 5.1), or both.

Given these two components of the misfit, two simple types of dislocations have been considered as candidates for plastic relaxation (Fig. 1B and C), namely straight edge dislocations parallel to the NW axis, relieving in-plane tangential misfit (Gutkin et al., 2000; Ovid'ko and Sheinerman, 2006; Raychaudhuri and Yu, 2006a,b) and edge dislocation loops normal to the NW axis (usually termed *prismatic loops*), relieving axial misfit (Aifantis et al., 2007; Colin, 2010; Haapamaki et al., 2012; Ovid'ko and Sheinerman, 2004, 2006; Raychaudhuri and Yu, 2006a,b). We note that a single straight dislocation breaks the rotational symmetry of the system and a single prismatic loop its translational symmetry.

As regards plastic relaxation, core–shell heterostructures present the same two basic features as axial heterostructures, namely a radius-dependent critical thickness and a critical radius, both dependent on misfit. Here, the critical radius is such that arbitrarily thick shells should grow without dislocations around any NW of radius $R \leq R_c$. The very existence of $R_c$ can be understood from simple dimensional arguments in line with those developed for axial heterostructures (Glas, 2006; see also Section 3.2.1). Actually, we have seen in Section 5.1 that, for elastically strained shells thicker than the core, the in-plane strain components extend only over a distance of order $R$ away from the interface (Fig. 7). In state Se, the total energy of a length $L$ of NW thus scales as $R^2$, independently of $H$ (assumed to be large). Conversely, in state Sd, the energy is surely larger than the energy of the dislocation core (since there is also a positive elastic energy), which, as a first approximation, for a dislocation along the NW axis, scales as $L$ and is independent of $R$ (Hirth and Lothe, 1982). This readily shows that, for narrow enough cores, Se has a lower energy than Sd. A similar argument can be developed for a single interfacial loop: if $L$ is now a typical axial length over which the loop relieves stress, introducing the loop will procure a decrease of elastic energy scaling as $LR^2$, whereas the dislocation core energy scales as $R$. What happens for cores wider than $2R_c$? Then, as in the planar and axial cases, there is a maximum shell thickness above which dislocations are favored. Of course, this critical thickness $H_c$ depends on $R$: it decreases from infinity at $R = R_c$ to the 2D critical thickness for very wide cores (Fig. 8). In addition, both critical dimensions are expected to decrease with increasing misfit (Figs. 8 and 9), since the elastic energy scales with the square of the misfit whereas the dislocation core energy is independent of it.

Whereas most models indeed predict the existence of a radius-dependent critical thickness and of a critical radius (Gutkin et al., 2000; Haapamaki et al., 2012; Nazarenko et al., 2013; Ovid'ko and Sheinerman, 2004; Raychaudhuri and Yu, 2006a,b; Salehzadeh et al., 2013; Trammell et al., 2008), quantitative estimates call for an evaluation of the energy of state Sd, taking into account that the dislocation modifies the strain and stress fields of state Se (Section 5.1). As in the axial case (Section 3.2.1), these modified fields are the sums of the Se fields and of those induced by the dislocation, which are again independent of the former but not of the system

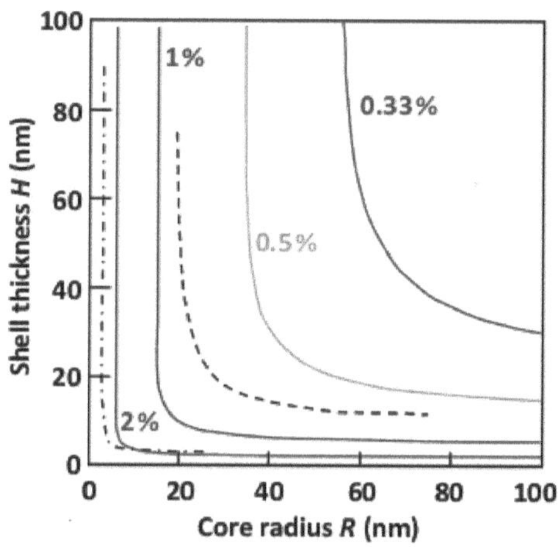


**Figure 8** Full lines: calculations by Raychaudhuri and Yu (2006b) of the critical shell thickness $H_c$ of GaN/$Al_xGa_{1-x}$N core–shell NWs as a function of core radius $R$. Each curve corresponds to a different shell composition-dependent relative misfit (taken as the average of axial and transverse misfits for c-oriented NWs), indicated in %. The asymptotes for $H \rightarrow \infty$ and $R \rightarrow \infty$ give respectively the critical radius $R_c$ and the 2D critical thickness. The diagram may also be read as a stability map in the plane (NW radius, shell thickness), where each curve separates the domain of elastic relaxation (state Se) at left and bottom from the domain of plastic relaxation (state Sd) at right and top. Dashed and dash-dotted lines: The same for WZ InAs/InP NWs (3.2% misfit) and WZ InAs/$Al_{0.15}In0_{0.85}$As NWs (1% misfit) as calculated by, respectively, Salehzadeh et al. (2013) and Haapamaki et al. (2012).

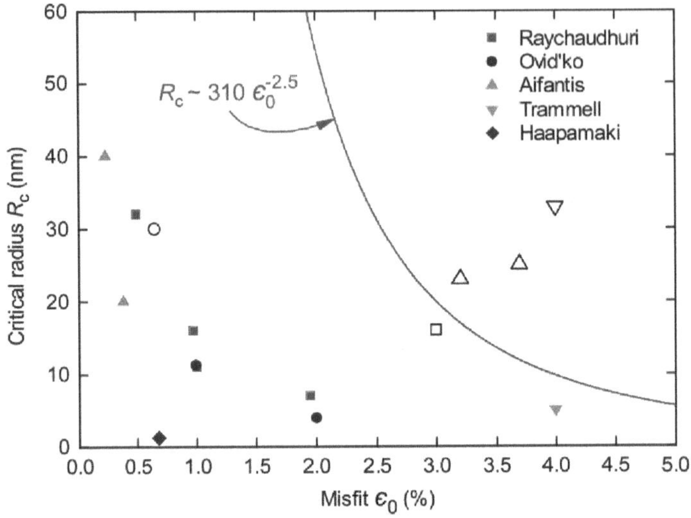


**Figure 9** Full symbols and line: critical radii calculated as a function of misfit $\epsilon_0$ by Raychaudhuri and Yu (2006b), Ovid'ko and Sheinerman (2004), Aifantis et al. (2007), Trammell et al. (2008), Haapamaki et al. (2012) (symbols), and Salehzadeh et al. (2013) (line) for various material systems (see text). In the case of Raychaudhuri and Yu, the misfit is the average of the axial and transverse values and data for two nitride systems are included. The open symbols give upper bounds of $R_c$ estimated by us from the observation of dislocations by Perillat-Merceroz et al. (2012) (○), Dayeh et al. (2013) (□), Salehzadeh et al. (2013) (▽), and Conesa-Boj et al. (2014) (△).

geometry. The total energy consists of the three same terms as before (elastic energy in state Se, dislocation self-energy, and interaction energy). As an approximation, one may sum the dislocation self-energy and an elastic energy corresponding to a reduced misfit, allowing for partial accommodation by the dislocation. However, this reduced misfit is not any easier to define than in the axial case.

### 5.2.2 Calculations

To evaluate the geometry-dependent self-energy of any of the dislocations considered, one must know their strain fields in a generic NW. As regards straight dislocations parallel to the NW axis, the full calculation of the elastic energy was performed by Gutkin et al. (2000) for isotropic materials with identical elastic constants (strain fields were not given). Raychaudhuri and Yu (2006a,b) include elastic anisotropy but their calculations appear oversimplified (reduced misfit approximation, with homogeneous strains in core and shell; dislocation strain fields for infinite media, neglecting relaxation at the sidewalls). As regards prismatic loops, Raychaudhuri and Yu (2006a,b) use the same crude approximations and Haapamaki et al. (2012) also rely on the strain field in an infinite medium. Before that, Ovid'ko and Sheinerman (2004) had however calculated the elastic fields of the loop in an elastically isotropic cylinder, and Cai and Weinberger (2009) had given a semi-analytical estimate of the corresponding energy. Chu et al. (2011) considered loops tilted with respect to the axis of an elastically isotropic circular NW, which therefore adopt an elliptical shape in the interface, taking into account the free surface by an image force method. This may look a more realistic configuration than circular loops since glide planes may not be normal to the NW axis (see Section 5.2.3). The authors carried out extensive numerical calculations but, strangely, this lead them to predict an increase of the critical shell thickness with core radius (and therefore no critical radius). The reasons for these discrepancies are not clear and this model will not be discussed further.

### *5.2.3 Which Dislocations May Actually Form?*

Comparing the energies of states Se and Sd results in an equilibrium criterion which does not specify which dislocations may actually form. This question has two aspects: (i) what are the energetically more favorable dislocations? (ii) Can these actually form in NWs? As for the first point, there exist speculations about which types of dislocations are possible and consequently about which part of the misfit might be relieved first. The predictions are however somewhat conflicting. Ovid'ko and Sheinerman (2004) and Nazarenko et al. (2013) predict, by simply comparing the total energies of the NW with either type of dislocations, and leaving aside crystallography, that loops are generally less costly to form than axial dislocations. Salehzadeh et al. (2013) also argue that loops should appear first because they relieve axial strain which is present in both core and shell, whereas radial and in-plane tangential strains are much weaker in the core (Section 5.1, Figs. 6 and 7). Raychaudhuri and Yu (2006a) point a difference between fcc and hexagonal crystals: in the former, dislocations along the $\langle 111 \rangle$ NW axis are not stable, according to Frank's criterion (Hirth and Lothe, 1982), so that stress relief should occur first via loops, whereas in the latter, dislocations along **c** (with in-plane Burgers vectors along **a**) are possible. These authors also find the critical dimensions relative to the axial dislocations to be always lower than those relative to loops, so that the former should appear first, in WZ crystals.

Beyond these energy-based arguments, the identification of the possible glide planes has recently emerged as a key question. The core being usually free of dislocation before shell growth, it is expected that dislocations form as half-loops at the surface of the shell and move toward the interface by glide or climb; dislocation reactions have also been invoked. It is thus capital to decide if a given dislocation may actually form by glide from the sidewalls. Dayeh et al. (2013) stressed that, in cubic NWs, loops relieving axial misfit cannot extend by glide in the (111) plane normal to the NW axis since their Burgers vectors necessarily have components out of this plane. The efficient loops should rather lie in the other, steeply inclined, {111} planes. Dayeh et al. (2013) performed dislocation dynamics simulations that describe the glide of these defects. This is a rare (if not unique) instance of the use in the field of NWs of this well-established technique, which allows one to go beyond the "equilibrium" approach to plastic relaxation. Reasoning on WZ NWs with m-plane facets, Perillat-Merceroz et al. (2012) similarly argue that glide on the basal **c** plane is not allowed, which rules out the direct formation of the prismatic loops considered in the simple models (Fig. 1C). They note moreover that the activated glide planes should depend on the location around the NW. Along most of each m facet, the stress state is nearly biaxial; glide of axial dislocations is then possible in prismatic planes (those planes that contain the NW axis) not normal to the facet. Near the vertical facet intersects, stress is not biaxial anymore and other slip systems may be activated, possibly facilitating plastic relaxation.

### *5.2.4 Results*

These considerations stress the limitations of the simple models that include prismatic loops and axial dislocations, and only these defects. We nevertheless summarize some quantitative results of these models, in part to indicate the problems that they raise. First, the predicted critical shell thickness $H_c(R)$ usually transits from a value close to the 2D critical thickness to infinity over a small range of decreasing core radii. Except for small misfits, the critical shell thickness is thus either very small or quasi-infinite (Fig. 8). Second, various authors find rather different values of critical dimensions for a given misfit (Figs. 8 and 9). A first illustration of such discrepancies is given in Fig. 8, where one sees that the critical thicknesses calculated by Haapamaki et al. (2012) for a misfit of 1% are much lower, at given core radius, than those calculated by Salehzadeh et al. (2013) for a misfit of 3.2%, although the reverse would have been expected. The critical core radii calculated in the framework of isotropic elasticity by Ovid'ko and

Sheinerman (2004) and Aifantis et al. (2007) and for nitrides by Raychaudhuri and Yu (2006b) are of the same order whereas those of Salehzadeh et al. (2013) may be 10 times larger (for both non-nitride WZ and ZB III–V compounds). The critical radii found by Trammell et al. (2008) lie somewhere in between whereas those of Haapamaki et al. (2012) (InAs/$Al_xIn_{1-x}As$) are extremely small and the value given by Nazarenko et al. (2013) extremely large. These discrepancies have not been much commented so far, although they affect key portions of the effectively accessible parameter space. For instance, it is only for misfits less than about 1% that the first groups find critical radii approaching those of the narrowest existing NWs whereas, for Salehzadeh et al. (2013), who claim that their calculations reproduce well the experimental results, the critical radii are still of about 20 nm for a misfit of 3% in the ZB structures.

Hence, beyond the general ability that NWs have to relax strain thanks to their free surfaces, there is an important difference between core–shell and axial heterostructures. In the latter, even for misfits of several %, the NW radius may be below the critical radius. In the former, the calculated critical radii are small (probably up to 10 times smaller for a given misfit; compare Figs. 5 and 9) and reach the order of typical NW radii only at low misfit (less than 1%), except for a few authors (Nazarenko et al. 2013; Salehzadeh et al. 2013). An obvious reason for this relatively inefficient strain relaxation in core–shell structures is that the core, being entirely surrounded by the shell, has little latitude to relax laterally, all the more so that the shell thickens.

### *5.2.5 Experiments*

There are convincing reports of misfitting core–shell heterostructures free of interfacial dislocations with shell thicknesses exceeding the 2D critical thickness in systems such as Ge/(Si,Ge) (Dayeh et al., 2013; Lu et al., 2005; Varahramyan et al., 2009), ZB (Ga,In)As/GaAs (Nazarenko et al., 2013), or InAs/GaSb (Rieger et al., 2015), WZ InAs/InP (Salehzadeh et al., 2013), GaN/AlN (Hestroffer et al., 2010; Rigutti et al., 2011), or ZnO/(Zn,-Mg)O (Perillat-Merceroz et al., 2012). The absence of dislocations for core–shell dimensions $(R, H)$ means that $H_c(R) > H$, whereas the presence of dislocations implies that $R_c < R$ and $H_c(R) < H$. As in the case of axial heterostructures (Section 3.2.2), determining the critical core radius is *a priori* difficult, since it requires growing very thick shells, that hinder the observation of the interface. In practice, we might satisfy ourselves with the growth of shells with thicknesses of the order of the core diameter, since the outer portion of thicker shells is effectively free of in-plane strain while most of the axial misfit is accommodated by the core (Fig. 7). The first reports of plastic relaxation in these structures only appeared recently but it does not seem that any measurement of the critical core radius has been claimed so far.

Dislocations possibly corresponding to the simple kinds discussed above have been reported by several authors. Extra (111) or **c** half-planes were observed by TEM in core–shell structures such as Ge/(Si,Ge) (Goldthorpe et al., 2009), InAs/GaAs (Kavanagh et al., 2012), GaAs/Si (Conesa-Boj et al., 2014), or GaN/AlN (Arbiol et al., 2012). The observations of extra half-planes normal to the axis being made edge-on, it is not easy to decide if a full loop is present [although this has been suggested on the basis of a comparison with strain calculations by Conesa-Boj et al. (2014)] or if only an interfacial segment, with terminations threading through the shell, is present. In addition to such half-planes, Popovitz-Biro et al. (2011) have observed axial lines in the same NWs consisting of high misfit (about 8%) WZ InAs/GaAs heterostructures.

In two different sets of experiments, it is clear that such loops in basal planes did not appear. In low misfit ZnO/(Zn,Mg)O WZ NWs, Perillat-Merceroz et al. (2012) observed two families of half-loops lying, respectively, in prismatic and pyramidal planes and presenting segments threading through the shell. The loops of the first family relieve in-plane misfit and might later evolve into the axial dislocations of the models (Fig. 1C). The loops of the second family, which surround the core only partially, have Burgers vectors with both in-plane and axial components and therefore relieve simultaneously axial and in-plane tangential misfits. A quasi-periodic distribution of similar loops relieving both misfit components has been observed in cubic Ge/(Si,Ge) heterostructures by Dayeh et al. (2013). These loops lie in a $\{111\}$ plane not normal to the NW axis and have a standard $\frac{a}{2}\langle 110\rangle$ Burgers vector with both in-plane and axial components. In this case, the close correspondence between the half-planes located in the two diametrically opposed portions of the shell observable in a single TEM image, with an axial shift corresponding to the tilt of the glide plane, strongly suggests that these are closed, and entirely located in the interface; they are likely to represent a more mature stage of plastic relaxation.

If instead of an infinite core–shell structure, the misfitting part of the core is limited in height (as in the case of axial quantum disks, barriers, or SLs inserted in a NW; see e.g. Fig. 2C), the formation of full loops by glide in steeply inclined $\{111\}$ planes is expected to be unfavorable, since the projection of such a loop along the axis would generally be taller than the insertion, so that part of it would actually not lie in the core–shell interface.

Dislocations have been observed in such mixed axial/core–shell structures, namely GaN disks in AlN NWs (Bougerol et al., 2009), or (In,Ga)N disks in GaN NWs (Roshko et al., 2014) but apparently, as expected, only on one side of the structure.

# 6. OTHER POSSIBLE INSTANCES OF STRAIN RELAXATION IN NWs

## 6.1 Augmented Strain Relaxation via Morphological Changes

In the context of 2D growth, it is well known that a misfitting layer deposited on a planar substrate may also accommodate the epitaxial stress by developing a nonplanar surface. If this starts via a smooth surface corrugation, the phenomenon is usually acknowledged as an instance of the Asaro–Tiller–Grinfeld (ATG) morphological instability of the planar free surface of a solid submitted to uniaxial or biaxial stress (Asaro and Tiller, 1972; Grinfel'd, 1986; Srolovitz, 1989). For an alloy layer, this morphological instability may be coupled to a compositional instability (Glas, 1997). At high misfit, islands may form without any initial planar growth stage (Volmer–Weber growth) or after a thin planar wetting layer has grown (Stranski–Krastanow growth). In any case, the development of the nonplanarity requires no extended defect and its driving force is an increased elastic relaxation (with respect to the 2D case) at the flanks of the corrugation or at the island facets.

In the case of NWs, since a similar lateral relaxation is at the root of the efficient accommodation of stress (Sections 3–5), one may however wonder if any significant extra elastic relaxation is to be expected from a morphological instability. This however seems to be the case, both for axial and core–shell heterostructures, in which cases the altered morphology corresponds respectively to a nonplanar top surface and to a noncircular section or nonplanar sidewall facets.

The possibility of the spontaneous formation of a nonuniformly thick axial deposit was investigated by Glas and Daudin (2012). The authors considered the formation of a cylindrical island that is narrower than the NW stem. To this end, they calculated numerically the total energy of the system (elastic and surface contributions) for a given deposit volume and determined the optimal shape of the deposit as a function of misfit, NW radius $R$, and nominal deposit thickness. For given values of any two of these parameters, there is a critical value of the third one above which the formation of an island narrower than the stem is favored compared to that of a disk of equal volume covering uniformly the top facet. These critical values are well within the ranges of current NW systems. At high misfit, islanding may reduce the total energy by more than 50%, provided $R$ is not too small. In most cases of instability, the optimal island is well defined, in that its radius is well below $R$ (Glas and Daudin, 2012). We believe that, for semiconductor NWs, such islanding is likely to occur primarily in the case of catalyst-free growth since, in the case of VLS growth, each new solid monolayer (ML) tends to nucleate at the triple phase line (Glas et al., 2007) or in its vicinity, before rapidly spreading over the whole top facet of the NW. It is possible that the spontaneous formation of an In-rich core surrounded by a Ga-rich shell which has been observed when an (In,Ga)N alloy is deposited on top of a GaN NW (Chang et al., 2010; Tourbot et al., 2011) is triggered by the formation of such an island (in this case, the shell would form after the core). The formation of faceted (In,Ga)N islands with bases narrower than the NW, observed in the same system, might be an even more convincing indication of the mechanism (Tourbot et al., 2012). However, core–shell structures might also form if all successive MLs adopted the NW radius but a radially modulated composition, as studied theoretically by Niu et al. (2012), a mechanism in which strain also plays a significant part.

In turn, the spontaneous corrugation of the outer surface of a core–shell NW with circular section was studied theoretically by Schmidt et al. (2008). The authors performed a linear stability analysis modeled on those developed for the usual ATG instability (Spencer et al., 1992; Srolovitz, 1989). They tested the stability of the cylindrical outer surface against an elementary corrugation with joint sinusoidal profiles along the NW axis and around it. The authors find that core–shell NWs may indeed be morphologically unstable. They also note that surface stress (Müller and Saúl, 2004), which relates to the change of energy due to straining the surface (as opposed to that due to a change of area, described by surface tension), must be taken into account in such nanostructures. This induces a dependence on the sign of the misfit. The morphological instability of NWs was studied theoretically by other authors (Duan et al., 2008; Li and Yang, 2014; Wang et al., 2008). Some experiments do indeed reveal surface islanding in strained core–shell Si–Ge NWs, which, as in the case of the ATG instability, might represent a more mature stage of the process; the islands positions along the axis are in antiphase for opposite facets (Pan et al., 2005), which might correspond to the second harmonic of the theory of Schmidt et al. (2008). InAs islands have also been grown on the side facets and at the edges of the side facets of AlAs-capped GaAs NWs (Uccelli et al., 2010) or GaAs NWs (Yan et al., 2015) and

GaN islands at the edges of AlN-capped GaN NWs (Arbiol et al., 2012), although it is not clear if strain plays an important part in these cases.

## 6.2 Stacking Faults, Twins, and Polytypism

We recalled in Section 1.3 that semiconductor NWs most often grow along the [111] (cubic) or ⟨0001⟩ (hexagonal) directions. Along these directions, all successive MLs are the same but they can occupy three distinct lateral positions, commonly noted A, B, and C. Cubic stacking corresponds to repeated ABC sequences, whereas the hexagonal structure consists in the repeat of only two ML positions, e.g., ABABAB... (Glas, 2008). Stacking on top of each other two MLs at the same lateral position (e.g., AA) costs such a large energy that it is not observed, at least in NWs, nor usually considered in the discussions. An error in the stacking sequence creates a basal stacking fault (e.g., ABCBCABC... in ZB) or a twin plane (ABCBABC...). Such faults transform locally one of the two structures into the other one (Glas, 2008; Yeh et al., 1992). Depending on material and growth conditions, these defects can be quite frequent, in particular in III–V NWs, where mixed ZB/WZ structures are also often observed (this is usually termed *polytypism*).

Because the stacked MLs are identical (in terms of in-plane atomic arrangement), it is at first not expected that a basal fault will induce any appreciable strain in the NW or conversely that it could release a sizeable portion of the strain generated by a misfitting inclusion. Single faults have however been mentioned as sources of strain by some authors (Shimamura et al., 2013), whereas others consider them as "strain-free" (Corfdir et al., 2014). On the other hand, the insertion of a several-ML-long block of one phase into the other might have a larger effect. The ZB and WZ structures of a given material may indeed have different lattice parameters. Unfortunately, the theoretical and experimental evidences for such polytypism-induced strain are somewhat inconclusive. Taking GaAs as an example, the early *ab initio* calculations of Yeh et al. (1992) predict relative WZ/ZB differences of in-plane and axial lattice parameters of about −2.1% and −1.3%, for bulk crystals. The XRD measurements of McMahon and Nelmes (2005) on polycrystalline WZ GaAs however give much lower values (−0.2% and +0.55%). As regards NWs, Zardo et al. (2009) have studied individual polytypic GaAs NWs by space-resolved Raman spectroscopy (Section 2.2) and noted shifts of some Raman peaks that they correlate with the varying abundance of ZB and WZ along the NW axis and therefore attribute to different lattice parameters of these two phases. On the other hand, in their detailed study of strain in individual faulted NWs by coherent XRD, Favre-Nicolin et al. (2010) do not consider such planar faults as a source of strain but only as an extra source of scattering that may hinder the analysis of strain (induced by the insertion of InSb sections in their InP NWs). We can only conclude that more studies are called for on this delicate topic.

### 6.3 Sidewall-Induced and Edge-Induced Strains

In addition, there are also calculations showing that, in the absence of any compositional or structural inhomogeneity, the NW sidewall facets and their edges may act as sources of strain. It is indeed well known that, even in the absence of reconstruction, the atoms at a semiconductor surface may be considerably displaced from their bulk positions because, due to their incomplete environment, they are submitted to forces different from those experienced by the internal atoms. The strain thus created can of course propagate inside the NW (for a review of elastic effects in surface physics, see Müller and Saúl, 2004). Yuan et al. (2012) have performed molecular dynamics calculations of this effect for GaAs NWs. They find that the GaAs nearest neighbor distances may be modified by several percents for the surface atoms but the effect dampens over only a few atomic distances away from the free surface.

The same group has recently shown that surface stress may mediate interactions between MLs (Yuan and Nakano, 2013). Using again molecular dynamics, they demonstrate that it is the atoms at the sidewall corners that are most displaced in a twin plane and that the displacements at a given edge tend to be opposite for two successive twins. This interaction has the effect that if a twin has appeared, the formation of a second twin is favored within a range of distances from the first one that depends on NW diameter. The authors propose that this may generate self-replicating twin SLS in NWs, an explanation alternative to that proposed before by Algra et al. (2008), based on the distortion of the catalyst droplet in VLS and not involving strain.

## 7. SUMMARY AND CONCLUSIONS

By choosing appropriate material combinations, growth methods, and growth conditions, it is possible to fabricate axial or core–shell heterostructures (and combinations thereof) in semiconductor NWs. It is particularly in this case that strain may manifest itself in these nanostructures. However, the NW geometry offers great flexibility for associating in such heterostructures materials with markedly different lattice constants, without introducing extended defects. This stems mainly from the ease with which the internal misfit stress may relax elastically at the NW sidewalls, which constitute a free boundary of the system. This opens the possibility of combining materials that cannot be assembled without introducing crystalline defects not only in planar heterostructures but even in the more favorable configuration of quantum dots on a substrate.

We have seen that, for both core–shell and axial heterostructures, there exists a specific critical deposit thickness under which no interfacial dislocation should form. Although this is also the case for planar deposits, the values for a NW are systematically larger, and in practice often much larger, for a given misfit. Even more striking, and specific to NWs, is the existence of a critical NW radius for each type of heterostructure, under which arbitrarily thick deposits may be fabricated without introducing interfacial defects. The basic reason for this is that a thick deposit, be it axial or in the form of a shell, is strained only over a distance from the interface of the order of the NW radius, the rest of it (if any) being effectively free of strain. Critical thicknesses and critical radii decrease when misfit increases. Experiments do confirm that thicker axial or shell layers may be grown without defects on a NW than in the case of a planar heterostructure.

Beyond these similarities, axial and core–shell heterostructures present important differences. For axial heterostructures, the extant calculations of critical thicknesses and critical radii yield values of the same order, which seem broadly confirmed by the experiments. For core–shell heterostructures, the values found by various authors may differ widely. As yet, the reasons for these discrepancies are not clear; they cannot be explained solely by differences of the material systems considered. Hence, whereas there are converging evidences that axial structures release strain so efficiently that one can assemble highly misfitting materials in NWs with diameters of several tens of nanometers, the same cannot be asserted of core–shell structures. Unless the core radius is very small, they present a high risk of plastic relaxation. In the core–shell geometry, since the misfit must be relieved along the axial and in-plane tangential directions of the interface, it is often assumed that two different types of dislocations of edge character (axial lines and basal loops) must form. Considering NW geometry and possible slip systems for a given crystal structure indicates that the actual situation might be more complex. The few experiments on plastic relaxation indeed show the "textbook" situation in some cases whereas in others, tilted dislocation loops that relieve simultaneously both misfit components are observed. Especially for core–shell structures, plastic relaxation thus calls for more theory and experiments.

This chapter did not deal with the important topic of how strain affects the electronic, optical, or magnetic properties of semiconductor NWs. However, the studies that we reviewed provide useful guidelines for selecting a geometry and material parameters likely to produce heterostructures with specific strain-related physical properties, without compromising the physical properties of the structure by introducing extended defects.

## REFERENCES

Aifantis, K.E., Kolesnikova, A.L., Romanov, A.E., 2007. Nucleation of misfit dislocations and plastic deformation in core/shell nanowires. Philos. Mag. 87, 4731–4757.

Alarcón-Lladó, E., Conesa-Boj, S., Wallart, X., Caroff, P., Fontcuberta i Morral, A., 2013. Raman spectroscopy of self-catalyzed $GaAs_{1-x}Sb_x$ nanowires grown on silicon. Nanotechnology 24, 405707.

Algra, R.E., Verheijen, M.A., Borgström, M.T., Feiner, L.-F., Immink, G., van Enckevort, W.J.P., Vlieg, E., Bakkers, E.P.A.M., 2008. Twinning superlattices in indium phosphide nanowires. Nature 456, 369–372.

Arbiol, J., Magen, C., Becker, P., Jacopin, G., Chernikov, A., Schfer, S., Furtmayr, F., Tchernycheva, M., Rigutti, L., Teubert, J., Chatterjee, S., Morante, J.R., Eickhoff, M., 2012. Self-assembled GaN quantum wires on GaN/AlN nanowire templates. Nanoscale 4, 7517–7524.

Asaro, R.J., Tiller, W.A., 1972. Interface morphology development during stress corrosion cracking: Part I. Via surface diffusion. Metall. Trans. 3, 1789–1796.

Barton, M.V., 1941. The circular cylinder with a band of uniform pressure on a finite length of the surface. J. Appl. Mech. 8, A97–A104.

Béché, A., Bougerol, C., Cooper, D., Daudin, B., Rouvière, J.L., 2011. Measuring two dimensional strain state of AlN quantum dots in GaN nanowires by nanobeam electron diffraction. J. Phys. Conf. Ser. 326, 012047.

Bellet-Amalric, E., Elouneg-Jamroz, M., Bougerol, C., Den Hertog, M., Genuist, Y., Bounouar, S., Poizat, J., Kheng, K., André, R., Tatarenko, S., 2010. Epitaxial growth of ZnSe and ZnSe/CdSe nanowires on ZnSe. Phys. Status Solidi C 7, 1526–1529.

Björk, M.T., Ohlsson, B.J., Sass, T., Persson, A.I., Thelander, C., Magnusson, M.H., Deppert, K., Wallenberg, L.R., Samuelson, L., 2002a. One-dimensional heterostructures in semiconductor nanowhiskers. Appl. Phys. Lett. 80, 1058–1060.

Björk, M.T., Ohlsson, B.J., Sass, T., Persson, A.I., Thelander, C., Magnusson, M.H., Deppert, K., Wallenberg, L.R., Samuelson, L., 2002b. One-dimensional steeplechase for electrons realized. Nano Lett. 2, 87–89.

Bougerol, C., Songmuang, R., Camacho, D., Niquet, Y.M., Mata, R., Cros, A., Daudin, B., 2009. The structural properties of GaN insertions in GaN/AlN nanocolumn heterostructures. Nanotechnology 20, 295706.

Bougerol, C., Songmuang, R., Camacho, D., Niquet, Y.M., Daudin, B., 2010. Structural properties of GaN nanowires and GaN/AlN insertions grown by molecular beam epitaxy. J. Phys. Conf. Ser. 209, 012010.

Boxberg, F., Søndergaard, N., Xu, H.Q., 2012. Elastic and piezoelectric properties of zinc blende and wurtzite crystalline nanowire heterostructures. Adv. Mater. 24, 4692–4706.

Cai, W., Weinberger, C.R., 2009. Energy of a prismatic dislocation loop in an elastic cylinder. Math. Mech. Solids 14, 192.

Camacho Mojica, D., Niquet, Y.-M., 2010. Stark effect in GaN/AlN nanowire heterostructures: influence of strain relaxation and surface states. Phys. Rev. B 81, 195313.

Camacho, D., Niquet, Y.M., 2010. Application of Keating's valence force field model to non-ideal wurtzite materials. Phys. E 42, 1361–1364.

Chang, Y.-L., Wang, J.L., Li, F., Mia, Z., 2010. High efficiency green, yellow, and amber emission from InGaN/GaN dot-in-a-wire heterostructures on Si(111). Appl. Phys. Lett. 96, 013106.

Chu, H.J., Wang, J., Zhou, C.Z., Beyerlein, I., 2011. Self-energy of elliptical dislocation loops in anisotropic crystals and its application for defect-free core/shell nanowires. Acta Mater. 59, 7114–7124.

Chuang, L.C., Moewe, M., Chase, C., Kobayashi, N.P., Chang-Hasnain, C., Crankshaw, S., 2007. Critical diameter for III–V nanowires grown on lattice-mismatched substrates. Appl. Phys. Lett. 90, 043115.

Cirlin, G.E., Dubrovskii, V.G., Soshnikov, I.P., Sibirev, N.V., Samsonenko, Y.B., Bouravleuv, A.D., Harmand, J.C., Glas, F., 2009. Critical diameters and temperature domains for MBE growth of III–V nanowires on lattice mismatched substrates. Phys. Status Solidi Rapid Res. Lett. 3, 112.

Clark, T.E., Nimmatoor, P., Lew, K.-K., Pan, L., Redwing, J.M., Dickey, E.C., 2008. Diameter dependent growth rate and interfacial abruptness in vapor-liquid-solid $Si/Si_{1-x}Ge_x$ heterostructure nanowires. Nano Lett. 8, 1246–1252.

Colin, J., 2010. Prismatic dislocation loops in strained core-shell nanowire heterostructures. Phys. Rev. B 82, 054118.

Conesa-Boj, S., Dunand, S., Russo-Averchi, E., Heiss, M., Rüffer, D., Wyrsch, N., Ballif, C., Fontcuberta i Morral, A., 2013. Hybrid axial and radial Si-GaAs heterostructures in nanowires. Nanoscale 5, 9633–9639.

Conesa-Boj, S., Boioli, F., Russo-Averchi, E., Dunand, S., Heiss, M., Rüffer, D., Wyrsch, N., Ballif, C., Miglio, L., Fontcuberta i Morral, A., 2014. Plastic and elastic strain fields in GaAs/Si core-shell nanowires. Nano Lett. 14, 1859–1864.

Consonni, V., 2013. Self-induced growth of GaN nanowires by molecular beam epitaxy: a critical review of the formation mechanisms. Phys. Status Solidi Rapid Res. Lett. 7, 699–712.

Consonni, V., Feuillet, G. (Eds.), 2014. Wide Band Gap Semiconductor Nanowires for Optical Devices. Wiley-ISTE, London.

Corfdir, P., Hauswald, C., Zettler, J.K., Flissikowski, T., Lähnemann, J., Fernández-Garrido, S., Geelhaar, L., Grahn, H.T., Brandt, O., 2014. Stacking faults as quantum wells in nanowires: density of states, oscillator strength, and radiative efficiency. Phys. Rev. B 90, 195309.

Cros, A., Mata, R., Hestroffer, K., Daudin, B., 2013. Ultraviolet Raman spectroscopy of GaN/AlN core-shell nanowires: core, shell, and interface modes. Appl. Phys. Lett. 102, 143109.

Dayeh, S.A., Tang, W., Boioli, F., Kavanagh, K.L., Zheng, H., Wang, J., Mack, N.H., Swadener, G., Huang, J.Y., Miglio, L., Tu, K.-N., Picraux, S.T., 2013. Direct measurement of coherency limits for strain relaxation in heteroepitaxial core/shell nanowires. Nano Lett. 13, 1869–1876.

de la Mata, M., Magén, C., Caroff, P., Arbiol, J., 2014. Atomic scale strain relaxation in axial semicondctor III–V nanowire heterostructures. Nano Lett. 14, 6614–6620.

Dillen, D.C., Varahramyan, K.M., Corbet, C.M., Tutuc, E., 2012. Raman spectroscopy and strain mapping in individual $Ge\text{-}Si_xGe_{1-x}$ core-shell nanowires. Phys. Rev. B 86, 045311.

Duan, H.L., Weissmüller, J., Wang, Y., 2008. Instabilities of core-shell heterostructured cylinders due to diffusions and epitaxy: spheroidization and blossom of nanowires. J. Mech. Phys. Solids 56, 1831–1851.

Ertekin, E., Greaney, P.A., Sands, T.D., Chrzan, D.C., 2002. Equilibrium analysis of lattice-mismatched nanowire heterostructures. In: Klimov, V., Buriak, J.M., Wayner, D.D.M., Priolo, F., White, B., Tsybeskov, L. (Eds.), Quantum Confined Semiconductor Nanostructures. In: Mat. Res. Soc. Proc., vol. 737. Materials Research Society, Pittsburgh, p. F10.4.

Ertekin, E., Greaney, P.A., Chrzan, D.C., 2005. Equilibrium limits of coherency in strained nanowire heterostructures. J. Appl. Phys. 97, 114325.

Eymery, J., Rieutord, F., Favre-Nicolin, V., Robach, O., Niquet, Y.-M., Fröberg, L., Mårtensson, T., Samuelson, L., 2007. Strain and shape of epitaxial InAs/InP nanowire superlattice measured by grazing incidence X-ray techniques. Nano Lett. 7, 2596–2601.

Favre-Nicolin, V., Eymery, J., Koester, R., Gentile, P., 2009. Coherent diffraction imaging of single nanowires of diameter 95 nanometers. Phys. Rev. B 79, 195401.

Favre-Nicolin, V., Mastropietro, F., Eymery, J., Camacho, D., Niquet, Y.M., Borg, B.M., Messing, M.E., Wernersson, L.-E., Algra, R.E., Bakkers, E.P.A.M., Metzger, T.H., Harder, R., Robinson, I.K., 2010. Analysis of strain and stacking faults in single nanowires using Bragg coherent diffraction imaging. New J. Phys. 12, 035013.

Ferrand, D., Cibert, J., 2014. Strain in crystalline core-shell nanowires. Eur. Phys. J. Appl. Phys. 67, 30403.

Fitzgerald, E.A., 1991. Dislocations in strained-layer epitaxy: theory, experiment, and applications. Mater. Sci. Rep. 7, 87–142.

Furtmayr, F., Teubert, J., Becker, P., Conesa-Boj, S., Morante, J.R., Chernikov, A., Schäfer, S., Chatterjee, S., Arbiol, J., Eickhoff, M., 2011. Carrier confinement in GaN/$Al_xGa_{1-x}N$ nanowire heterostructures ($0 \leq x(1$). Phys. Rev. B 84, 205303.

Geng, H., Yan, X., Zhang, X., Li, J., Huang, Y., Ren, X., 2012. Analysis of critical dimensions for axial double heterostructure nanowires. J. Appl. Phys. 112, 114307.

Glas, F., 1997. Thermodynamics of a stressed alloy with a free surface: coupling between the morphological and compositional instabilities. Phys. Rev. B 55, 11277–11286.

Glas, F., 2006. Critical dimensions for the plastic relaxation of strained axial heterostructures in free-standing nanowires. Phys. Rev. B 74, 121302.

Glas, F., 2008. A simple calculation of energy changes upon stacking fault formation or local crystalline phase transition in semiconductors. J. Appl. Phys. 104, 093520.

Glas, F., 2013. Heterostructures and strain relaxation in semiconductor nanowires. In: Wang, S. (Ed.), Lattice Engineering: Technologies and Applications. Pan Stanford, Singapore, pp. 189–228.

Glas, F., Daudin, B., 2012. Stress-driven island growth on top of nanowires. Phys. Rev. B 86, 174112.

Glas, F., Guille, C., Hénoc, P., Houzay, F., 1987. TEM study of the molecular beam epitaxy island growth of InAs on GaAs. In: Cullis, A.G., Augustus, P.D. (Eds.), Microscopy of Semiconducting Materials 1987. In: Inst. Phys. Conf. Ser., vol. 87. The Institute of Physics, Bristol, pp. 71–76.

Glas, F., Gors, C., Hénoc, P., 1990. Diffuse scattering, size effect and alloy disorder in ternary and quaternary III/V compounds. Philos. Mag. B 62, 373–394.

Glas, F., Harmand, J.C., Patriarche, G., 2007. Why does wurtzite form in nanowires of III-V zinc blende semiconductors? Phys. Rev. Lett. 99, 146101.

Goldthorpe, I.A., Marshall, A.F., McIntyre, P.C., 2009. Inhibiting strain-induced surface roughening: dislocation-free Ge/Si and Ge/SiGe core-shell nanowires. Nano Lett. 9, 3715–3719.

Grinfel'd, M.A., 1986. Instability of the separation boundary between a nonhydrostatically stressed elastic body and a melt. Sov. Phys. Dokl. 31, 831–834.

Grönqvist, J., Søndergaard, N., Boxberg, F., Guhr, T., Åberg, S., Xu, H.Q., 2009. Strain in semiconductor core-shell nanowires. J. Appl. Phys. 106, 053508.

Guo, Y.N., Zou, J., Paladugu, M., Wang, H., Gao, Q., Tan, H.H., Jagadish, C., 2006. Structural characteristics of GaSb/GaAs nanowire heterostructures grown by metal-organic chemical vapor deposition. Appl. Phys. Lett. 89, 231917.

Gutkin, M.Y., Ovid'ko, I.A., Sheinerman, A.G., 2000. Misfit dislocations in wire composite solids. J. Phys. Condens. Matter 12, 5391–5401.

Haag, S.T., Richard, M.-I., Favre-Nicolin, V., Welzel, U., Jeurgens, L.P.H., Ravy, S., Richter, G., Mittemeijer, E.J., Thomas, O., 2013. In situ coherent X-ray diffraction of isolated core-shell nanowires. Thin Solid Films 530, 113–119.

Haapamaki, C.M., Baugh, J., LaPierre, R.R., 2012. Critical shell thickness for InAs-$Al_xIn_{1-x}P$ core-shell nanowires. J. Appl. Phys. 112, 124305.

Hanke, M., Eisenschmidt, C., Werner, P., Zakharov, N.D., Syrowatka, F., Heyroth, F., Schäfer, P., Konovalov, O., 2007. Elastic strain relaxation in axial Si/Ge whisker heterostructures. Phys. Rev. B 75, 161303.

Hestroffer, K., Mata, R., Camacho, D., Leclere, C., Tourbot, G., Niquet, Y.M., Cros, A., Bougerol, C., Renevier, H., Daudin, B., 2010. The structural properties of GaN/AlN core-shell nanocolumn heterostructures. Nanotechnology 21, 415702.

Hirth, J.P., Lothe, J., 1982. Theory of dislocations, second ed. Wiley, New York.

Hiruma, K., Murakoshi, H., Yasawa, M., Katsuyama, T., 1996. Self-organized growth of GaAs/InAs heterostructure nanocylinders by organometallic vapor phase epitaxy. J. Cryst. Growth 163, 226–231.

Hocevar, M., Immink, G., Verheijen, M., Akopian, N., Zwiller, V., Kouwenhoven, L., Bakkers, E., 2012. Growth and optical properties of axial hybrid III-V/silicon nanowires. Nat. Commun. 3, 1286.

Hocevar, M., Giang, L.T.T., Songmuang, R., den Hertog, M., Besombes, L., Bleuse, J., Niquet, Y.-M., Pelekanos, N.T., 2013. Residual strain and piezoelectric effects in passivated GaAs/AlGaAs core-shell nanowires. Appl. Phys. Lett. 102, 191103.

Hÿtch, M.J., Snoeck, E., Kilaas, R., 1998. Quantitative measurement of displacement and strain fields from HREM micrographs. Ultramicroscopy 74, 131–146.

Jalabert, D., Curé, Y., Hestroffer, K., Niquet, Y.M., Daudin, B., 2012. Strain state of GaN nanodisks in AlN nanowires studied by medium energy ion spectroscopy. Nanotechnology 23, 425703.

Jeppsson, M., Dick, K.A., Wagner, J.B., Caroff, P., Deppert, K., Samuelson, L., Wernersson, L.-E., 2008. GaAs/GaSb nanowire heterostructures grown by MOVPE. J. Cryst. Growth 310, 4115–4121.

Kaganer, V.M., Belov, A.Y., 2012. Strain and X-ray diffraction from axial nanowire heterostructures. Phys. Rev. B 85, 125402.

Kästner, G., Gösele, U., 2004. Stress and dislocations at cross-sectional heterojunctions in a cylindrical nanowire. Philos. Mag. 84, 3803–3824.

Kavanagh, K.L., Saveliev, I., Blumin, M., Swadener, G., Ruda, H.E., 2012. Faster radial strain relaxation in InAs-GaAs core-shell heterowires. J. Appl. Phys. 111, 044301.

Keating, P.N., 1966. Effect of invariance requirements on the elastic strain energy of crystals with application to the diamond structure. Phys. Rev. 145, 637–645.

Kehagias, T., Dimitrakopulos, G.P., Becker, P., Kioseoglou, J., Furtmayr, F., Koukoula, T., Husler, I., Chernikov, A., Chatterjee, S., Karakostas, T., Solowan, H.M., Schwarz, U.T., Eickhoff, M., Komninou, P., 2013. Nanostructure and strain in InGaN/GaN superlattices grown in GaN nanowires. Nanotechnology 24, 435702.

Keplinger, M., Mårtensson, T., Stangl, J., Wintersberger, E., Mandl, B., Kriegner, D., Holý, V., Bauer, G., Deppert, K., Samuelson, L., 2009. Structural investigations of core-shell nanowires using grazing incidence X-ray diffraction. Nano Lett. 9, 1877–1882.

Kloeffel, C., Trif, M., Loss, D., 2014. Acoustic phonons and strain in core/shell nanowires. Phys. Rev. B 90, 115419.

Kremer, P.E., Dada, A.C., Kumar, P., Ma, Y., Kumar, S., Clarke, E., Gerardot, B.D., 2014. Strain-tunable quantum dot embedded in a nanowire antenna. Phys. Rev. B 90, 201408.

Landré, O., Camacho, D., Bougerol, C., Niquet, Y.M., Favre-Nicolin, V., Renaud, G., Renevier, H., Daudin, B., 2010. Elastic strain relaxation in GaN/AlN nanowire superlattice. Phys. Rev. B 81, 153306.

Laneuville, V., Demangeot, F., Péchou, R., Salles, P., Ponchet, A., Jacopin, G., Rigutti, L., de Luna Bugallo, A., Tchernycheva, M., Julien, F.H., March, K., Zagonel, L.F., Songmuang, R., 2011. Double strain state in a single GaN/AlN nanowire: probing the core-shell effect by ultraviolet resonant Raman scattering. Phys. Rev. B 83, 115417.

Larsson, M.W., Wagner, J.B., Wallin, M., Håkansson, P., Fröberg, L.E., Samuelson, L., Wallenberg, L.R., 2007. Strain mapping in freestanding heterostructured wurtzite InAs/InP nanowires. Nanotechnology 18, 015504.

Lauhon, L.J., Gudiksen, M.S., Wang, D., Lieber, C.M., 2002. Epitaxial core-shell and core-multishell nanowire heterostructures. Nature 402, 57–61.

Lauhon, L.J., Gudiksen, M.S., Lieber, C.M., 2004. Semiconductor nanowire heterostructures. Philos. Trans. R. Soc. Lond. A 362, 1247–1260.

Li, X., Yang, G., 2014. Modification of Stranski-Krastanov growth on the surface of nanowires. Nanotechnology 25, 435605.

Lim, S.K., Brewster, M., Qian, F., Li, Y., Lieber, C.M., Gradečak, S., 2009. Direct correlation between structural and optical properties of III–V nitride nanowire heterostructures with nanoscale resolution. Nano Lett. 9, 3940–3944.

Lu, W., Xiang, J., Timko, B.P., Wu, Y., Lieber, C.M., 2005. One-dimensional hole gas in germanium/silicon nanowire heterostructures. Proc. Natl. Acad. Sci. U.S.A. 102, 10046–10051.

Martin, R.M., 1970. Elastic properties of ZnS structure semiconductors. Phys. Rev. B 1, 4005–4011.

McMahon, M.I., Nelmes, R.J., 2005. Observation of a wurtzite form of gallium arsenide. Phys. Rev. Lett. 95, 215505.

Menéndez, J., Singh, R., Drucker, J., 2011. Theory of strain effects on the Raman spectrum of Si-Ge core-shell nanowires. Ann. Phys. (Berlin) 523, 145.

Montazeri, M., Fickenscher, M., Smith, L.M., Jackson, H.E., Yarrison-Rice, J., Kang, J.H., Gao, Q., Tan, H.H., Jagadish, C., Guo, Y., Zou, J., Pistol, M.-E., Pryor, C.E., 2010. Direct measure of strain and electronic structure in GaAs/GaP core-shell nanowires. Nano Lett. 10, 880–886.

Müller, P., Saúl, A., 2004. Elastic effects on surface physics. Surf. Sci. Rep. 54, 157–258.

Musin, R.N., Wang, X.-Q., 2005. Structural and electronic properties of epitaxial core-shell nanowire heterostructures. Phys. Rev. B 71, 155318.

Nazarenko, M.V., Sibirev, N.V., Ng, K.W., Ren, F., Ko, W.S., Dubrovskii, V.G., Chang-Hasnain, C., 2013. Elastic energy relaxation and critical thickness for plastic deformation in the core-shell InGaAs/GaAs nanopillars. J. Appl. Phys. 113, 104311.

Niquet, Y.M., 2006. Electronic and optical properties of InAs/GaAs nanowire superlattices. Phys. Rev. B 74, 155304.

Niquet, Y.M., 2007. Effects of a shell on the electronic properties of nanowire superlattices. Nano Lett. 7, 1105–1109.

Niquet, Y.-M., Camacho Mojica, D., 2008. Quantum dots and tunnel barriers in InAs/InP nanowire heterostructures: electronic and optical properties. Phys. Rev. B 77, 115316.

Niquet, Y.-M., Delerue, C., Krzeminski, C., 2012. Effects of strain on the carrier mobility in silicon nanowires. Nano Lett. 12, 3545–3550.

Niu, X., Stringfellow, G.B., Lee, Y.-J., Liu, F., 2012. Simulation of self-assembled compositional core-shell structures in $In_xGa_{1-x}N$ nanowires. Phys. Rev. B 85, 165316.

Ovid'ko, I.A., 2002. Relaxation mechanisms in strained nanoislands. Phys. Rev. Lett. 88, 046103.

Ovid'ko, I.A., Sheinerman, A.G., 2004. Misfit dislocation loops in composite nanowires. Philos. Mag. 84, 2103–2118.

Ovid'ko, I.A., Sheinerman, A.G., 2006. Misfit dislocation in nanocomposites with quantum dots, nanowires and their ensembles. Adv. Phys. 55, 627–689.

Pan, L., Lew, K.-K., Redwing, J.M., Dickey, E.C., 2005. Stranski-Krastanow growth of germanium on silicon nanowires. Nano Lett. 5, 1081–1085.

Perillat-Merceroz, G., Thierry, R., Jouneau, P.-H., Ferret, P., Feuillet, G., 2012. Strain relaxation by dislocation glide in ZnO/ZnMgO core-shell nanowires. Appl. Phys. Lett. 100, 173102.

Poole, P.J., Lefebvre, J., Fraser, J., 2003. Spatially controlled nanoparticle-free growth of InP nanowires. Appl. Phys. Lett. 83, 2055–2057.

Popovitz-Biro, R., Kretinin, A., Von Huth, P., Shtrikman, H., 2011. InAs/GaAs core-shell nanowires. Cryst. Growth Des. 11, 3858–3865.

Priante, G., Harmand, J.-C., Patriarche, G., Glas, F., 2014. Random stacking sequences in III–V nanowires are correlated. Phys. Rev. B 89, 241301.

Rajadell, F., Royo, M., Planelles, J., 2012. Strain in free standing CdSe/CdS core-shell nanorods. J. Appl. Phys. 111, 014303.

Ramdani, M.R., Harmand, J.C., Glas, F., Patriarche, G., Travers, L., 2013. Arsenic pathways in self-catalyzed growth of GaAs nanowires. Cryst. Growth Des. 13, 91–96.

Raychaudhuri, S., Yu, E.T., 2006a. Calculation of critical dimensions for wurtzite and cubic zinc blende coaxial nanowire heterostructures. J. Vac. Sci. Technol. B 24, 2053–2059.

Raychaudhuri, S., Yu, E.T., 2006b. Critical dimensions in coherently strained coaxial nanowire heterostructures. J. Appl. Phys. 99, 114308.

Rieger, T., Grutzmacher, D., Lepsa, M.I., 2015. Misfit dislocation free InAs/GaSb core-shell nanowires grown by molecular beam epitaxy. Nanoscale 7, 356–364.

Rigutti, L., Jacopin, G., Largeau, L., Galopin, E., De Luna Bugallo, A., Julien, F.H., Harmand, J.-C., Glas, F., Tchernycheva, M., 2011. Correlation of optical and structural properties of GaN/AlN core-shell nanowires. Phys. Rev. B 83, 155320.

Roshko, A., Geiss, R.H., Schlager, J.B., Brubaker, M.D., Bertness, K.A., Sanford, N.A., Harvey, T.E., 2014. Characterization of InGaN quantum disks in GaN nanowires. Phys. Status Solidi C 11, 505–508.

Sadd, M.H., 2005. Elasticity. Elsevier, Amsterdam.

Salehzadeh, O., Kavanagh, K.L., Watkins, S.P., 2013. Geometric limits of coherent III–V core/shell nanowires. J. Appl. Phys. 114, 054301.

Sburlan, S., Nakano, A., Dapkus, P.D., 2012. Effect of substrate strain on critical dimensions of highly lattice mismatched defect-free nanorods. J. Appl. Phys. 111, 054907.

Schmidt, V., McIntyre, P.C., Gösele, U., 2008. Morphological instability of misfit-strained core-shell nanowires. Phys. Rev. B 77, 235302.

Shimamura, K., Yuan, Z., Shimojo, F., Nakano, A., 2013. Effects of twins on the electronic properties of GaAs. Appl. Phys. Lett. 103, 022105.

Singh, R., Dailey, E.J., Drucker, J., Menéndez, J., 2011. Raman scattering from Ge-Si core-shell nanowires: validity of analytical strain models. J. Appl. Phys. 110, 124305.

Søndergaard, N., He, Y., Fan, C., Han, R., Guhr, T., Xu, H.Q., 2009. Strain distributions in lattice-mismatched semiconductor core-shell nanowires. J. Vac. Sci. Technol. B 27, 827–830.

Spencer, B.J., Tersoff, J., 2000. Dislocation energetics in epitaxial strained islands. Appl. Phys. Lett. 77, 2533–2535.
Spencer, B.J., Voorhees, P.W., Davis, S.H., 1992. Morphological instability in epitaxially strained dislocation-free solid films: linear stability theory. J. Appl. Phys. 73, 4955–4970.
Srolovitz, D.J., 1989. On the stability of surfaces of stressed solids. Acta Metall. 37, 621–625.
Svensson, C.P.T., Seifert, W., Larsson, M.W., Wallenberg, L.R., Stangl, J., Bauer, G., Samuelson, L., 2005. Epitaxially grown $GaP/GaAs_{1-x}P_x/GaP$ double heterostructure nanowires for optical applications. Nanotechnology 16, 936–939.
Swadener, J.G., Picraux, S.T., 2009. Strain distributions and electronic property modifications in Si/Ge axial nanowire heterostructures. J. Appl. Phys. 105, 044310.
Thillosen, N., Sebald, K., Hardtdegen, H., Meijers, R., Calarco, R., Montanari, S., Kaluza, N., Gutowski, J., Lüth, H., 2006. The state of strain in single GaN nanocolumns as derived from micro-photoluminescence measurements. Nano Lett. 6, 704–708.
Timoshenko, S.P., Goodier, J.N., 1951. Theory of Elasticity. McGraw-Hill, New York.
Tourbot, G., Bougerol, C., Grenier, A., Den Hertog, M., Sam-Giao, D., Cooper, D., Gilet, P., Gayral, B., Daudin, B., 2011. Structural and optical properties of InGaN/GaN nanowire heterostructures grown by PA-MBE. Nanotechnology 22, 075601.
Tourbot, G., Bougerol, C., Glas, F., Zagonel, L.F., Mahfoud, Z., Meuret, S., Gilet, P., Kociak, M., Gayral, B., Daudin, B., 2012. Growth mechanism and properties of InGaN insertions in GaN nanowires. Nanotechnology 23, 135703.
Trammell, T.E., Zhang, X., Li, Y., Chen, L.-Q., Dickey, E.C., 2008. Equilibrium strain-energy analysis of coherently strained core-shell nanowires. J. Cryst. Growth 310, 3084–3092.
Uccelli, E., Arbiol, J., Morante, J.R., Fontcuberta i Morral, A., 2010. InAs quantum dot arrays decorating the facets of GaAs nanowires. ACS Nano 4, 5985–5993.
Varahramyan, K.M., Ferrer, D., Tutuc, E., Banerjee, S.K., 2009. Band engineered epitaxial $Ge\text{-}Si_xGe_{1-x}$ core-shell nanowire heterostructures. Appl. Phys. Lett. 95, 033101.
Verheijen, M.A., Immink, G., de Smet, T., Borgström, M.T., Bakkers, E.P.A.M., 2006. Growth kinetics of heterostructured GaP-GaAs nanowires. J. Am. Chem. Soc. 128, 1353–1359.
Wang, H., Upmanyu, M., Ciobanu, C.V., 2008. Morphology of epitaxial core-shell nanowires. Nano Lett. 8, 4305–4311.
Warwick, C.M., Clyne, T.W., 1991. Development of composite coaxial cylinder stress analysis model and its application to SiC monofilament systems. J. Mater. Sci. 26, 3817–3827.
Wölz, M., Ramsteiner, M., Kaganer, V.M., Brandt, O., Geelhaar, L., Riechert, H., 2013. Strain engineering of nanowire multi-quantum well demonstrated by Raman spectroscopy. Nano Lett. 13, 4053–4059.
Wu, Y., Fan, R., Yang, P., 2002. Block-by-block growth of single-crystalline Si/SiGe superlattice nanowires. Nano Lett. 2, 83–86.
Yan, X., Zhang, X., Li, J., Cui, J., Ren, X., 2015. Fabrication and optical properties of multishell InAs quantum dots on GaAs nanowires. J. Appl. Phys. 117, 054301.
Ye, H., Yu, Z., 2014. Plastic relaxation of mixed dislocation in axial nanowire heterostructures using Peach-Koehler approach. Phys. Status Solidi Rapid Res. Lett. 8, 445–448.
Ye, H., Lu, P., Yu, Z., Han, L., 2009a. Critical lateral dimension for a nanoscale patterned heterostructure using the finite element method. Semicond. Sci. Technol. 24, 025029.
Ye, H., Lu, P., Yu, Z., Song, Y., Wang, D., Wang, S., 2009b. Critical thickness and radius for axial heterostructure nanowires using finite-element method. Nano Lett. 9, 1921–1925.
Yeh, C.-Y., Lu, Z.W., Froyen, S., Zunger, A., 1992. Zinc-blende-wurtzite polytypism in semiconductors. Phys. Rev. B 46, 10086.
Yuan, Z., Nakano, A., 2013. Self-replicating twins in nanowires. Nano Lett. 13, 4925–4930.

Yuan, Z., Nomura, K.-i., Nakano, A., 2012. A core/shell mechanism for stacking fault generation in GaAs nanowires. Appl. Phys. Lett. 100, 163103.
Zagonel, L.F., Mazzucco, S., Tenc, M., March, K., Bernard, R., Laslier, B., Jacopin, G., Tchernycheva, M., Rigutti, L., Julien, F.H., Songmuang, R., Kociak, M., 2011. Nanometer scale spectral imaging of quantum emitters in nanowires and its correlation to their atomically resolved structure. Nano Lett. 11, 568–573.
Zardo, I., Conesa-Boj, S., Peiro, F., Morante, J.R., Arbiol, J., Uccelli, E., Abstreiter, G., Fontcuberta i Morral, A., 2009. Raman spectroscopy of wurtzite and zinc-blende GaAs nanowires: polarization dependence, selection rules, and strain effects. Phys. Rev. B 80, 245324.
Zhang, G., Tateno, K., Gotoh, H., Sogawa, T., Nakano, H., 2010. Structural, compositional, and optical characterizations of vertically aligned AlAs/GaAs/GaP heterostructure nanowires epitaxially grown on Si substrate. Jpn. J. Appl. Phys. 49, 015001.
Zhang, X., Dubrovskii, V.G., Sibirev, N.V., Ren, X., 2011. Analytical study of elastic relaxation and plastic deformation in nanostructures on lattice mismatched substrates. Cryst. Growth Des. 11, 5441–5448.
Zhong, Z., Sun, Q.P., 2002. Analysis of a transversely isotropic rod containing a single cylindrical inclusion with axisymmetric eigenstrains. Int. J. Solids Struct. 39, 5763–5765.

CHAPTER THREE

# van der Waals Heteroepitaxy of Semiconductor Nanowires

**Young Joon Hong*[,1], Chul-Ho Lee[†]**
*Faculty of Nanotechnology & Advanced Materials Engineering, Graphene Research Institute, Sejong University, Seoul, South Korea
[†]KU-KIST Graduate School of Converging Science and Technology, Korea University, Seoul, South Korea
[1]Corresponding author: e-mail address: yjhong@sejong.ac.kr

## Contents



## 1. INTRODUCTION

### 1.1 Heteroepitaxy of Semiconductors on Atomic-Layered Materials (ALMs)

Epitaxy is an ideal route to monolithic integration of semiconductors on single-crystal substrates for diverse electronic and optoelectronic device applications (Ayers, 2007). Paradigm of the microelectronic substrates has been recently shifted from thick, rigid single-crystalline substrates to flexible

*Semiconductors and Semimetals*, Volume 93
ISSN 0080-8784
http://dx.doi.org/10.1016/bs.semsem.2015.07.002

and transparent soft films by a demand arising from lightweight, wearable, transparent, and transferable electronics. Mostly, the polymeric films have fulfilled such requirements for the emerging electronics (Park et al., 2008). However, the plastic or organic materials have no tolerance to high temperature process needed for semiconductor growth, and, as amorphous materials, provide no epitaxial relationship with inorganic semiconductors for the monolithic epitaxial integration. Thus, a new class of flexible and transparent inorganic substrates is promising for heteroepitaxial integration of semiconductor devices (Chung et al., 2010; Kobayashi et al., 2012; Lee et al., 2011a).

Two-dimensional atomic-layered materials (2d-ALMs; Butler et al., 2013; Geim and Grigorieva, 2013), such as graphene (Geim and Novoselov, 2007), hexagonal boron nitride (hBN; Kim et al., 2012; Song et al., 2010), transition metal dichalcogenides (e.g., $MoS_2$, $WS_2$, $MoSe_2$, and $WSe_2$; Wang et al., 2012), topological insulators (e.g., $Bi_2Se_3$, $Bi_2Te_3$, $Sb_2Te_3$, and BiSb; Zhang et al., 2009), have been attracting much attention, owing to their excellent electrical, thermal, mechanical, and optical properties. The 2d-ALMs have potential advantages to be used as a semiconductor substrate due to the following reasons. Since the typical 2d-ALMs have the crystalline surface with triangular lattices (or honeycomb atomic arrangement), they can be structurally compatible with many semiconductors of zinc blende (ZB), wurtzite (WZ), and diamond crystal structures (Hong et al., 2012; Kim et al., 2009a; Kobayashi et al., 2012; Yoo et al., 2012). Most of the 2d-ALMs exhibit sufficiently high melting (or decomposition) temperature and chemical/mechanical stability so that can be endurable for many microfabrication processes. Since the atomically thin 2d-ALMs of graphene, hBN, and $MoS_2$ have inherently semimetallic, insulating (or topologically insulating), and semiconducting properties, respectively, the combination of semiconductors and 2d-ALMs can create diverse hybrid junctions with the tailored characteristics (Chandramohan et al., 2012; Geim and Grigorieva, 2013). More importantly, the electrical conductivities of 2d-ALMs are amphoterically modulated by chemical and physical doping (Giovannetti et al., 2008; Guo et al., 2010; Lee et al., 2013; Liu et al., 2011a; Shi et al., 2010) or applying electric field (Yu et al., 2009), and bandgap of the 2d-ALMs is also tunable by forming alloy (e.g., hBN–graphene; Gong et al., 2014), $MoS_2$–$WS_2$–$WSe_2$, $Bi_2Se_3$–$Bi_2Te_3$–$Sb_2Te_3$ (Cha et al., 2012; Ko et al., 2013; Kong et al., 2011), or lateral heterostructures (Levendorf et al., 2012; Liu et al., 2013), all of which enables to form either Ohmic or Schottky junctions with diverse n- and p-type semiconductors on demands (Chandramohan et al., 2012; Jo et al., 2012; Tongay et al., 2012). High optical transmittance of few-atomic-layer-thick 2d-ALMs is also advantageous for high-efficiency optoelectronic devices with high light absorption or extraction. Furthermore, wafer-sized synthesis of single-crystalline 2d-ALMs (Lee et al., 2014) with precisely controlled layer thickness (Li et al., 2009) implies that the use of 2d-ALM wafers is possibly compatible with large-area semiconductor process. Accordingly, the heteroepitaxial integration of semiconductors on 2d-ALMs substrate is expected to bring significant advances in fabrication, design, and performance for new functional semiconductor electronics and optoelectronics (Munshi and Weman, 2013; Park et al., 2011). Although the 2d-ALMs have great potential as a substrate, the problems associated with chemical stability of a surface without dangling bonds make it difficult to deposit inorganic materials by conventional epitaxy that requires covalent bonding between substrate and overlayer.

## 1.2 van der Waals Epitaxy (Versus Covalent Epitaxy)

In general, covalent heteroepitaxy of inorganic thin films is driven by unoccupied bonds on the surface of a bulk crystalline substrate: the unstable dangling bonds promote a consecutive "adatom–nucleation–growth" process to form the overlayer via covalent bonding for substantial energy minimization (Ayers, 2007). Due to the strong covalent bonds, even small lattice misfit produces extended crystal defects such as threading dislocations at the heterointerface (Fig. 1A): it is well known that the threading dislocations deteriorate the performance of semiconductor devices by acting as carrier leakage paths (or carrier trap sites) and nonradiative recombination centers. Thus, only limited combination of an overlayer and a substrate could yield high-quality heteroepitaxial structures because the length and angle of the covalent bonds are not easily altered.

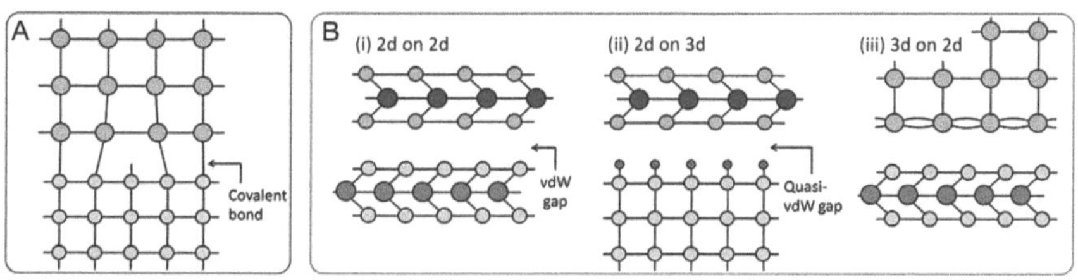


**Figure 1** Schematics showing heterointerfaces formed by (A) covalent heteroepitaxy and (B) vdW heteroepitaxy.

In 1984, Koma et al. demonstrated unconventional, noncovalent heteroepitaxy, and named van der Waals (vdW) epitaxy for the first time (Koma et al., 1984). The epitaxy is driven by vdW attraction between overlayer precursors and an inactive substrate without surface dangling bonds (or terminated with suitable atoms). Then, the heterojunction of overlayer/substrate is adhered by the weak vdW interaction (Fig. 1B). Noticeably, in contrast to the covalent epitaxy, since the heterointerfacial stress caused by lattice mismatch can be sufficiently released through the noncovalent heterointerface gap, the vdW heteroepitaxy typically yields a single-crystalline overlayer with reduced threading dislocation density (Koma, 1992, 1999). Previous literatures have also demonstrated that the vdW epitaxy produces heteroepitaxial overlayers with the well-defined epitaxial relationship and the abrupt clean heterointerface (Koma and Yoshimura, 1986). Accordingly, owing to the choice of substrate free from severe lattice-matching conditions, the vdW heteroepitaxy is regarded as a promising route to integrate high-quality semiconductor thin films and nanowires on arbitrary 2d-ALM surfaces with highly incommensurate heteroepitaxial system.

The vdW heteroepitaxy technique has so far been typically used for preparing 2d-ALM films of transition metal dichalcogenides (Shi et al., 2012), topological insulators (Peng et al., 2012), and atomic-layered semiconductors (Schlaf et al., 1999a) on other 2d-ALMs (Schlaf et al., 1994; Wisotzki et al., 2000), hydrogenated silicon (Tada et al., 1992), sulfur-terminated GaAs (Ueno et al., 1990), and so on. However, the vdW epitaxy has versatile capability of growing various crystalline materials with two-dimensional (2d) or three-dimensional (3d) crystal structures (Koma et al., 1991): Fig. 1B depicts the feasible heterogeneous combinations of 2d-ALMs on 2d-ALMs, 2d-ALMs on surface-passivated (or -reconstructed) 3d semiconductors, and $sp^3$-hybridized (3d) semiconductors on 2d-ALMs fabricated by vdW heteroepitaxy. Since typical semiconductor (groups IV, III–V, and II–VI) nanowires consist of $sp^3$ bonds, the vdW heteroepitaxy of semiconductor nanowires can also be achieved on 2d-ALMs. In this chapter, we present recent progresses and opportunities in vdW heteroepitaxy of semiconductor nanowires on 2d-ALMs.

## 2. VAN DER WAALS (vdW) HETEROEPITAXY OF SEMICONDUCTOR NANOWIRES

One-dimensional (1d) semiconductor nanostructures (i.e., nanowires, nanorods, nanopillars, nanotubes, etc.) have attracted much attention as a promising material for high-performance future electronics and optoelectronics (Heiss et al., 2013; Hyun et al., 2013; Kanungo et al., 2013; Reimer et al., 2012; Tomioka et al., 2012; Wallentin et al., 2013; Xiang et al., 2006), due to several important structural advantages over conventional planar films: (i) high surface-to-volume ratio for enlarged active layer area (Tian et al., 2007) in an equivalent dice surface area; (ii) high aspect ratio for high photon extraction (or absorption) as a waveguide (An et al., 2008; Kim et al., 2004); (iii) small footprint for the high-quality heteroepitaxial integration on highly lattice-mismatched substrates (Park et al., 2002; Tomioka et al., 2008); and (iv) structural versatility of designing 3d heterostructured architectures with compositional modulation along the axial or radial direction (Gudiksen et al., 2002; Hong et al., 2011; Lauhon et al., 2002; Park et al., 2003). Hence, the integration of semiconductor nanowires on 2d-ALMs is expected to create novel properties for new functional device applications (Munshi and Weman, 2013; Park et al., 2011). In this section, the vdW epitaxy of vertical nanowires on 2d-ALMs, nanowire alloying, nanowire heterostructures, use of ultrathin 2d-ALM substrates, and the fabrication of semiconductor/2d-ALM/semiconductor double heterostructures are discussed.

## 2.1 Vertically Aligned Nanowires on 2d-ALMs

Semiconductor nanowire arrays that are vertically aligned on a substrate offer suitable geometry for device processing and applications (Kanungo et al., 2013; Lee et al., 2009; Tomioka et al., 2012). The aligned growth of nanowires along the preferential direction requires a well-defined epitaxial relationship. To date, various semiconductor nanowires were successfully grown on the layered substrates by the vdW epitaxy as shown in Fig. 2. In 2010, Xiong and coworkers firstly demonstrated the vdW epitaxial growth of vertically aligned semiconductor nanowires (Utama et al., 2011), proposing a new strategy for nanowire integration on 2d-ALMs, as displayed in scanning electron microscopy (SEM) images of Fig. 2A. Cadmium chalcogenides (CdS, CdSe, and CdTe) nanowires with uniform diameters and lengths were grown on freshly cleaved (001) muscovite mica ($K_2O \cdot 3Al_2O_3 \cdot 6SiO_2 \cdot 2H_2O$) substrate via a vapor transport method. The growth direction of CdS and CdSe nanowires was [0001] of WZ and that of CdTe nanowires was [111] of ZB, all of which corresponds to the preferential growth direction of WZ and ZB.

Pan et al. demonstrated vertically aligned $CdS_xSe_{1-x}$ nanowire arrays grown on the muscovite mica substrate by vdW epitaxy (Pan et al.,

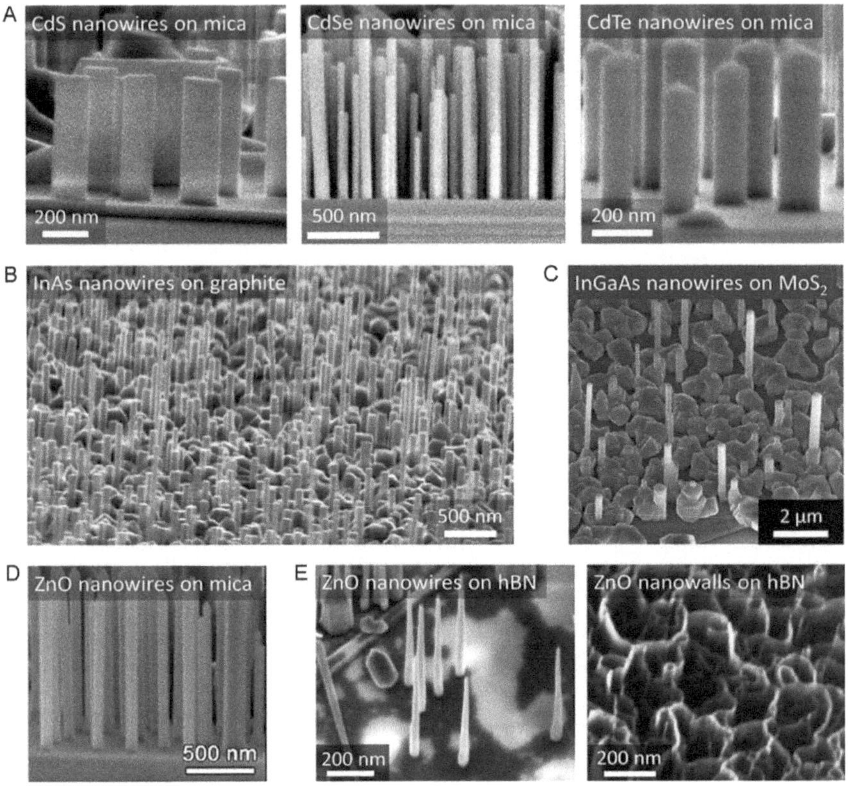


**Figure 2** SEM images of diverse vertically aligned semiconductor nanowire arrays grown by vdW epitaxy. (A) Cadmium chalcogenides nanowires grown on muscovite mica. (B) InAs nanowires on graphite. (C) InGaAs nanowires on $MoS_2$. (D) ZnO nanowires on muscovite mica. (E) ZnO nanowires and nanowalls on hBN. *Panel (A): Reproduced with permission from Utama et al. (2011). Copyright 2011 American Chemical Society. Panel (B): Adapted with permission from Hong and Fukui (2011). Copyright 2011 American Chemical Society. Panel (C): Reproduced with permission from Mohseni et al. (2013). Copyright 2013 American Chemical Society. Panel (D): Adapted with permission from Utama et al. (2012a). Copyright 2012 American Chemical Society. Panel (E): Reprinted with permission from Oh et al. (2014). Copyright 2014 Nature Publishing Group.*

2012). The composition of the alloyed nanowire was completely modulated in the full range of $0 < x < 1$ by changing the ratio of S:Se source powders, which was thoroughly confirmed by X-ray diffraction, transmission electron microscopy (TEM), and optical spectroscopies of UV–Vis absorption and photoluminescence. Hence, the vdW epitaxy was demonstrated to yield single-crystalline CdS and CdSe nanowires with complete composition tunability and sufficiently high optical quality so that those nanowires can be used for high-performance nanowire optoelectronic device applications.

Hong and Fukui reported that the vdW epitaxy enabled the growth of vertically aligned InAs nanowires on graphite and graphene layers using metal–organic vapor-phase epitaxy (MOVPE; Fig. 2B; Hong and Fukui, 2011). Vertical InAs nanowires and nanoislands were concurrently formed on the cleaved graphite (or graphene) by vdW epitaxy, as shown in Fig. 3A. The nanowires exhibited the hexagonal prismatic morphology, and the neighboring nanowires have an identical hexagonal in-plane alignment within a single domain of graphite, suggesting well-defined heteroepitaxial relationship of InAs/graphene (inset of Fig. 3A): $(\bar{1}\bar{1}\bar{1})[\bar{2}11]_{InAs} \| (0001)[1\bar{2}10]_{graphite}$ with in-plane misfit of ~0.5% (see Section 3.1). The vdW growth produced high-density InAs nanoislands (with line number density of $\sim 3 \times 10^4\ cm^{-1}$) specifically along the step edges of the freshly cleaved graphite, as marked with wedges in Fig. 3B, while vertical nanowires were mostly formed on the flat terrace region. Because the height of step edges was typically ~0.4–2 nm and root mean square (rms) surface roughness of the graphite terrace was as low as $0.08 \pm 0.05$ nm, measured by atomic force microscopy (AFM), the vdW growth of either nanowires or nanoislands is presumably determined by the surface morphology of graphite. It is noted that the step edge, which is a continuous atomic imperfection, plays an important role in forming the high-density InAs nanoislands (see also Section 4.1).

Li and coworkers demonstrated InGaAs nanowires grown vertically on $MoS_2$ substrate by vdW epitaxy using MOVPE (Fig. 2C; Mohseni et al., 2013). The $MoS_2$, a semiconducting 2d-ALM, was shown to be capable of accommodating vdW epitaxy as a substrate for growing vertical

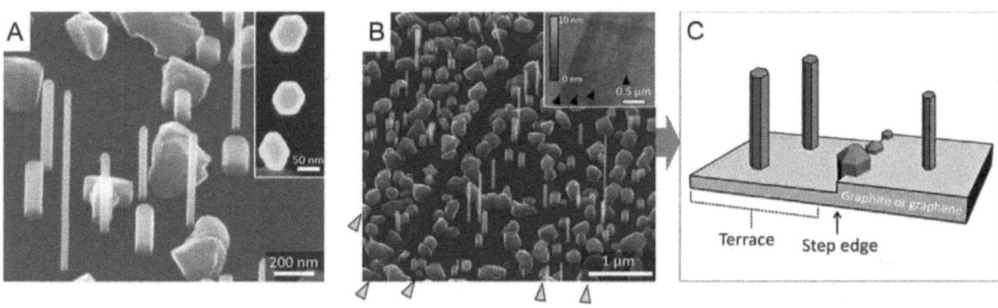


**Figure 3** InAs nanostructures grown on graphite by vdW heteroepitaxy. (A) Bird's eye-view SEM image. The inset is a top-view image. (B) Low-magnification SEM image showing that the nanoislands are favorably grown along a specific direction. According to the AFM surface topography of graphite (inset), the nanoislands were found to preferentially form along the step edges. (C) Schematic illustrating the general morphology of InAs nanostructures grown on graphite.

semiconductor nanowires. The homogeneous composition of InGaAs was observed along the whole length of a nanowire as growing the nanowires on $MoS_2$. However, in contrast to the $MoS_2$ substrate, the use of the graphene substrate yielded InGaAs/InAs (shell/core) nanowire heterostructures during the vdW epitaxial growth of InGaAs nanowires, which is further discussed in the next section.

Xiong and coworkers reported incommensurate vdW epitaxy of vertical ZnO nanowires on a mica substrate (Fig. 2D; Utama et al., 2012a). Despite highly anisotropic in-plane lattice mismatch of 64.0% for the $[11\bar{2}0]_{ZnO} \| [020]_{mica}$ and −8.36% for the $[1\bar{1}00]_{ZnO} \| [200]_{mica}$ orientation, the ZnO nanowires were vertically grown on mica(001) with a singular heteroepitaxial relation of $(0001)[11\bar{2}0]_{ZnO} \| (001)[010]_{mica}$. At this highly incommensurate heterojunction of the nanowire/mica, no threading dislocations were generated owing to nearly complete lattice relaxation at the nanowire/substrate heterointerface, thus signifying the unnecessity of lattice matching for defect-free epitaxial nanowires on arbitrary substrates. This also implies that the vdW heteroepitaxy is suitable for fabricating high-performance semiconductor electronics and optoelectronics on 2d-ALMs.

The vdW heteroepitaxy of ZnO on hBN using MOVPE yielded two distinct nanostructure morphologies (Fig. 2E). Yi and coworkers demonstrated that either ZnO nanoneedles or nanowalls are selectively grown on the hBN substrate depending on surface treatment conditions (Oh et al., 2014). Under the same MOVPE growth batch, the freshly cleaved or pristine chemical vapor deposition (CVD)-grown hBN surface yielded vertical ZnO nanowire arrays (left panel of Fig. 2E), whereas the plasma-treated hBN surface resulted in vertical nanowalls (right panel of Fig. 2E). Because the plasma damages and etches the surface layer of hBN, the atomic step edges and kinks formed by the plasma treatment are thought to play a crucial role in determining the favorable growth sites of ZnO nanowalls. The plausible scenario is that the atomic step edges promote the high-density heterogeneous vdW nucleation–growth of ZnO along them, which leads to formation of laterally interconnected nanowires that are nanowalls. This rationale for preferential nucleation–growth sites can be extended to control the position and shape of nanostructures by forming artificially designed atomic steps on 2d-ALMs substrates. In Sections 4.1 and 4.2, we further discuss how the position and exterior shape of semiconductor nanostructures are controlled in a designed way.

In addition to vertically aligned nanowires, the vdW epitaxy could be extended to fabricate the complex nanoarchitectures on 2d-ALMs

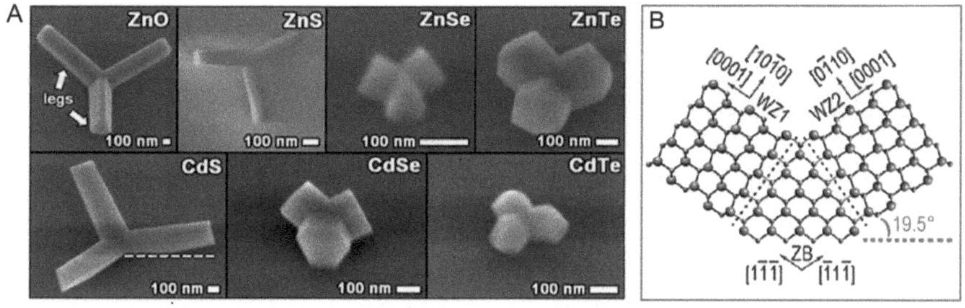


**Figure 4** II–VI semiconductor nanotripod grown by vdW epitaxy. (A) A series of SEM images of Zn–VI and Cd–VI nanotripods. (B) Cross-sectional side view of atomic configuration of a nanotripod. Inclined WZ legs were grown on hemioctahedral ZB seed crystal. *Adapted with permission from Utama et al. (2012b). Copyright 2012 American Chemical Society.*

substrates. Utama et al. reported the epitaxial growth of nonplanar nanotripods of II–VI compounds (ZnO, ZnS, ZnSe, ZnTe, CdS, CdSe, and CdTe) on the muscovite mica substrate as shown in Fig. 4A (Utama et al., 2012b). Regardless of the choice of materials, the threefold symmetrically branched legs of a tripod were preferentially oriented on the substrate without forming a buffer layer or any intermediated structures, strongly indicating the vdW epitaxial growth of the tripods. To explain the formation of the tripods, the polytypic growth mechanism was proposed as depicted in Fig. 4B, where the legs in the hexagonal WZ phase were growing on the +{111} surfaces of the octahedral seed in the cubic ZB phase. The epitaxial relationship of $(\bar{1}1\bar{1})\langle\bar{1}\bar{1}0\rangle_{ZB} \| (0001)\langle 2\bar{1}\bar{1}0\rangle_{WZ}$ between the seed and the legs was confirmed by TEM analyses. This work has proven the general potential of vdW epitaxy to fabricate the ordered arrays of more complex nanoarchitectures from a wide range of materials toward various electronic and optoelectronic device applications.

## 2.2 Nanowire Heterostructure

Followed by the epitaxial growth of homogeneous nanowires, the bandgap engineering and doping in the heterostructured nanowires should be achieved for practical device applications. Hence, in this section, we briefly review the fabrication of nanowire heterostructures on 2d-ALMs substrates.

Mohseni et al. reported both axial and radial InAs/InGaAs nanowire heterostructures grown on graphene substrates by the vdW heteroepitaxy (Mohseni et al., 2013, 2014). The sequential growth of InAs and InGaAs segments using metal–organic chemical vapor deposition (MOCVD)

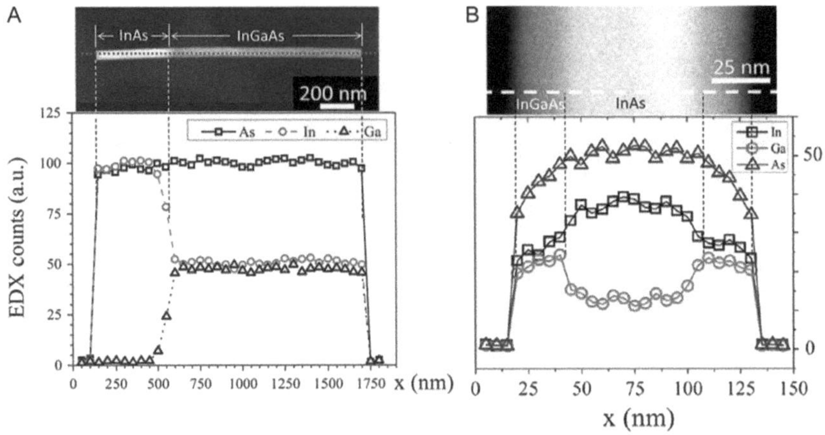


**Figure 5** III–As nanowire heterostructures grown by vdW epitaxy. HAADF-STEM image and radial EDX linescan data of As, In, and Ga of (A) axial and (B) radial InAs/InGaAs nanowire heterostructures, obtained along the dashed lines in the STEM images. *Reproduced with permission from Mohseni et al. (2013, 2014) for panels (A) and (B), respectively. Copyright 2014 John Wiley & Sons, Inc. and 2013 American Chemical Society.*

enabled to fabricate the axial InAs/InGaAs heterostructure without tapering on the graphene substrate. The high-angle annular dark-field (HAADF) scanning transmission electron microscopy (STEM) image and the energy-dispersive X-ray (EDX) spectroscopy line profile in Fig. 5A clearly demonstrate the uniform diameter along the growth direction and abrupt modulation in composition at the heterointerface. In addition, they found that the well-defined InAs/InGaAs core/shell heterostructure can be formed through spontaneous phase separation during the InGaAs growth on the graphene substrate. Figure 5B shows the composition profile of the core/shell heterostructure. Such phase separation was interpreted that the binary InAs component that has small lattice mismatch of ~0.5% with graphene can be self-assembled into the core segment at the early growth stage because strain sharing is not permitted in the case of vdW epitaxy on an inert substrate. It should be noted that such intriguing growth behavior was not observed on $MoS_2$ that has a lattice constant of 3.16 Å larger than graphene by 28%. These experimental demonstrations for simple heterostructures suggest great potential to fabricate the more complicated heterostructures embedded with the multiple quantum wells and/or quantum dots for various sophisticated device applications, which will be further discussed in the latter part (see Section 5).

## 2.3 vdW Epitaxial Nanowires on a Monoatomic Layer Substrate

The 2d-ALMs have been proven to accommodate the substantial vdW attraction for epitaxy of semiconductor nanowires, as described in the previous sections. Among various 2d-ALMs, graphene is a group IV elemental 2d-ALM with superb electrical and mechanical, and thermal properties, such as extremely high electron mobility of 200,000 $cm^2\ V^{-1}\ s^{-1}$ (Bolotin et al., 2008), high elastic modulus of 1 TPa (Lee et al., 2008), and thermal conductivity (Balandin et al., 2008; Seol et al., 2010), respectively. In particular, due to high optical transmittance and chemical stability, monoatomic thin graphene is regarded as an attractive building block for transparent electronics and optoelectronics (Bae et al., 2010; Jo et al., 2012; Kim et al., 2009b). However, electrostatic force microscopy revealed that surface potential significantly decreases with reducing the graphene to monoatomic thickness (Datta et al., 2009), which may provide insufficient vdW attraction for heteroepitaxy. Therefore, it is meaningful to investigate how thin 2d-ALMs can be exploited as epitaxial substrates for the validity of vdW heteroepitaxy.

The vdW heteroepitaxy of InAs nanowires was attempted using a series of graphene substrate with different thicknesses: single-layer graphene (SLG; Li et al., 2009), few-layer graphene (FLG; Chen et al., 2011), and multilayer graphene (MLG) were synthesized by CVD using copper, copper–nickel alloy, and nickel foils, respectively, and were then transferred onto a $SiO_2$/Si substrate (Suk et al., 2011). For a control experiment, the bare $SiO_2$/Si substrate was used.

The SEM images of Fig. 6 display that the vdW epitaxy produced vertically well-aligned InAs nanowires on the SLG substrate, whereas the use of

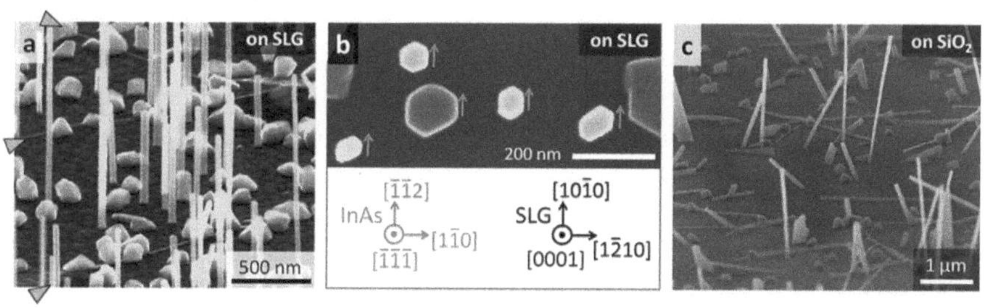


**Figure 6** SEM images of InAs nanowires grown on (A and B) SLG and (C) $SiO_2$ substrates. Yellow (gray in the print version) wedges in (A) indicate the wrinkles of graphene. Plan-view SEM image of (B) shows the homogeneous in-plane alignment of hexagonal faceted InAs nanowires and nanoislands. The corresponding heteroepitaxial relationship of InAs/SLG is depicted in the lower panel of (B).

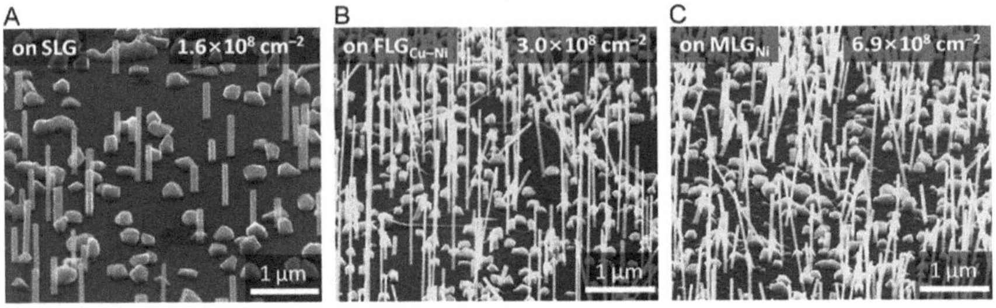


**Figure 7** Tilt-view SEM images of InAs nanowires grown on (A) SLG, (B) FLG, and (C) MLG substrates. The vertical alignment of nanowires grown by vdW epitaxy is deteriorated due to rough surface of thicker graphene substrate.

$SiO_2$ substrate resulted in randomly oriented nanowires even in the same MOVPE batch, due to the lack of epitaxial relationship between InAs and amorphous $SiO_2$ (Hong et al., 2012). The number density of nanowires was measured to be ca. $10^8$ and $10^7$ cm$^{-2}$ on SLG and $SiO_2$ substrates, respectively. Higher density formation of nanowires on SLG/$SiO_2$/Si than $SiO_2$/Si substrate implies that the electrostatic attractions from the underlying $SiO_2$/Si layer (across the SLG) are almost negligible for the vdW epitaxy. Importantly, the excellent vertical and in-plane sidewall facet alignment of the nanowires on SLG suggests the heteroepitaxial relationship between InAs and SLG.

Intriguingly, the vdW epitaxy yielded InAs nanowires with different vertical alignment and number density on the SLG, FLG, and MLG substrates. Figure 7 displays that the vertical alignment of nanowires grown by the vdW epitaxy was gradually deteriorated when using thicker graphene: the portion of nanowire vertical alignment was measured to be 91.0, 71.3, and 55.2% for the use of SLG, FLG, and MLG substrates, respectively (Table 1). Only the nanowires with standing angles of $90\pm5°$ from the substrate surface were counted to be "vertically aligned." In addition, the number density of InAs nanowires exhibited increasing trend on thicker graphene substrate: the SLG resulted in nanowire number density of $1.6\times10^8$ cm$^{-2}$, and the increased number density of $3.0\times10^8$ and $6.9\times10^8$ cm$^{-2}$ for the FLG and MLG, respectively. These results imply that the vertical alignment and number density of nanowire were related to the thickness of graphene. However, these trends are not resulted simply from the graphene thickness but presumably from the surface roughness of graphene substrates.

Figure 8 shows the photomicrographs and AFM surface topographies of SLG and FLG films transferred onto $SiO_2$/Si substrates. The SLG presented

**Table 1** Morphological Information of Graphene and InAs Nanowires Grown on the Graphene

| | Freshly Cleaved Graphene | $SLG_{Cu}$ | $FLG_{Cu-Ni}$ | $MLG_{Ni}$ |
|---|---|---|---|---|
| rms surface roughness (nm) | – | $2.4 \pm 0.6$ | $12.0 \pm 8.8$ | $20.5 \pm 13.8$ |
| rms surface roughness of wrinkle-free area (nm) | $0.08 \pm 0.05$ | $0.5 \pm 0.2$ | $2.9 \pm 0.7$ | $6.8 \pm 2.2$ |
| Nanowire number density ($\times 10^8$ cm$^{-2}$) | $0.9 \pm 0.5$ | $1.6 \pm 0.7$ | $3.0 \pm 0.8$ | $6.9 \pm 1.1$ |
| Portion of vertical nanowires (%) | >99 | 91.0 | 71.3 | 55.2 |

The subscripts of Cu, Cu–Ni, and Ni denote the catalyst foils used for the synthesis of graphene films. The AFM characterizations were performed after transferring the graphene films on $SiO_2$/Si substrate. Partly adapted with permission from Hong et al. (2012). Copyright 2012 American Chemical Society.

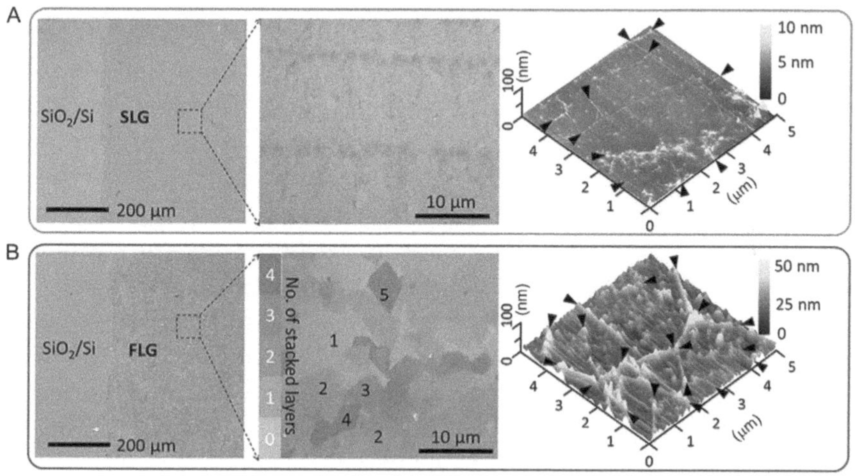


**Figure 8** Single- and few-layer graphene used as substrate for vdW epitaxy of InAs nanowires. Photomicrographs and AFM surface topographies of (A) SLG and (B) FLG. *Reproduced with permission from Hong et al. (2012). Copyright 2012 American Chemical Society.*

the uniform color and contrast over the entire surface in the optical microscopic image, implying uniform thickness of the SLG film (Fig. 8A). In contrast, the FLG or MLG showed lots of dappled areas with uneven contrasts, indicating that those films consisted of partially thickly formed graphene layers (Fig. 8B). It is well known that the number of stacked layers in the graphene films increases locally during CVD for FLG and MLG synthesis, producing atomic steps (Fig. 8B).

The surface topographies scanned by AFM in Fig. 8A and B clearly show higher surface roughness as well as higher density wrinkles of FLG (or MLG) than those of SLG. Since the difference of thermal expansion coefficients between metal foil and graphene causes the graphene wrinkles to release interfacial thermal stress during quenching, higher growth temperature of the FLG and MLG films results in denser and taller wrinkles than those in the SLG films. Noticeably, the surface roughness of the wrinkle-free area presented the increasing trend with increase of thickness of graphene films (Table 1), implying that the ledges and kinks are sources of surface roughness.

According to the comparative study on the AFM surface topographies of SLG and FLG (or MLG), it is envisioned that the steps and kinks determine the nucleation density and growth direction of InAs nanowires (this will be discussed in Section 4.1). As summarized in Table 1, the vertical alignment was monotonically deteriorated as a function of the rms surface roughness of graphene. In addition, the number density of nanowires increased as graphene surface roughness increased. These two correlations demonstrate that rough graphene surface, caused by surface imperfections, facilitated the heterogeneous nucleation and growth of InAs nanowires but highly rough surface deteriorated the vertical growth of nanowires due to weakly bound vdW heterojunctions between InAs and graphene. Since the graphene wrinkles did not cause preferential nucleation and growth of InAs nanowires or nanoislands, as indicated with wedges in Fig. 6A, the wrinkles are no longer considered as a nucleation–growth site for the vdW epitaxy. This is quite distinct from the high-density formation of InAs nanoislands along the step edges of cleaved graphene (Fig. 3B). It is also noticeable that extremely flat graphite (or graphene) exfoliated from the bulk graphite resulted in InAs nanowire arrays with nearly 100% vertical alignment, irrespective of the substrate thickness, supporting that the graphene thickness does not simply determine the vertical alignment of InAs nanowires.

Regarding the nucleation and growth of InAs on graphene, it is surmised that not only the graphene ledges (or kinks) but also the vacancies or point defects in the SLG acted as heterogeneous nucleation sites for the vdW epitaxy because the interplanar crystal imperfections can promote the consecutive process of adatom–vacancy (Hashimoto et al., 2004; or adatom–ledge, adatom–kink) pair formation, heterogeneous nucleation, and heteroepitaxial growth of nanowires. Thus, it is expected that single-crystalline 2d-ALMs with well-controlled surface ledges can be readily exploited for fabricating the nanowire device arrays with designed spatial arrangement, which will be discussed in Section 4.2.

Graphene has great advantages for nanowire heteroepitaxy. First, due to the limited coalescence size of InAs on graphene, it is easy to obtain the vertical nanowire arrays with mean diameter of $40 \pm 14$ nm without the use of the mask patterns. Second, the vdW epitaxy yields exclusively vertical InAs nanowires with preferential growth direction of $[111]_{ZB}$ even on damaged graphene surface. This is distinguishable from the covalent heteroepitaxy of InAs nanowires on Si: the topmost surface of $Si^{1+}$ or $Si^{3+}$ on Si should have either As- or In-termination that determines the InAs nanowire growth direction (Tomioka et al., 2008). In particular, the CVD-grown SLG enabled growing nanowire arrays with the vertical yield higher than 90%, signifying that one-atom-thick 2d-ALMs can be exploited as epitaxial substrates for the vdW epitaxy.

## 2.4 vdW Epitaxial Double Heterostructure: InAs/Graphene/InAs

2d-ALM-based vertical multiheterostructures provided an important platform for a new class of electronics and optoelectronics. The multiheterostructures have been typically fabricated using the peel/lift layer-by-layer stacking technique (Britnell et al., 2012), which yielded vdW heterojunctions with undefined in-plane crystallographic orientations with respect to underlying lattices. The layer-by-layer assembly method enables high-performance heterostructure electronic and optoelectronic devices in which less layer-to-layer electronic coupling inflicts only a slight effect on the innate sublattice symmetry at the heterojunctions (Georgiou et al., 2013). However, it has been recently reported that new different electronic structures can arise from occurrence of heterojunction sublattices, depending on the heterointerfacial lattice alignment (Dean et al., 2013; Hunt et al., 2013; Ponomarenko et al., 2013). Hence, there has remained a challenge of developing a new method for fabricating semiconductor/2d-ALM multiheterostructures with homogeneous in-plane epitaxial relationship. This section shows that the vdW heteroepitaxy is a facile route to fabricate vdW epitaxial 3d–2d–3d double heterostructures.

The basic approach for achieving the epitaxial semiconductor/2d-ALM/semiconductor double heterostructure is to grow the InAs nanostructures (nanowires and nanoislands) on both sides of suspended SLG (Hong et al., 2013). Prior to the growth, suspended SLG substrates were prepared by transferring the SLG onto perforated-hole-patterned $SiN_x$ membrane/Si substrates, as shown in Fig. 9A. The vapor-phase transport growth (i.e., MOVPE, CVD, etc.) with low reactor pressure and high gas velocity is suitable to grow the semiconductor nanostructures concurrently on/underneath the suspended 2d-ALM for double-side growth because the precursors are supplied more rapidly to front and backside of the suspended 2d-ALM across thinner boundary layer under faster laminar precursor flow ($\delta \propto u_0^{-1/2}$ from the Blasius solution, where $\delta$ is the boundary layer thickness and $u_0$ is the free-stream velocity of laminar flow; Blasius, 1908). As the laminar vapor stream is slower, the boundary layer becomes thicker so that the precursors are not sufficiently diffused to the backside of the suspended 2d-ALM. There are two precursor supply paths to grow the InAs nanostructures on the backside of the suspended SLG: i.e., the gap between the suspended SLG and the susceptor or the perforated membrane holes not covered with the SLG.

The low pressure and high velocity precursor transport conditions of 76 Torr and $\sim$1 m $s^{-1}$, respectively, enabled the double-side growth of InAs nanostructures on/underneath the suspended SLG (Fig. 9B and C). The suspended SLG was transparent enough to examine the InAs nanostructures formed on the opposite side of the SLG under typical SEM conditions. The SEM images of Fig. 10A and B demonstrate that the InAs nanostructures including vertical nanowires and nanoislands were grown on both sides of the suspended SLG, signifying that suspended SLG, a freestanding monoatom-thick layer, provides substantial vdW attractions for the

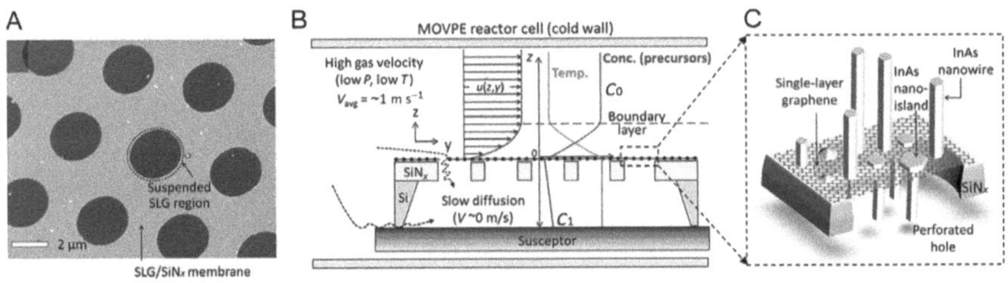


**Figure 9** Substrate and MOVPE reactor used for fabricating vdW epitaxial InAs/SLG/InAs double heterostructure. (A) SEM image of SLG suspended on perforated-hole-patterned $SiN_x$/Si substrate. (B) Illustration depicting substrate and reactor of vapor transport vdW heteroepitaxy. (C) Schematic of substrate after the growth of InAs nanostructures on both sides of suspended SLG. *Reproduced with permission from Hong et al. (2013). Copyright 2013 John Wiley & Sons, Inc.*

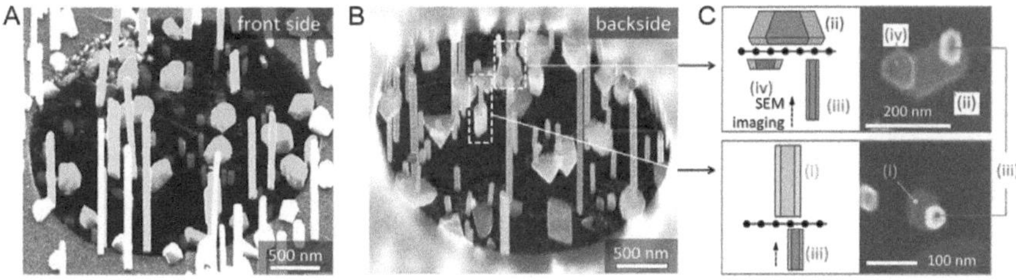


**Figure 10** False-colored SEM images of InAs nanostructures grown on both sides of suspended SLG, taken from (A) front side and (B) backside of suspended SLG region. (C) High-magnification plan-view image of InAs/SLG/InAs double heterostructures, taken from backside of the membrane. Yellow (light gray in the print version) (i) and green (gray in the print version) (ii) objectives are InAs nanowire and nanoisland grown on front side of suspended SLG, respectively; red (gray in the print version) (iii) and purple (gray in the print version) (iv) are nanowire and nanoisland formed on the backside, respectively. *Reproduced with permission from Hong et al. (2013). Copyright 2013 John Wiley & Sons, Inc.*

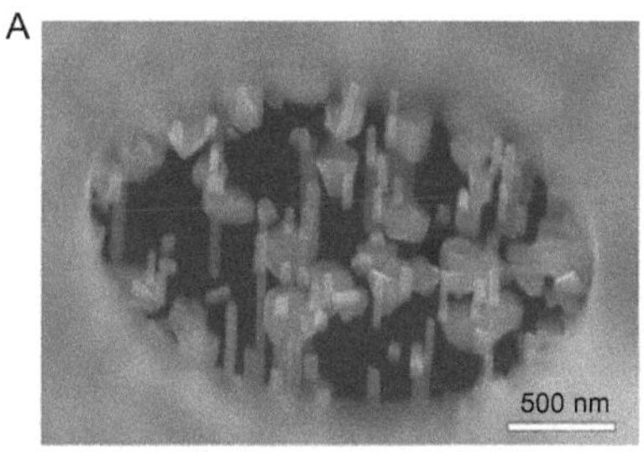


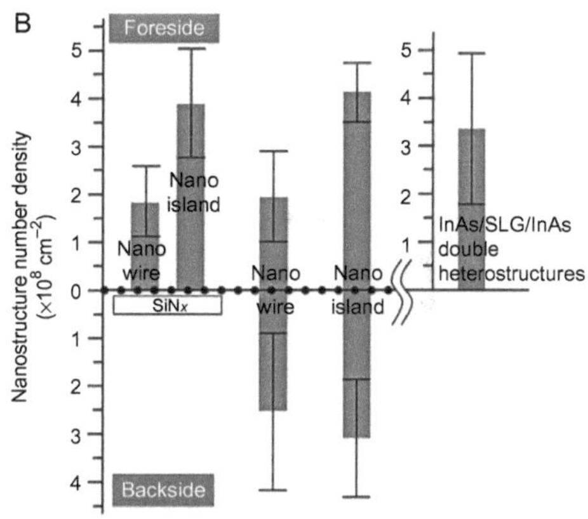


C

| | Island /SLG /wire | Island /SLG /island | Wire /SLG /wire | Total |
|---|---|---|---|---|
| Number density ($\times10^8$ cm$^{-2}$) | 1.4 ± 0.5 | 1.8 ± 1.3 | $<5\times10^{-2}$ | 3.2 ± 1.5 |

**Figure 11** Statistical information on number density of InAs nanostructures on suspended SLG. (A) SEM images of InAs nanostructures grown on both sides of suspended SLG, taken from backside of the membrane. (B) Histogram of number density of InAs nanostructures formed on foreside and backside of SLG and SLG/$SiN_x$. (C) Table listing the number density of InAs/SLG/InAs double heterostructures.

heteroepitaxial integration (see also noncolored SEM image of Fig. 11A). Interestingly, the SEM inspection clearly revealed that the vdW epitaxy enabled the formation of InAs/SLG/InAs double heterostructures in which the InAs nanostructures are diametrically formed at the same position on both sides of the suspended SLG (Fig. 10C). The homogeneous sixfold rotational in-plane ordering of InAs nanostructures on/underneath the suspended SLG implies that those are epitaxial double heterostructures with excellent lattice and symmetry matching. In terms of morphology, these epitaxial double heterostructures can be classified into three architectures of nanowire/SLG/nanowire, nanowire/SLG/nanoisland, and nanoisland/SLG/nanoisland, as displayed in schematics and SEM images of Fig. 10C.

The SEM image of Fig. 11A, taken on the backside of the suspended SLG, displays that the high-density InAs nanostructures were formed on the backside of the SLG, and considerable portion of these nanostructures was found to be double heterostructures. The number density values of InAs nanostructures formed on both sides of the suspended SLG and SLG/$SiN_x$ were counted as shown in Fig. 11B. The number density values of nanowires and nanoislands were ca. $2 \times 10^8$ and $4 \times 10^8$ $cm^{-2}$, respectively, and were almost the same with acceptable error ranges, regardless of the substrate area. This outcome implies that the vdW heterogeneous nucleation density of InAs was not virtually influenced by the different precursor concentration on/underneath the suspended SLG but presumably determined by the homogeneous density of atomic steps (or interplanar vacancies) of SLG. It is also noted that the InAs nanostructures on SLG/$SiN_x$ exhibited little difference in number density with those on/underneath the suspended SLG. This suggests that the $SiN_x$ sublayers under the SLG provide little electrostatic attraction across the SLG for the vdW nucleation–growth of InAs, but vdW attraction from the SLG sufficiently contributed to the epitaxy of InAs.

Among the high-density InAs nanostructures on either sides of the suspended SLG (~$6 \times 10^8$ $cm^{-2}$), half of those was found to be double heterostructures (~$3 \times 10^8$ $cm^{-2}$), suggesting high yield of forming the double heterostructures (Fig. 11B and C). Such high yield is surmised to be correlated with the preferential vdW nucleation–growth of InAs on atomic steps of SLG. Typically, the CVD-grown SLG has a few kinds of structural imperfections, such as steps (or kinks) along periphery of unwantedly formed bilayer, vacancy, etc. Specifically, the interplanar vacancy of the suspended SLG provides a preferential nucleation site with the deep potential well that drives the diffuse-in of precursor adatoms on both sides during the vapor transport growth, which accommodated the diametrical growth of InAs nanostructures at the same position of the suspended SLG. This implies that the control of surface imperfections of the suspended SLG can be extended to fabricate the double heterostructures in a designed way.

# 3. vdW HETEROEPITAXIAL RELATIONSHIP AND HETEROINTERFACE OF NANOWIRE/2D-ALM

This section covers the heteroepitaxial relationship and heterointerface of the vdW epitaxial semiconductor nanostructure/2d-ALM. Nearly commensurate and highly incommensurate vdW heteroepitaxial systems are discussed with a few examples of the InAs nanowire on graphene and ZnO nanostructures on hBN and mica, respectively, based on TEM observations. Theoretical computational analyses on the stable vdW heterointerface and vdW cohesion binding energy are also discussed.

## 3.1 Nearly Commensurate System: InAs/Graphene

As shown in Sections 2.1 and 2.3, InAs nanowires vertically grown on SLG presented the homogeneous hexagonal sidewall facet alignment, signifying the excellent matching of in-plane lattices and symmetries between InAs(111) and carbon honeycomb lattice. This section discusses how the InAs is epitaxially grown on graphene by means of vdW attraction and why the epitaxial double heterostructures of InAs/SLG/InAs are readily formed by vdW epitaxy.

The vdW heteroepitaxial relationship of InAs/SLG was investigated by plan-view TEM and selected area electron diffraction (SAED) analyses (Fig. 12). The nanostructures grown on the suspended 2d-ALMs facilitated the plan-view TEM observations without further sampling process. The plan-view image presents that the neighboring InAs nanowires have an identical in-plane sixfold rotational symmetry (Fig. 12A). Figure 12B and C are the SAED patterns obtained from the InAs nanowire and its vicinal SLG area, respectively. As marked in Fig. 12B, the SAED patterns show that InAs nanowire was composed of mixed WZ/ZB phases and grown along $[\bar{1}\,\bar{1}\,\bar{1}]_{ZB}$ or $[000\bar{1}]_{WZ}$. The diffraction spots of InAs and SLG were not sufficiently well resolved to distinguish them because of small difference of $d$-spacing of InAs and graphene: $d_{\bar{2}20}$ of ZB-InAs (or $d_{110}$ of WZ-InAs) and $d_{100}$ of graphene have only a small difference of $\sim$0.5%. The SAED pattern of Fig. 12C, which was obtained exclusively from the SLG, clearly presents that the (100) diffraction spot nearly coincided with SAED $(\bar{2}20)_{ZB}$ spot of InAs (or $(110)_{WZ}$ of InAs).

Figure 12D and E are the corresponding plan-view atomic configurations of InAs and graphene induced from the SAED patterns of

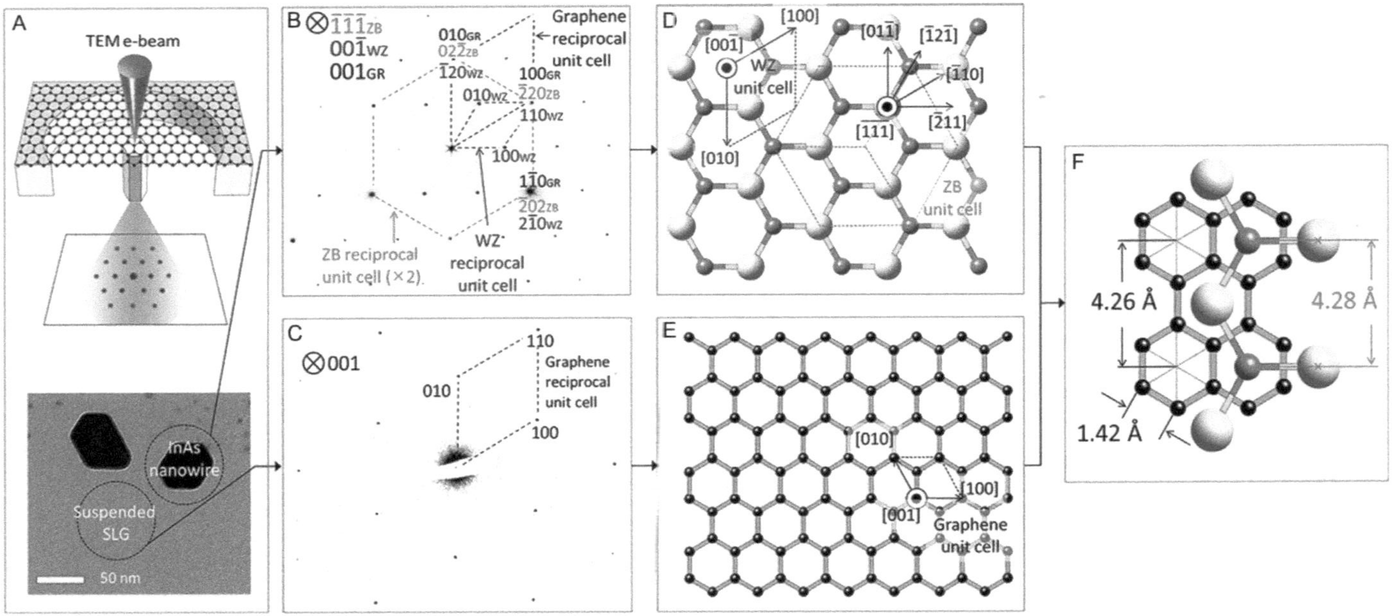


**Figure 12** Nearly commensurate vdW heteroepitaxial relationship of InAs/SLG observed by plan-view TEM. (A) Schematic depicting the plan-view TEM observation (upper panel) and TEM image of InAs nancwires grown on the backside of suspended SLG. SAED patterns of (B) InAs/suspended SLG and (C) suspended SLG obtained from the enclosed area marked in (A). (D and E) The corresponding lattices obtained from the SAED patterns by Fourier transforms. (F) Schematic of atomic configuration for the vdW heteroepitaxial InAs/graphene.

Fig. 12B and C via Fourier transforms, respectively. The InAs/SLG heterostructure of which two atomic planes are merged into one shows nearly commensurate heteroepitaxial relationship of ZB-InAs $(\bar{1}\,\bar{1}\,\bar{1})$ $[\bar{2}11]\,\|\,$SLG $(001)[\bar{1}2\bar{1}0]$ (or ZB-InAs $(\bar{1}\,\bar{1}\,\bar{1})[\bar{2}11]\,\|\,$SLG $(001)[100]$) with a small lattice misfit (Fig. 12F): the distance of the nearest carbon honeycomb–honeycomb along $[10\bar{1}0]$ is 4.26 Å, and has only a small difference of less than 0.5% with the nearest In–In (or As–As) distance of 4.28 Å along $[01\bar{1}]_{ZB}$. This represents that the basal plane of the InAs nanowire is structurally compatible with graphene for the vertical alignment of nanowires.

Cross-sectional STEM and EDX spectroscopic analyses clarified the growth directions and crystal structures of InAs nanowires and nanoislands grown on SLG. The HAADF-STEM and EDX atomic-column mapping images (Fig. 13A) taken from the InAs nanowire exhibit the As-polarized preferential growth direction (i.e., $[\bar{1}\,\bar{1}\,\bar{1}]_{ZB}$ or $[00\bar{1}]_{WZ}$) of the nanowire with mixed ZB/WZ crystal phases due to rotational twin stacking faults. The InAs nanoislands also presented the same preferred crystallographic growth direction of $[\bar{1}\,\bar{1}\,\bar{1}]_{ZB}$ (Fig. 13B), but ZB single crystallinity was observed from nanoislands (diameter $> \sim 150$ nm) presumably because of the size-dependent thermodynamic stability (Akiyama et al., 2006; Johansson et al., 2010).

It is necessary to clarify why the InAs nanostructures were exclusively grown along As-terminated polar direction of $[\bar{1}\,\bar{1}\,\bar{1}]_{ZB}$ or $[0001]_{WZ}$ on the nonpolar graphene. Such preferred growth orientation is thought to be determined at the early growth stage, so that the adatom process must

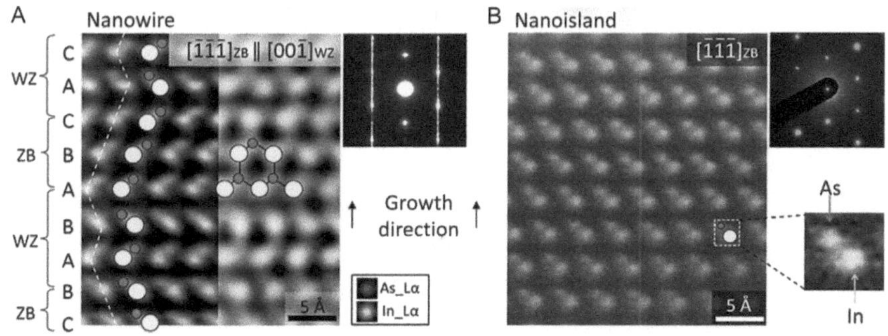


**Figure 13** Crystal structure and preferential growth direction of InAs nanostructures grown on graphene. (A) Cross-sectional atomic-resolution HAADF-STEM image (left) and EDX atomic-column mapping image (right) of InAs nanowire. (B) HAADF-STEM image of InAs nanoisland. Upper right panels are the corresponding SAED patterns. *Adapted with permission from Hong et al. (2013). Copyright 2013 John Wiley & Sons, Inc.*

be primarily considered. The total energy values for the adsorption of In and As adatoms were calculated with respect to the graphene surfaces using First-principle calculations. As depicted in Fig. 14A, In or As adatoms must occupy one of three plausible adatom sites of "H," "T," or "B" on graphene, which stand for the center hollow of the carbon honeycomb, on-top above a carbon atom, and the above of carbon–carbon bond bridge, respectively. Chan et al. and Liu et al. reported that In adatoms prefer to occupying the "H" sites rather than "T" or "B" for energy minimization with higher coordination (Chan et al., 2008; Liu et al., 2011b). On the "H" sites, the adsorption energy of In–SLG was calculated to be −0.746 eV at an equilibrium distance of 2.5 Å, which is much lower than that of As–SLG of −0.125 eV at an equilibrium distance of 3.0 Å (Fig. 14B), suggesting that the In adatoms are initially incorporated to form vdW heterointerfacial layer with lattitude closer to the SLG surface than As adatoms. As a result of the adatom process, the InAs nanostrutures are grown exclusively along the As-terminated direction, irrespective of nanostructure morphology. The atomic-resolution STEM data shown in Fig. 13 conversely corroborate the calculation results.

The equilibrium heterointerface gap of the vdW heteroepitaxial InAs/graphene was measured to be 0.31 ± 0.02 nm by the cross-sectional high-resolution TEM (HRTEM), as shown in Fig. 15. This spacing is distinct from $d_{002}$ of graphite (0.34 nm) and $d_{111}$ of ZB-InAs (0.35 nm). It is notable that the vdW heteroepitaxial gap of InAs/graphene (0.31 nm) is smaller than the vdW homoepitaxial gap (0.34 nm) of graphene–graphene (graphite). This observation suggests that the vdW heteroepitaxial bonds may be stronger than those of graphite (Benedict et al., 1998), which is discussed in the following.

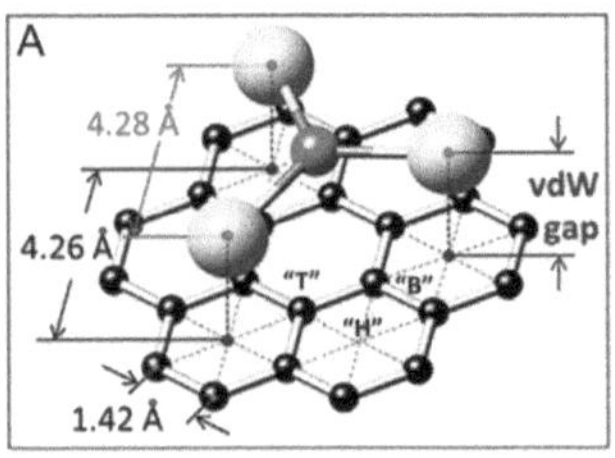


B

| Adatom on "H" site of SLG | Adsorption energy (eV) | Adatom–SLG distance (Å) |
|---|---|---|
| In | −0.746 | 2.5 |
| As | −0.125 | 3.0 |

**Figure 14** (A) Schematic of atomic configuration of the vdW epitaxial InAs/graphene heterostructure. There are three plausible adatom sites denoted as "H," "T," and "B," representing the In or As atoms on the hollow of a honeycomb, the top of C atom, and the bond bridge of C–C, respectively. (B) Table of adhesion energy and equilibrium distance of adatoms to SLG.

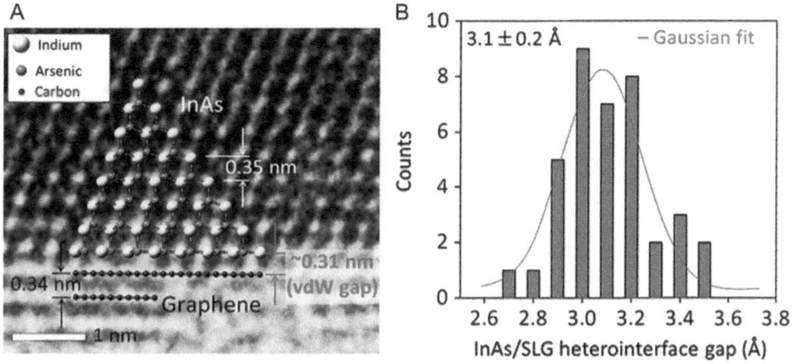


**Figure 15** vdW gap of InAs/graphene. (A) Cross-sectional HRTEM image of InAs/graphene heterointerface. (B) Histogram of the vdW gap of InAs/graphene counted from HRTEM analyses. *Adapted with permission from Hong et al. (2013). Copyright 2013 John Wiley & Sons, Inc.*

The stable crystal and electronic structure of the vdW epitaxial InAs/SLG heterostructure was simulated using the first-principle calculations of total energy, vdW binding energy, and electronic structure based on vdW density-functional (DF) method (Blöchl, 1994; Kresse and Furthmüller, 1996; Kresse and Joubert, 1999). For the calculations, InAs(111)A with the clean free surface and reconstructed surface was approached to the SLG to form the epitaxial heterostructure. Interestingly, it was calculated that the total energy values of the reconstructed surface with the In vacancy in every $(2 \times 2)$ InAs (111)A are considerably lower than those of the nonreconstructed InAs(111)A by 1.151 eV. Moreover, the same trend was also observed in the vdW epitaxial InAs/SLG heterostructures (Fig. 16A and B): the total energy of the InAs(111)/SLG heterostructure with the same reconstruction was computed to be lower than that of the unreconstructed heterostructure by 0.931 eV. Previous literatures already showed InAs(111)A surface reconstruction with the vacancy and buckling using reflection high-energy electron diffraction (Ohtake et al., 2000) and scanning tunneling microscopy (Taguchi and Kanisawa, 2006) under free surface vacuum conditions. Meaningfully, the simulation data were in good agreement with those previous observations, and further the same reconstruction was simulated to be stable when forming the vdW heterostructure, implying that the graphene has no chemical links with bottommost InAs(111)A plane in the heterostructure. While the reconstruction was induced in the formation of vdW epitaxial

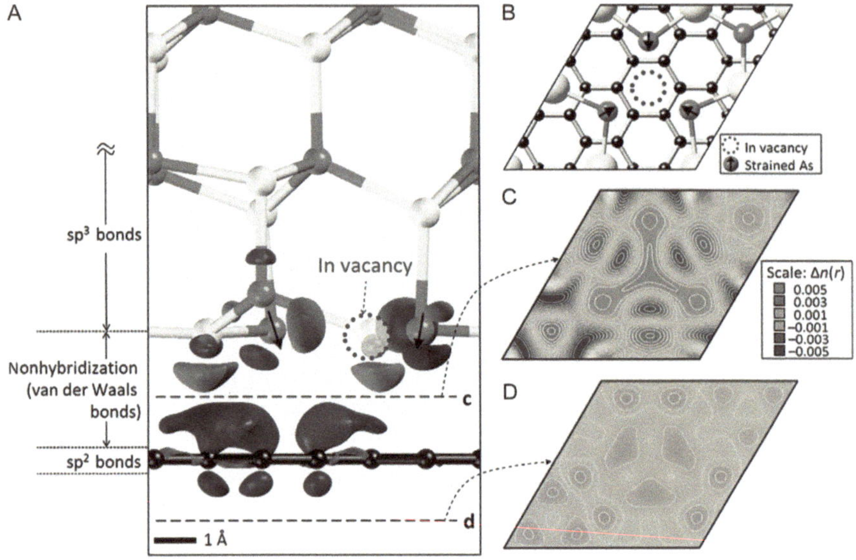


**Figure 16** Theoretically simulated vdW epitaxial InAs/SLG heterostructure in equilibrium. (A) Cross-sectional view crystal structure of an InAs/SLG supercell with EDD isosurfaces contoured at the level of $\pm 0.005e$ Å$^{-3}$. The red (gray in the print version) and blue (dark gray in the print version) EDD isosurfaces represent electron accumulation and depletion, respectively. Plan-view of (B) atomic configuration of the heterostructure with graphene and InAs (111)A basal plane and (C and D) contour map of EDD simulated at a distance of $\pm 1.55$ Å (at midpoint of the equilibrium vdW gap) from the SLG, as marked in (A). The diamond-shaped domain in (B–D) corresponds to the supercell of InAs (111)–($2 \times 2$)/graphene (0001)–($2\sqrt{3} \times 2\sqrt{3}$). *Reproduced with permission from Hong et al. (2013). Copyright 2013 John Wiley & Sons, Inc.*

heterostructure, the graphene exhibited no clear buckling or lattice distortion, due to no structural hybridization with the InAs bottommost plane.

Electron density difference (EDD) contour map of Fig. 16C shows no chemical bonds with valence electrons being shared or donated to form the cohesion between InAs and SLG, signifying the vdW binding feature with less electron orbital hybridization: the green (light gray in the print version) color represents the neutral region while red (light gray in the print version) and blue (gray in the print version) correspond to electron accumulation and depletion, respectively. More intriguingly, the EDD map of the SLG backside (Fig. 16D) shows the nearly charge neutral feature, similar to the bare graphene surface.

Figure 17A displays the binding potential curve of the vdW epitaxial InAs/SLG heterostructure plotted as a function of gap spacing (blue (light

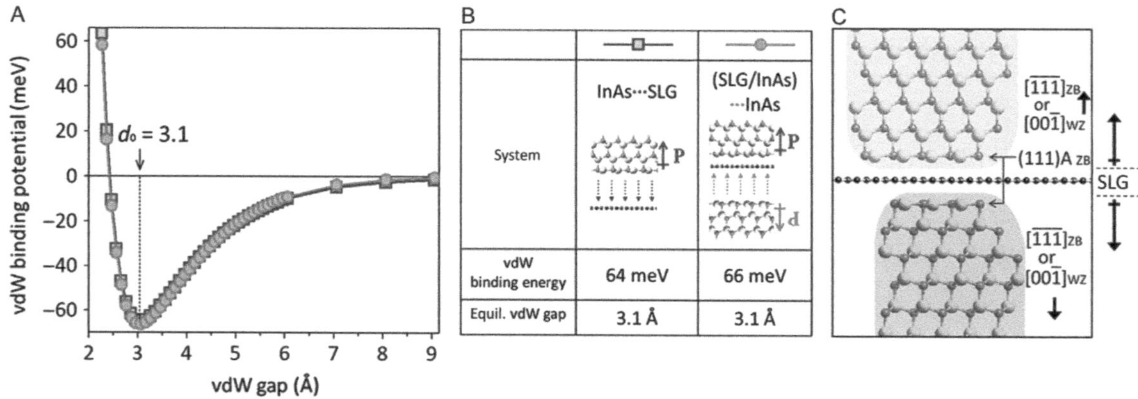


**Figure 17** vdW binding and equilibrium gap for epitaxial InAs/SLG heterostructure and InAs/SLG/InAs double heterostructure. (A) vdW binding energy curves of epitaxial InAs/SLG heterostructure (blue (gray in the print version) squares) and InAs/SLG/InAs double heterostructure (red (light gray in the print version) circles) as a function of the vdW gap. (B) Table summarizing the vdW binding energy and vdW gap of the InAs/SLG and InAs/SLG/InAs. (C) Schematic of atomic configuration for InAs/SLG/InAs double heterostructure with polarity inversion. The electric polarization of InAs is marked with sign of "+→" according to Lewis notation. *Reproduced with permission from Hong et al. (2013). Copyright 2013 John Wiley & Sons, Inc.*

gray in the print version) squares) calculated by the vdW–DF method, revealing that the equilibrium vdW gap distance of 3.1 Å and binding energy of −64 meV with respect to the vacuum level, which is consistent with the HRTEM observation shown in Fig. 15. It should be noted that the binding energy of −64 meV per carbon atom corresponds to 2.43 eV $nm^{-2}$, greater than the interplanar vdW binding energy of graphite (24–35 meV), presumably due to a structural reason. This value is much smaller than that of covalent or ionic bonds (typically 1–3 eV) by two orders of magnitude. The reconstructed bottommost InAs (111)A has flatter surface than the nonreconstructed InAs (111)A with a zigzag In–As dumbbell configuration, thus both the In and As atoms at the same latitude of the flatly reconstructed (111)A can contribute to stronger vdW cohesion on SLG (64 meV at a small vdW gap of 3.1 Å) than the nonreconstructed one (calculated to be 46 meV at the same gap).

Regarding the formation of epitaxial double heterostructures, the theoretical calculations revealed that the vdW binding energy between InAs/SLG and the backside InAs as a function of gap distances was almost similar to that between InAs and SLG. The vdW binding energy of InAs/SLG–InAs was calculated to be 66 meV per carbon atom at the equilibrium heterointerface gap of 3.1 Å. Because there is little charge transfer from/into the foreside InAs at the vdW heterointerface (Fig. 16A), EDD maps of the SLG backside, shown in Fig. 16D, are almost the same as that of a pristine SLG surface. This implies that the backside InAs is hardly affected by the foreside InAs with the distance of 6.2 Å, due to quite weak vdW attraction at the distance and excellent charge-screening capability of the conductive graphene. Hence, the electronically decoupled backside surface of InAs/SLG could promote the formation of [111]B-oriented InAs, yielding the thermodynamically stable InAs/SLG/InAs double heterostructures with polarity inversion, as marked in table of Fig. 17B. Noticeably, the InAs/SLG/InAs is the double heterostructure of $sp^3/sp^2/sp^3$ type, dissimilar to the 2d-based epitaxial double heterostructure fabricated by vdW epitaxy (Schlaf et al., 1999b).

## 3.2 Highly Incommensurate System: ZnO/hBN and ZnO/Mica

The major advantage of the vdW heteroepitaxy is that high-quality single-crystalline overlayers are easily obtained despite highly lattice-mismatched heteroepitaxial system, owing to the weakly bound heterojunction (Koma, 1992). In this section, highly incommensurate systems of ZnO/hBN and ZnO/mica, already shown in Fig. 2E and D, respectively, are discussed based on HRTEM observations and theoretical calculations.

As already shown in Section 2.1, the ZnO nanowires and nanowalls were vertically grown on hBN by vdW epitaxy (Fig. 2E; Oh et al., 2014). The vdW heterointerface and crystallographic growth direction of the ZnO/hBN were examined by cross-sectional HRTEM. The TEM images of Fig. 18A show that a single-crystalline ZnO nanowire with $d_{0002}$ of 0.26 nm grew vertically on hBN ($d_{0002}$ of 0.34 nm). The atomically abrupt, clean heterointerface with no extended crystal defects (i.e., threading dislocation) was observed by HRTEM and two-beam microscopic conditions, presumably attributable to sufficient stress relaxation through the noncovalent vdW heterointerface. A series of the diffraction patterns obtained through fast Fourier transform (FFT) process shown in Fig. 18A demonstrated that the heteroepitaxial relationship of ZnO $(000\bar{1})$ $[10\bar{1}0] \parallel$ hBN $(0001)[10\bar{1}0]$.

The convergent beam electron diffraction (CBED) analyses showed disk patterns with asymmetric stripy contrasts (Fig. 18B): the ZnO nanowires were found to grow along the oxygen-polar $[000\bar{1}]$ direction by vdW epitaxy on hBN, which was in close accordance with those simulated with the same TEM condition. ZnO nanowalls grown on hBN also exhibited the

**Figure 18** Heterointerface of vdW epitaxial ZnO/hBN observed by cross-sectional TEM. (A) HRTEM lattice images of the ZnO nanowire/hBN heterointerface. The right insets are the SAED patterns obtained through a FFT process of the HRTEM images marked with enclosed areas. (B) Experimental and simulated CBED patterns of ZnO nanoneedles preferentially grown along the O-polar $(000\bar{1})$ direction. *Adapted with permission from Oh et al. (2014). Copyright 2014 Nature Publishing Group.*

same growth orientation (not shown here), indicating that the ZnO nanostructures grew exclusively along the oxygen-polar *c*-axis direction, as a result of vdW epitaxy on hBN.

The heteroepitaxial relationship of ZnO/hBN was further determined by the plan-view TEM observation. The low-magnification TEM image of Fig. 19A shows the neighboring nanowires exhibit a hexagonal prismatic morphology with six $\{10\overline{1}0\}_{WZ}$ sidewall facets and a fairly uniform sixfold rotational in-plane alignment as marked with arrows. The HRTEM image clearly displays the hexagonal lattices of ZnO and hBN with the same crystallographic alignments. Notably, though the lattice mismatch of ZnO/hBN is −30.0%, the ZnO/hBN showed the domain-aligned epitaxial relationship of $(000\overline{1})[10\overline{1}0]_{ZnO} \| (0001)[10\overline{1}0]_{hBN}$. It is well known that large lattice misfits in hexagonal heteroepitaxial systems (e.g., ZnO, GaN, and AlN on *c*-$Al_2O_3$) yield the heterostructures of $(0001)[11\overline{2}0]_{overlayer} \| (0001)[10\overline{1}0]_{substrate}$, in which the overlayer is crystallographically rotated by 30° with respect to the substrate lattices to reduce the dangling bond density through covalent epitaxial links at the heterointerface (Grandjean et al., 1997; Narayan et al., 1998; Ponce et al., 1994). Such in-plane rotation also occurs for the covalent epitaxy of nanowires with small footprints even on 2d-ALMs (Munshi et al., 2012). Hence, the observed domain-aligned epitaxial relationship of ZnO/hBN is thought to be resulted from the noncovalent heterointerfaces.

Due to the large misfit of ZnO/hBN, triangular Moiré fringes with periodicity of ~1.0 nm were observed by plan-view TEM (Fig. 19B). The periodicity corresponds to a triple of the $d_{100}$ of ZnO (3.25 Å) and a quadruple of the $d_{100}$ of hBN (2.50 Å), but these values still have a large misfit of 2.5%.

Based on the plan-view TEM observations, a ball-and-stick model could be simulated with an assumption that the ZnO/hBN interfacial layers are fully relaxed through weakly bound vdW heterointerface (Fig. 19C). The low-magnification ball-and-stick image of inset of Fig. 19C clearly reveals the regular triangular Moiré patterns with periodicity of ~1.0 nm, in good agreement with the plan-view TEM observation shown in Fig. 19B. In a unit Moiré pattern marked with a triangle in Fig. 19C, there must exist two distinct heteroepitaxial atomic configurations, as depicted in Fig. 19D: i.e., ball-to-ball (b–b) and ball-to-hollow (b–h) atomic configurations. Due to the large lattice misfit of ZnO/hBN, a few epitaxial arrangements with different b–b and b–h combinations are occurred as displayed in Fig. 19E. Then, two supercells of $(3 \times 3)$-ZnO/$(4 \times 4)$-hBN can be

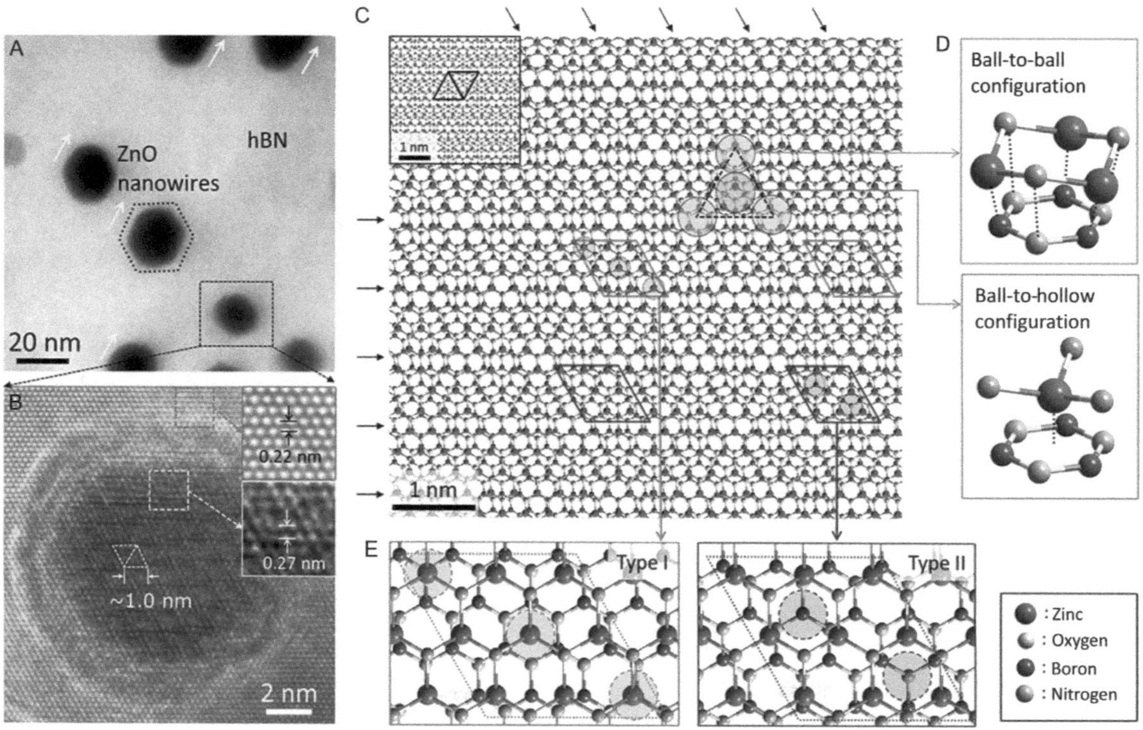


**Figure 19** vdW heteroepitaxial relationship of ZnO/hBN investigated by plan-view TEM. (A) Low-magnification TEM image and (B) HRTEM lattice image of ZnO nanowire grown on few-layer hBN. Triangular Moiré patterns are repeatedly appeared every 1 nm in the high-resolution lattice image. (C) Ball-and-stick model for ZnO/hBN. (D) Two heterointerfacial atomic epitaxial arrangement of ball-to-ball (upper panel) and ball-to-hollow (bottom panel) configurations appearing in a Moiré pattern. (E) Two representative supercell structures of ball-top-matching (Type I) and hollow-matching (Type II) $(3 \times 3)$-ZnO/$(4 \times 4)$-hBN heterojunctions. *Reproduced with permission from Oh et al. (2014). Copyright 2014 Nature Publishing Group.*

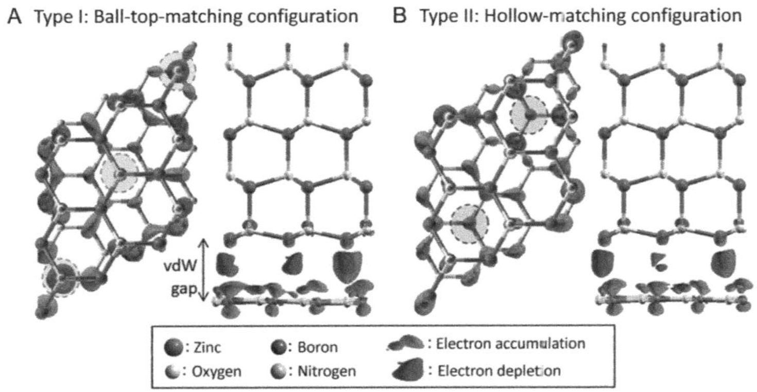


C

| | Before epitaxy (electrons) | vdW epitaxy (electrons) |
|---|---|---|
| ZnO | 657 | 656.79 (type I)<br>656.85 (type II) |
| hBN | 128 | 128.21 (type I)<br>128.15 (type II) |

D

| | ΔE (eV) | ΔE/unit area (eV nm$^{-2}$) | vdW gap (Å) |
|---|---|---|---|
| Type I | −1.894 | −2.170 | 3.4 |
| Type II | −1.890 | −2.165 | 3.4 |

**Figure 20** Electronic structure and cohesion energy of the vdW epitaxial ZnO/hBN heterointerface computed by the DF method. Atomic model structures and electron density difference isosurfaces at the periphery of the heterointerfaces with (A) ball-top-matching and (B) hollow-matching configurations of ZnO/hBN. The electron density was contoured with an isovalue of $\pm 0.0002e$ Å$^{-3}$. (C) The electron populations in the ZnO and hBN parts of the supercells before and after the vdW epitaxy, calculated by Bader analysis. (D) vdW hetero-interfacial adhesion energy and equilibrium vdW distance values for type I and II supercells. *Reprinted with permission from Oh et al. (2014). Copyright 2014 Nature Publishing Group.*

represented for the highly incommensurate vdW epitaxial ZnO/hBN system: i.e., a ball-top-matching configured supercell in which two Zn atoms are on top of B and/or N atoms (type I, Fig. 19E) and a hollow-matching supercell consisting of two hollows of ZnO hexagon lattices on top of B or N atoms (type II).

The vdW epitaxial ZnO/hBN heterointerfaces of these two supercells (type I and II) were theoretically simulated using DF calculations of total energy and electronic structure. As a consequence, the theoretical computations for type I and II supercell heterointerfaces showed almost negligible disparities between them in terms of electronic structure, binding energy, and equilibrium vdW distance. The electron density maps of these supercells that are contoured with an isovalue of $\pm 0.0002e$ Å$^{-3}$ showed few electron exchanges between ZnO and hBN (Fig. 20A and B), featuring the vdW binding. Bader analysis supports little electron orbital hybridization across the vdW heterointerface (Fig. 20C). Little buckling (or distortion) was observed at the heterointerfaces, implying weakly hybridized heterointerface.

The binding energy ($\Delta E = E_{total} - E^{ZnO} - E^{hBN}$) values of both type I and II supercells were calculated to be 2.170 and 2.165 eV nm$^{-2}$ with quite small difference of 5 meV nm$^{-2}$, and the equilibrium vdW gaps were calculated to be 3.4 Å (Fig. 20D). Thus, the homogeneous heterointerface feature (i.e., binding force and gap distance) is expected over the whole vdW heterointerface. The binding energy values correspond to ~59 meV per B and N atom, which is almost comparable to 64 meV per carbon atom (2.43 eV nm$^{-2}$) for InAs/SLG shown in the previous section.

Utama et al. found out that ZnO nanowires can be heteroepitaxially grown on the muscovite mica substrate via vdW epitaxy despite the large crystal symmetry and lattice mismatches (Table 2; Utama et al., 2012a): ZnO nanowires were grown on the mica substrate with excellent vertical alignment, and the hexagonal sidewall facets presented an in-plane alignment between the nanowires. As shown in Fig. 21, heterointerface quality and epitaxial relation were thoroughly investigated by HRTEM analyses along the two orthogonal zone axes. The heterointerface from either direction is atomically abrupt without observation of a buffer layer. Additionally, analyses of the diffraction patterns obtained through FFT process revealed the heteroepitaxial relation of $(0001)[11\bar{2}0]_{ZnO} \| (001)[010]_{mica}$ and $(0001)[1\bar{1}00]_{ZnO} \| (001)[100]_{mica}$. This is remarkable, considering that ZnO/mica system presents highly anisotropic lattice mismatches of 64.0% in

**Table 2** Crystallographic Information of the vdW Heteroepitaxial Relationship of InAs/Graphene, ZnO/hBN, and ZnO/Muscovite Mica

| vdW Heteroepitaxial System | Lattice Constant | Heteroepitaxial Relationship | In-Plane Lattice Mismatch |
|---|---|---|---|
| InAs/graphene | InAs: $a = c = 6.06$ Å Graphite: $a = 2.46$ Å and $c = 6.71$ Å | $(\bar{1}\,\bar{1}\,\bar{1})[\bar{2}11]_{ZB\text{-}InAs} \parallel (001)[\bar{1}2\bar{1}0]_{SLG}$ | 0.5% |
| ZnO/hBN | ZnO: $a = 3.25$ Å and $c = 5.20$ Å hBN: $a = 2.50$ Å and $c = 6.66$ Å | $(000\bar{1})[10\bar{1}0]_{ZnO} \parallel (0001)[10\bar{1}0]_{hBN}$ | −30.0% |
| ZnO/muscovite mica | $KAl_2(Si_3Al)O_{10}(OH)_2$ for 2 $M_1$: $a = 5.18$ Å, $b = 8.99$ Å, and $c = 20.07$ Å (monoclinic $\beta = 95.7°$) | $(0001)[11\bar{2}0]_{ZnO} \parallel (001)[010]_{mica}$ | 64.0% along $[11\bar{2}0] \parallel [020]$; −8.36% along $[1\bar{1}00] \parallel [200]$ |

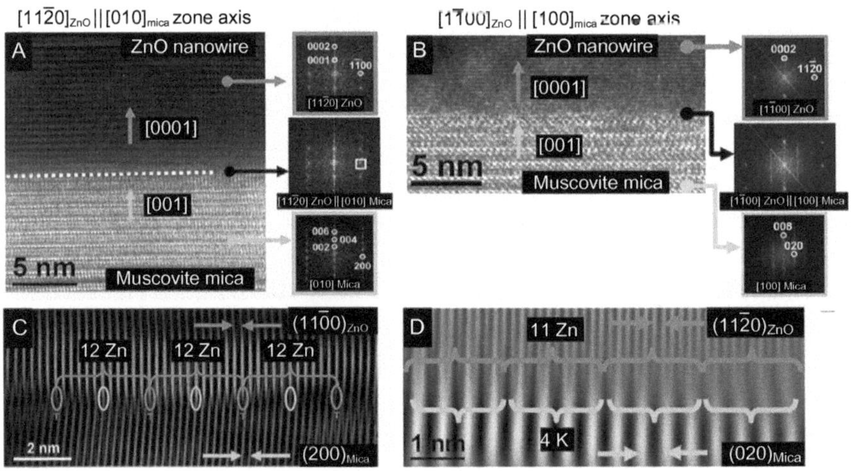


**Figure 21** HRTEM observation and analyses of the vdW epitaxial ZnO/mica heterointerface (A and C) along $[11\bar{2}0]_{ZnO} \parallel [010]_{mica}$ zone axis and (B and D) along the $[1\bar{1}00]_{ZnO} \parallel [100]_{mica}$ zone axis. In each zone axis, the analyses include the HRTEM imaging and electron diffraction patterns, and Fourier filtering. *Adapted with permission from Utama et al. (2012a). Copyright 2012 American Chemical Society.*

$[11\bar{2}0]_{ZnO} \| [020]_{mica}$ direction and −8.36% in $[1\bar{1}00]_{ZnO} \| [100]_{mica}$ direction. Moreover, threading dislocations were not observed at the heterointerface. All of these observations suggest that the vdW heteroepitaxy can be adopted to create high-quality semiconductor overlayers on arbitrary substrates especially for a highly incommensurate system.

# 4. CONTROLLED vdW EPITAXY OF SEMICONDUCTOR NANOWIRES

The vdW epitaxy of nanowires or nanostructures on 2d-ALMs is driven by the weak vdW attraction, resulting in low-density random nucleation–growth. This section reviews the comprehensive study on correlation of nanostructure density and the surface of 2d-ALMs, then provides a general scheme to control the position of semiconductor nanowires or the shape of nanostructures. We also discuss how the chemical stability of 2d-ALMs can be circumvented for fabricating epitaxial semiconductor films on 2d-ALMs by the vdW heteroepitaxy.

## 4.1 Nucleation and Growth on 2d-ALM with Surface Imperfections

Figure 22 displays the cross-sectional TEM images showing the vdW heterointerfaces of InAs nanowires and nanoislands on graphite. For both the nanowires and nanoislands, an obtuse contact angle of ~150° was observed, as marked with "$\theta$" in Fig. 22A, implying the high chemical stability of the graphite surface with very few unoccupied electron pairs (Wang et al., 2009). According to Young's equation (de Gennes, 1985), the high interfacial tension between InAs/graphite exists thus that makes a small contact area with a small coalescence size of InAs. This leads to the diameter-limited formation of fine nanowires with mean diameters of $42 \pm 16$ nm. The obtuse contact angle of InAs/graphene is in contrast to the small acute contact angle at the bottom of InAs nanowires grown on the InAs (111)B substrate by covalent homoepitaxy (Dayeh et al., 2007).

Even on the same graphene surface, some of InAs was grown to be nanowires while the others became nanoislands. Specifically, the HRTEM image reveals an isolated graphite atomic step edge at the center of the nanowire bottom and high-density graphite atomic steps under the entire bottom of a nanoisland, as marked with wedges in Fig. 22B and C, respectively. This

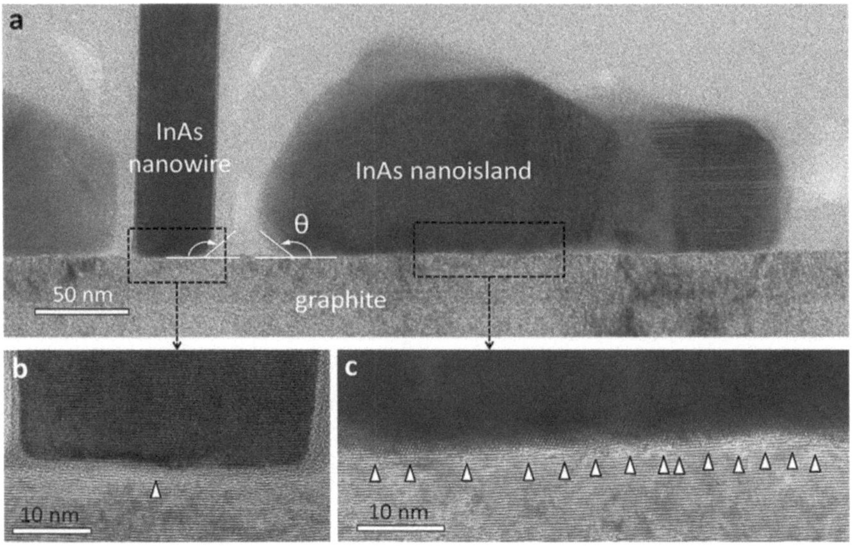


**Figure 22** Heterointerface of InAs nanostructures/graphite observed by cross-sectional TEM. (A) Low-magnification image of InAs nanowire and nanoisland grown on graphite. (B and C) HRTEM images obtained from the enclosed box areas marked in (A). The wedges indicate the atomic steps of graphite.

observation implies that an atomic step provided a preferential site for InAs nucleation, and when the spacings of a series of steps are smaller than the coalescence size of 42 ± 16 nm, the dense InAs nuclei on the steps coalesce one another, which forms the nanoisland. Hence, a scenario for vdW heteroepitaxy of semiconductor nanostructures on 2d-ALMs can be envisioned as follows: the atomic step (or ledge) of 2d-ALMs has infinitesimal radius of curvature with the deep potential well on which the precursors are crowded by diffusion (Fig. 23A). The deep potential enables long residential time of precursors around the steps, increasing the impingement rate of adatoms. As a result, the adatoms possibly agglomerate to form an embryo for the nanowire. Thus, the deep potential at the ledges overcomes the chemically stable, nonwettable 2d-ALM surface for the heterogeneous nucleation of the semiconductor overlayer. It should be noted that the atomic steps do not provide chemical binding sites because of stable hydrogenation for both graphene and hBN because the hydrogenation process was conducted just before vdW growth: according to theoretical computation, the binding energies of H–graphene (Sofo et al., 2007) and H–hBN (Wang, 2010) are calculated to be 6.5 and 1.04 eV $\text{atom}^{-1}$, respectively, suggesting that the

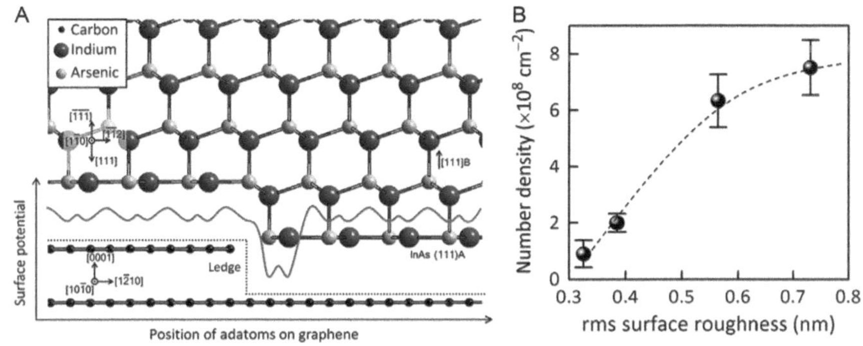


**Figure 23** vdW nucleation–growth on atomic step edge of 2d-ALM. (A) Schematic representations of vdW epitaxial InAs/graphene heterostructure. The red (gray in the print version) curves are the surface potential for precursor adatoms plotted as a function of position on graphene surface. (B) Number density of InAs nanowires as a function of rms surface roughness of graphite.

hydrogenation is stable and can readily passivate the ledges even at the high temperature process of MOVPE of InAs and ZnO nanostructures.

The density of surface structural imperfections of 2d-ALMs, such as steps, ledges, and kinks, can be artificially engineered using diverse surface treatments. One of the simple methods is plasma treatment or reactive ion etching, which damages the surface atoms of 2d-ALMs (Lu et al., 1999). The rms surface roughness of the graphite terrace was gradually increased by increasing the plasma treatment time because of the formation of more steps, ledges, and kinks. On these roughened graphite surfaces, more InAs nanowires were grown, as shown in Fig. 23B, demonstrating the correlation between vdW epitaxial nanostructure density and roughness of 2d-ALMs.

Figure 24 displays another example that demonstrates the correlation of nanostructure density and plasma treatment time of the 2d-ALM substrate. On untreated pristine hBN surfaces, ZnO nanowires are grown with low number density of $\sim 10^8$ cm$^{-2}$. Once the plasma treatment was performed on the hBN surface, the vdW epitaxy yielded the ZnO nanowalls with higher density. It is surmised that ZnO nanowires were grown on very tiny atomic edges of hBN while the nanowalls are grown along the long steps. Accordingly, the proportional relationship between plasma treatment time of hBN and line density of ZnO nanowalls (Fig. 24B) suggests that the plasma treatment systematically increases the density of atomic step edges leading to nanowall formation.

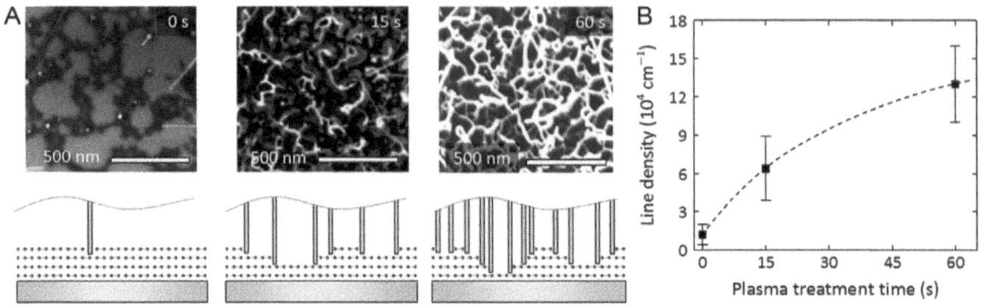


**Figure 24** Control of nanostructure number density using plasma-treated 2d-ALM substrate. (A) A series of top-view SEM images of ZnO nanowalls grown on hBN treated by plasma with different exposure times of 0, 15, and 60 s. (B) Plot of line density of ZnO nanowalls as a function of plasma treatment time.

## 4.2 Position- and Shape-Controlled vdW Epitaxy

Precise control of the spatial arrangement and shape of semiconductor nanostructures is necessary for various device applications (Hong et al., 2009; Kanungo et al., 2013; Tomioka et al., 2012; Wallentin et al., 2013). As discussed above, the surface atomic steps of 2d-ALMs provided energetically favorable nucleation sites. Thus, once the positions of atomic steps are artificially defined, the position of nanostructures can be determined. Motivated by the observation that the plasma treatment formed the atomic steps, the position of atomic steps could be controlled by conducting the plasma treatment on the predetermined positions or areas through the mask layer with the designed openings.

Figure 25 presents the position- and shape-controlled semiconductor nanostructures grown by vdW epitaxy on atomic-step-patterned 2d-ALMs. The approach for the position-controlled InAs nanowire arrays was the site-selective vdW epitaxy using a patterned growth-mask layer (Fig. 25A). For the patterned mask layer, the $SiO_x$ layer was deposited, followed by patterning the hole openings using electron-beam lithography and selective etching. Then, to form the surface imperfections on the opened graphite area, plasma treatment or reactive ion etching was performed. Without this process, hole openings hardly yielded nanowires. Since the mask layer covers the naturally formed step edges of graphene and graphite, only the hole opening with surface steps facilitated the vdW epitaxy of InAs nanowires. As shown in Fig. 25A, the diameter of nanowires was not determined by patterned hole-opening diameter, dissimilar to the selective-area covalent epitaxy of InAs nanowires on Si or III–V substrates (Tomioka et al., 2007, 2008). Such hole-size-independent growth can be explained by the coalescence-size-limited growth of vdW epitaxy on

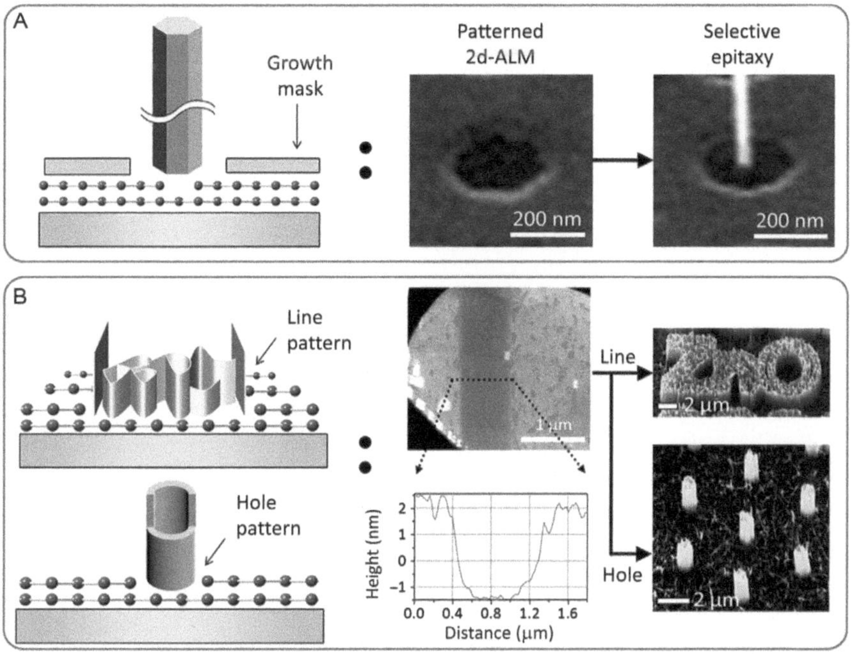


**Figure 25** Position- and shape-controlled vdW heteroepitaxy of nanostructures on patterned 2d-ALMs. (A) Single InAs nanowire grown on graphite covered with hole-opening-patterned growth mask. (B) Shape-controlled ZnO nanowalls and nanotubes grown on intaglio-patterned hBN by mask-free vdW epitaxy. *Reprinted with permission from Oh et al. (2014). Copyright 2014 Nature Publishing Group.*

2d-ALMs. Meanwhile, the position-controlled ZnO microrods were also implemented by seedless solution-phase vdW epitaxy, suggesting that the vdW attraction between the adatoms and 2d-ALMs can promote the vdW epitaxy in a liquid phase (Zhu et al., 2013).

The preferred vdW epitaxy of ZnO nanowalls along the patterned hBN atomic steps enabled shape-designed semiconductor nanoarchitectures, as displayed in Fig. 25B. Specifically, after the formation of designed step edges, polymeric pattern-mask layer (i.e., poly(methyl methacrylate)) used for the electron-beam lithography was removed for mask-free selective epitaxy. Inside the etched area, the nanowalls were grown to form random networked morphology, but the intaglio atomic steps along the circumference of the hBN-etched area promoted the preferential growth of ZnO nanowalls along them, which determined the exterior shape of semiconductor nanostructures in a designed shape.

The enhanced nucleation–growth of ZnO along the atomic steps, confirmed by the shape-designed selective vdW epitaxy, indicates the presence of a reduced Ehrlich–Schwoebel energy barrier encountered by the precursors being diffused downward (Pierre-Louis et al., 1999; Trushin et al., 1997) from the ultraflat hBN terrace into the plasma-etched atomic steps. The reduced barrier possibly contributed to overcome the difficulties of heterogeneous nucleation at the specific sites caused by the weak vdW attraction. As a result, the shape-designed ZnO nanowalls could be fabricated by maskless vdW heteroepitaxy with the artificially formed atomic steps, which is distinguished from the conventional selective-area growth that necessarily employs amorphous growth-mask layers (Hong et al., 2009).

Substrate patterning by chemical surface modification can be another approach for the maskless selective vdW epitaxy on 2d-ALMs. Liu and coworkers employed selected-area oxygen plasma treatment on the fluorophlogopite mica substrates so as to define the inactive area where the vdW epitaxy is inhibited: only the intact area contributed to the selective vdW epitaxy. This method was successfully applied to the selective vdW epitaxy of semiconducting $In_2Se_3$ nanoflakes (Lin et al., 2013) and topological insulator $Bi_2Se_3$ and $Bi_2Te_3$ nanoplates (Li et al., 2012) with the desired spatial arrangement.

Meanwhile, it is expected that the site-selective vdW epitaxy technique can be utilized to grow the smooth thin film structure by combining the two techniques of position-controlled vdW epitaxy of nanostructures (Fig. 26A) and epitaxial lateral overgrowth on the nanostructures (Fig. 26B). If the

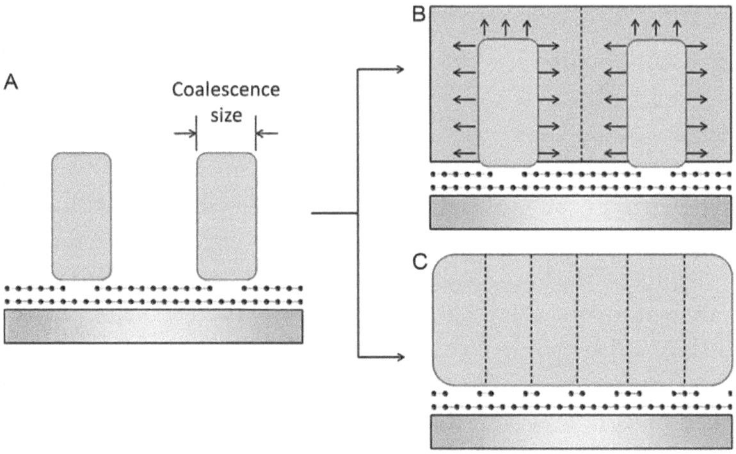


**Figure 26** Schematics of growth process to obtain smooth thin-film structures on 2d-ALMs. (A) Site-selective vdW epitaxy of semiconductor nanostructures with limited coalescence size. (B) Lateral homoepitaxial overgrowth. (C) High-density site-selective growth for thin film formation.

spacing of patterns is sufficiently smaller than the coalescence size of nanostructures, the high-density formation of nanostructures may coalesce to be the thin films, as depicted in Fig. 26C. For the thin film growth on 2d-ALMs, the crystalline quality of 2d-ALMs is crucial to obtain the smooth and single-crystalline film structure. Yoo et al. reported that the use of the CVD-grown graphene substrate produced highly *c*-axis-oriented textured GaN films with typical domain sizes of a few tens micrometer and in-plane domain misorientation angles of 3–5° because the size of the underlying graphene domain determines the overlayer grain sizes. Thus, it is necessary to use the single-domain 2d-ALMs to obtain single-crystalline overlayer films (Yoo et al., 2012). We believe that recent progresses on fabrication of the wafer-scale single-crystalline graphene (Lee et al., 2014) can be readily exploited for obtaining high-quality semiconductor thin films by the vdW heteroepitaxy.

## 5. OPTOELECTRONIC DEVICE APPLICATIONS

The semiconductor nanowires integrated on 2d-ALMs have great potentials for optoelectronic device applications due to high optical transmittance and electrical conductivity (or resistivity) of few-layer-thick 2d-ALMs. In this section, two different types of optoelectronic devices that exploited the inherent physical properties of conducting graphene and insulating hBN are reviewed.

Li and coworkers demonstrated a novel solar cell architecture utilizing the axial-and-radial heterostructured III–As nanowire arrays, monolithically fabricated on the graphene substrate via vdW heteroepitaxy (Mohseni et al., 2014). After the growth of the axial n-InAs/InGaAs heterostructure on graphene, the radial p–n junction of the InGaAs segment was fabricated through a two-step MOCVD process. Note that the n-InAs segment was introduced not only to prevent the preferential phase segregation during the direct growth of InGaAs on graphene but also to improve the electron collection from InAs with smaller band gap to the semimetallic graphene. Additionally, the p-type GaAs layer was coated on the core–shell structure as a passivation layer. To fabricate the device, then, the indium tin oxide (ITO) was deposited as a top electrode after filling the gap with the polymer (SU-8) for electrical insulation. In this device geometry illustrated in Fig. 27A, the graphene layer served as a bottom electrode as well as an epitaxial substrate. As shown in Fig. 27B, the fabricated solar cell exhibited the excellent device performances with the power conversion efficiency ($\eta$) of

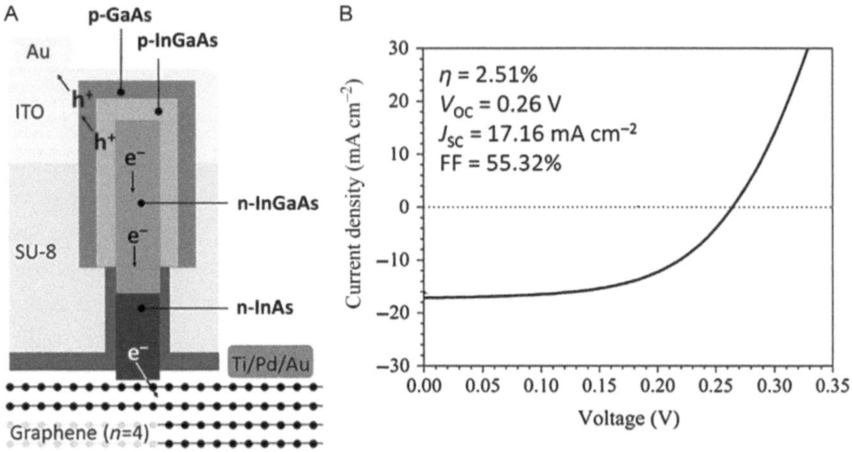


**Figure 27** III–As-based axial-and-radial nanowire heterostructure solar cells grown on graphene. (A) Schematic of solar cell structure. (B) *I–V* characteristic curve of axial-and-radial nanowire array-based solar cells, measured under AM 1.5 G illumination condition. *Reproduced with permission from Mohseni et al. (2014). Copyright 2014 John Wiley & Sons, Inc.*

2.51 %, the open-circuit voltage ($V_{oc}$) of 0.26 V, the short-circuit current ($J_{sc}$) of 17.16 mA $cm^{-2}$, and the fill factor (FF) of 55.32%. Those could be attributed to effective suppression of defect-mediated Schockley–Read–Hall (SRH) recombination processes in a core–shell structure with the additional surface passivation layer and efficient charge collection from the p–n junction to the graphene layer. Owing to the excellent electrical conductivity of graphene, the charge carriers could be injected into or extracted from the semiconductor overlayers for operating the semiconductor/graphene heterostructure devices.

The use of an insulating substrate is essential for individual operation of functional devices monolithically integrated on a substrate. The hBN, an electrical insulator 2d-ALM with wide direct band gap (5.2–5.9 eV), can satisfy such a requirement (Kubota et al., 2007; Watanabe et al., 2004). Moreover, the hBN exhibits high optical transparency (Zunger et al., 1976), chemical/thermal stability (Song et al., 2010), high thermal conductivity (Jo et al., 2013), and low dielectric constant (Wang et al., 2011), all of which are beneficial for optoelectronic device applications. The ultraviolet (UV) photosensor devices were fabricated using vdW epitaxial ZnO nanowall/hBN heterostructure, and the dark and photogenerated currents were measured with three different electrode configurations at an applied bias voltage of 1.0 V under duty-cycled on/off (20/20 s) UV irradiation

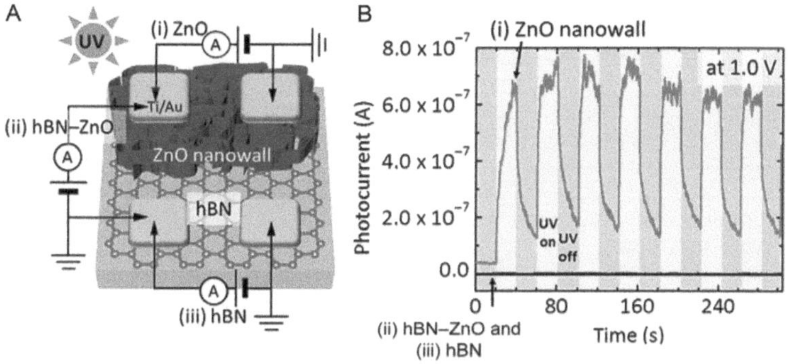


**Figure 28** UV photosensor of ZnO nanowalls grown on hBN. (A) Schematics of ZnO nanowall/hBN photoconductor device with three different electrode configurations for measuring the UV photocurrent through (i) ZnO nanowall networks, (ii) ZnO nanowall–hBN heterojunction, and (iii) bare hBN layer. (B) Photoresponse curve measured at bias voltage of 1.0 V under alternate on/off UV cycles. *Adapted with permission from Oh et al. (2014). Copyright 2014 Nature Publishing Group.*

(excitation light: Hg/Ar lamp, $<5$ W), as shown in Fig. 28. Among the three device configurations, only the electrode–ZnO nanowall–electrode configuration exhibited the sensitive UV detection performance with a mean on/off photocurrent-to-dark current ratio of $\sim$4–5 (Fig. 28B); the other device configurations with ZnO nanowall–hBN and bare hBN showed no clear photoresponses with a typical electrical leakage of $\sim 10^{-12}$ A, demonstrating that the hBN is a good wide-band-gap insulator sufficiently insensitive to the UV irradiation. Hence, it is expected that the hBN can be used as a vdW epitaxial substrate to integrate the nanodevice arrays that require good electrical insulation among neighboring devices.

## 6. CONCLUSIONS AND PERSPECTIVES

This chapter provided an overview of the recent efforts and advances on the vdW heteroepitaxy of semiconductor nanostructures on 2d-ALMs, covering the topics from growth to device applications. The recent works verified that weak vdW attraction facilitates the epitaxial integration of vertical nanowires on the 2d-ALM surfaces free of dangling bonds. A new class of epitaxial double heterostructures of $sp^3/sp^2/sp^3$ was also demonstrated by vapor-phase transport-based vdW epitaxy of InAs on the suspended SLG. Both nearly commensurate and highly incommensurate vdW heteroepitaxial systems exhibited well-defined heteroepitaxial relationships and atomically abrupt and clean heterointerfaces without threading dislocations,

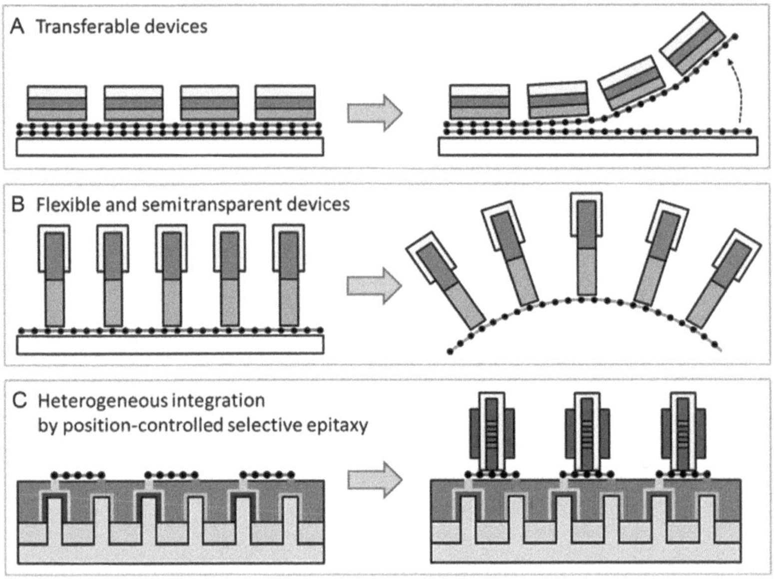


**Figure 29** Schematic illustrations representing various device applications of vdW epitaxial semiconductor nanostructure/2d-ALM heterostructure. (A) Transferable devices by exfoliation of 2d-ALM substrate. (B) Flexible and semitransparent devices. (C) Heterogeneous integration of nanowire devices on as-prepared semiconductor circuitry with 2d-ALM interconnect.

all of which are the notable features of vdW heteroepitaxy. Theoretical simulated vdW epitaxial heterostructures clarified stable vdW epitaxial relationship and heterointerface gap with little electron orbital hybridization, consistent with HRTEM observations.

Initiated by finding of the preferential nucleation–growth of semiconductors on the atomic steps of 2d-ALMs, the surface lithographic patterning of 2d-ALMs was adopted to create position-controlled and shape-designed semiconductor nanoarchitectures on 2d-ALMs. The vdW epitaxial semiconductor nanostructure/2d-ALM heterostructures were applied to optoelectronic devices, i.e., nanowire photovoltaic cells on graphene and nanowall UV photosensors on hBN, which exploited the electrical conducting and insulating properties of graphene and hBN, respectively. These experimental demonstrations presented their potentials for electronic and optoelectronic device applications.

We still have many challenges to the vdW epitaxy of semiconductors on 2d-ALMs for new functional electronics and optoelectronics that exploit the unique physical properties of 2d-ALMs. As schematically depicted in Fig. 29A, the semiconductor devices heteroepitaxially integrated on

2d-ALMs can be applied for transferring or stamp-printing the devices because the topmost layer of 2d-ALM is mechanically releasable and transferable (Chung et al., 2010; Kobayashi et al., 2012). This is expected to be achieved in a wafer scale in near future. By virtue of excellent mechanical properties of ultrathin 2d-ALMs, the spatially separated nanodevice arrays integrated on 2d-ALMs can be elastically tolerable to various mechanical deformations, such as bending, stretching, and rumpling (Fig. 29B; Choi et al., 2010; Lee et al., 2011b). Also, the ultrathin 2d-ALMs with high optical transmittance can be applied to fabricate semitransparent optoelectronics with high light-extraction or -absorption efficiency. As depicted in Fig. 29C, the site-selective vdW epitaxy can be useful for heterogeneous integration of the semiconductor/2d-ALM on as-prepared circuits, which would provide new electrical (Kim et al., 2011) and optical links. In addition to our perspectives, we believe that new findings and insights obtained through future works on vdW epitaxy would contribute to creating unconventional devices required for emerging functional electronics and optoelectronics.

## ACKNOWLEDGMENT

This work was supported by Basic Science Research Program through the National Research Foundation of Korea (NRF) funded from the Ministry of Education, Science, and Technology (MEST) (Grant No. NRF-2013R1A1A2058744; NRF-2014R1A1A2055112; 2010-0020207). Y.J.H. expresses his best gratitude to Prof. T. Fukui at Hokkaido University and Profs. Kwang S. Kim and R. S. Ruoff at UNIST for their supports and discussions.

## REFERENCES

Akiyama, T., et al., 2006. An empirical potential approach to wurtzite-zinc-blende polytypism in group III–V semiconductor nanowires. Jpn. J. Appl. Phys. 45, L275–L278.

An, S.J., et al., 2008. Enhanced light output of GaN-based light-emitting diodes with ZnO nanorod arrays. Appl. Phys. Lett. 92, 121108.

Ayers, J.E., 2007. Heteroepitaxy of Semiconductors: Theory, Growth, and Characterization. CRC Press, Boca Raton, FL.

Bae, S., et al., 2010. Roll-to-roll production of 30-inch graphene films for transparent electrodes. Nat. Nanotechnol. 5, 574–578.

Balandin, A.A., et al., 2008. Superior thermal conductivity of single-layer graphene. Nano Lett. 8, 902–907.

Benedict, L.X., et al., 1998. Microscopic determination of the interlayer binding energy in graphite. Chem. Phys. Lett. 286, 490–496.

Blasius, P.R.H., 1908. Grenzschichten in flüssigkeiten mit kleiner reibung. Z. Math. Phys. 56, 1–37.

Blöchl, P.E., 1994. Projector augmented-wave method. Phys. Rev. B 50, 17953–17979.

Bolotin, K.I., et al., 2008. Ultrahigh electron mobility in suspended graphene. Solid State Commun. 146, 351–355.

Britnell, L., et al., 2012. Field-effect tunneling transistor based on vertical graphene heterostructures. Science 335, 947–950.

Butler, S.Z., et al., 2013. Progress, challenges, and opportunities in two-dimensional materials beyond graphene. ACS Nano 7, 2898–2926.

Cha, J.J., et al., 2012. Weak antilocalization in Bi2(SexTe1−x)3 nanoribbons and nanoplates. Nano Lett. 12, 1107–1111.

Chan, K.T., et al., 2008. First-principles study of metal adatom adsorption on graphene. Phys. Rev. B 77, 235430.

Chandramohan, S., et al., 2012. Work-function-tuned multilayer graphene as current spreading electrode in blue light-emitting diodes. Appl. Phys. Lett. 100, 023502.

Chen, S., et al., 2011. Synthesis and characterization of large-area graphene and graphite films on commercial Cu-Ni alloy foils. Nano Lett. 11, 3519–3525.

Choi, D., et al., 2010. Fully rollable transparent nanogenerators based on graphene electrodes. Adv. Mater. 22, 2187–2192.

Chung, K., et al., 2010. Transferable GaN layers grown on ZnO-coated graphene layers for optoelectronic devices. Science 330, 655–657.

Datta, S.S., et al., 2009. Surface potentials and layer charge distributions in few-layer graphene films. Nano Lett. 9, 7–11.

Dayeh, S.A., et al., 2007. III-V nanowire growth mechanism: V/III ratio and temperature effects. Nano Lett. 7, 2486–2490.

Dean, C.R., et al., 2013. Hofstadter's butterfly and the fractal quantum Hall effect in moire superlattices. Nature 497, 598–602.

de Gennes, P.G., 1985. Wetting: statics and dynamics. Rev. Mod. Phys. 57, 827–863.

Geim, A.K., Grigorieva, I.V., 2013. Van der waals heterostructures. Nature 499, 419–425.

Geim, A.K., Novoselov, K.S., 2007. The rise of graphene. Nat. Mater. 6, 183–191.

Georgiou, T., et al., 2013. Vertical field-effect transistor based on graphene-$WS_2$ heterostructures for flexible and transparent electronics. Nat. Nanotechnol. 8, 100–103.

Giovannetti, G., et al., 2008. Doping graphene with metal contacts. Phys. Rev. Lett. 101, 026803.

Gong, Y., et al., 2014. Direct chemical conversion of graphene to boron- and nitrogen- and carbon-containing atomic layers. Nat. Commun. 5, 3193.

Grandjean, N., et al., 1997. Epitaxial relationships between GaN and $Al_2O_3$(0001) substrates. Appl. Phys. Lett. 70, 643–645.

Gudiksen, M.S., et al., 2002. Growth of nanowire superlattice structures for nanoscale photonics and electronics. Nature 415, 617–620.

Guo, B., et al., 2010. Controllable N-doping of graphene. Nano Lett. 10, 4975–4980.

Hashimoto, A., et al., 2004. Direct evidence for atomic defects in graphene layers. Nature 430, 870–873.

Heiss, M., et al., 2013. Self-assembled quantum dots in a nanowire system for quantum photonics. Nat. Mater. 12, 439–444.

Hong, Y.J., Fukui, T., 2011. Controlled van der Waals heteroepitaxy of InAs nanowires on carbon honeycomb lattices. ACS Nano 5, 7576–7584.

Hong, Y.J., et al., 2009. Shape-controlled nanoarchitectures using nanowalls. Adv. Mater. 21, 222–226.

Hong, Y.J., et al., 2011. Visible-color-tunable light-emitting diodes. Adv. Mater. 23, 3284–3288.

Hong, Y.J., et al., 2012. van der Waals epitaxy of InAs nanowires vertically aligned on single-layer graphene. Nano Lett. 12, 1431–1436.

Hong, Y.J., et al., 2013. Van der Waals epitaxial double heterostructure: InAs/single-layer graphene/InAs. Adv. Mater. 25, 6847–6853.

Hunt, B., et al., 2013. Massive dirac fermions and hofstadter butterfly in a van der waals heterostructure. Science 340, 1427–1430.

Hyun, J.K., et al., 2013. Nanowire heterostructures. Annu. Rev. Mater. Res. 43, 451–479.
Jo, G., et al., 2012. The application of graphene as electrodes in electrical and optical devices. Nanotechnology 23, 112001.
Jo, I., et al., 2013. Thermal conductivity and phonon transport in suspended few-layer hexagonal boron nitride. Nano Lett. 13, 550–554.
Johansson, J., et al., 2010. Diameter dependence of the wurtzite-zinc blende transition in InAs nanowires. J. Phys. Chem. C 114, 3837–3842.
Kanungo, P.D., et al., 2013. Selective area growth of III–V nanowires and their heterostructures on silicon in a nanotube template: towards monolithic integration of nano-devices. Nanotechnology 24, 225304.
Kim, H.M., et al., 2004. High-brightness light emitting diodes using dislocation-free indium gallium nitride/gallium nitride multiquantum-well nanorod arrays. Nano Lett. 4, 1059–1062.
Kim, Y.-J., et al., 2009a. Vertically aligned ZnO nanostructures grown on graphene layers. Appl. Phys. Lett. 95, 213101.
Kim, K.S., et al., 2009b. Large-scale pattern growth of graphene films for stretchable transparent electrodes. Nature 457, 706–710.
Kim, K., et al., 2011. A role for graphene in silicon-based semiconductor devices. Nature 479, 338–344.
Kim, K.K., et al., 2012. Synthesis and characterization of hexagonal boron nitride film as a dielectric layer for graphene devices. ACS Nano 6, 8583–8590.
Ko, W., et al., 2013. Atomic and electronic structure of an alloyed topological insulator, Bi1.5Sb0.5Te1.7Se1.3. Sci. Rep. 3, 2656.
Kobayashi, Y., et al., 2012. Layered boron nitride as a release layer for mechanical transfer of GaN-based devices. Nature 484, 223–227.
Koma, A., 1992. Van der Waals epitaxy—A new epitaxial growth method for a highly lattice-mismatched system. Thin Solid Films 216, 72–76.
Koma, A., 1999. Van der Waals epitaxy for highly lattice-mismatched systems. J. Cryst. Growth 201–202, 236–241.
Koma, A., Yoshimura, K., 1986. Ultrasharp interfaces grown with van der Waals epitaxy. Surf. Sci. 174, 556–560.
Koma, A., et al., 1984. Fabrication and characterization of heterostructures with sub-nanometer thickness. Microelectron. Eng. 2, 129–136.
Koma, A., et al., 1991. Heteroepitaxial growth by van der Waals interaction in one-dimensional, 2-dimensional and 3-dimensional materials. J. Cryst. Growth 111, 1029–1032.
Kong, D., et al., 2011. Ambipolar field effect in the ternary topological insulator (BixSb1-x)(2)Te-3 by composition tuning. Nat. Nanotechnol. 6, 705–709.
Kresse, G., Furthmüller, J., 1996. Efficient iterative schemes for *ab initio* total-energy calculations using a plane-wave basis set. Phys. Rev. B 54, 11169–11186.
Kresse, G., Joubert, D., 1999. From ultrasoft pseudopotentials to the projector augmented-wave method. Phys. Rev. B 59, 1758–1775.
Kubota, Y., et al., 2007. Deep ultraviolet light-emitting hexagonal boron nitride synthesized at atmospheric pressure. Science 317, 932–934.
Lauhon, L.J., et al., 2002. Epitaxial core-shell and core-multishell nanowire heterostructures. Nature 420, 57–61.
Lee, C., et al., 2008. Measurement of the elastic properties and intrinsic strength of monolayer graphene. Science 321, 385–388.
Lee, C.-H., et al., 2009. GaN/$In_{1-x}Ga_xN$/GaN/ZnO nanoarchitecture light emitting diode microarrays. Appl. Phys. Lett. 94, 213101.
Lee, C.-H., et al., 2011a. Flexible inorganic nanostructure light-emitting diodes fabricated on graphene films. Adv. Mater. 23, 4614–4619.

Lee, D.H., et al., 2011b. Flexible field emission of nitrogen-doped carbon nanotubes/reduced graphene hybrid films. Small 7, 95–100.

Lee, D.H., et al., 2013. Engineering electronic properties of graphene by coupling with Si-rich, two-dimensional islands. ACS Nano 7, 301–307.

Lee, J.-H., et al., 2014. Wafer-scale growth of single-crystal monolayer graphene on reusable hydrogen-terminated germanium. Science 344, 286–289.

Levendorf, M.P., et al., 2012. Graphene and boron nitride lateral heterostructures for atomically thin circuitry. Nature 488, 627–632.

Li, X., et al., 2009. Large-area synthesis of high-quality and uniform graphene films on copper foils. Science 324, 1312–1314.

Li, H., et al., 2012. Controlled synthesis of topological insulator nanoplate arrays on mica. J. Am. Chem. Soc. 134, 6132–6135.

Lin, M., et al., 2013. Controlled growth of atomically thin $In_2Se_3$ flakes by van der Waals epitaxy. J. Am. Chem. Soc. 135, 13274–13277.

Liu, H., et al., 2011a. Chemical doping of graphene. J. Mater. Chem. 21, 3335–3345.

Liu, X., et al., 2011b. Bonding and charge transfer by metal adatom adsorption on graphene. Phys. Rev. B 83, 235411.

Liu, Z., et al., 2013. In-plane heterostructures of graphene and hexagonal boron nitride with controlled domain sizes. Nat. Nanotechnol. 8, 119–124.

Lu, X., et al., 1999. Patterning of highly oriented pyrolytic graphite by oxygen plasma etching. Appl. Phys. Lett. 75, 193–195.

Munshi, A.M., Weman, H., 2013. Advances in semiconductor nanowire growth on graphene. Phys. Status Solidi RRL 7, 713–726.

Mohseni, P.K., et al., 2013. $In_xGa_{1-x}As$ nanowire growth on graphene: van der Waals epitaxy induced phase segregation. Nano Lett. 13, 1153–1161.

Mohseni, P.K., et al., 2014. Monolithic III–V nanowire solar cells on graphene via direct van der Waals epitaxy. Adv. Mater. 26, 3755–3760.

Munshi, A.M., et al., 2012. Vertically aligned GaAs nanowires on graphite and few-layer graphene: generic model and epitaxial growth. Nano Lett. 12, 4570–4576.

Narayan, J., et al., 1998. Defects and interfaces in epitaxial $ZnO/\alpha$-$Al_2O_3$ and $AlN/ZnO/\alpha$-$Al_2O_3$ heterostructures. J. Appl. Phys. 84, 2597–2601.

Oh, H., et al., 2014. Architectured van der Waals epitaxy of ZnO nanostructures on hexagonal BN. NPG Asia Mater. 6, e145.

Ohtake, A., et al., 2000. Strain relaxation in InAs/GaAs(111)A heteroepitaxy. Phys. Rev. Lett. 84, 4665–4668.

Pan, J., et al., 2012. Composition-tunable vertically aligned $CdS_xSe_{1-x}$ nanowire arrays via van der Waals epitaxy: investigation of optical properties and photocatalytic behavior. Adv. Mater. 24, 4151–4156.

Park, W.I., et al., 2002. Metalorganic vapor-phase epitaxial growth of vertically well-aligned ZnO nanorods. Appl. Phys. Lett. 80, 4232–4234.

Park, W.I., et al., 2003. Quantum confinement observed in ZnO/ZnMgO nanorod heterostructures. Adv. Mater. 15, 526–529.

Park, S.-I., et al., 2008. Theoretical and experimental studies of bending of inorganic electronic materials on plastic substrates. Adv. Funct. Mater. 18, 2673–2684.

Park, W.I., et al., 2011. Inorganic nanostructures grown on graphene layers. Nanoscale 3, 3522–3533.

Peng, H., et al., 2012. Topological insulator nanostructures for near-infrared transparent flexible electrodes. Nat. Chem. 4, 281–286.

Pierre-Louis, O., et al., 1999. Edge diffusion during growth: the kink Ehrlich-Schwoebel effect and resulting instabilities. Phys. Rev. Lett. 82, 3661–3664.

Ponce, F.A., et al., 1994. Crystalline structure of AlGaN epitaxy on sapphire using AlN buffer layers. Appl. Phys. Lett. 65, 2302–2304.

Ponomarenko, L.A., et al., 2013. Cloning of Dirac fermions in graphene superlattices. Nature 497, 594–597.
Reimer, M.E., et al., 2012. Bright single-photon sources in bottom-up tailored nanowires. Nat. Commun. 3, 737.
Schlaf, R., et al., 1994. Van der Waals epitaxy of thin InSe films on $MoTe_2$. Surf. Sci. 303, L343–L347.
Schlaf, R., et al., 1999a. Band lineup of layered semiconductor heterointerfaces prepared by van der Waals epitaxy: charge transfer correction term for the electron affinity rule. J. Appl. Phys. 85, 2732–2753.
Schlaf, R., et al., 1999b. Band lineup of a $SnS_2/SnSe_2/SnS_2$ semiconductor quantum well structure prepared by van der Waals epitaxy. J. Appl. Phys. 85, 6550–6556.
Seol, J.H., et al., 2010. Two-dimensional phonon transport in supported graphene. Science 328, 213–216.
Shi, Y., et al., 2010. Work function engineering of graphene electrode via chemical doping. ACS Nano 4, 2689–2694.
Shi, Y., et al., 2012. van der Waals epitaxy of $MoS_2$ layers using graphene as growth templates. Nano Lett. 12, 2784–2791.
Sofo, J.O., et al., 2007. Graphane: a two-dimensional hydrocarbon. Phys. Rev. B 75, 153401.
Song, L., et al., 2010. Large scale growth and characterization of atomic hexagonal boron nitride layers. Nano Lett. 10, 3209–3215.
Suk, J.W., et al., 2011. Transfer of CVD-grown monolayer graphene onto arbitrary substrates. ACS Nano 5, 6916–6924.
Tada, H., et al., 1992. Epitaxial growth of vanadyl-phthalocyanine ultrathin films on hydrogen-terminated Si(111) surfaces. Appl. Phys. Lett. 61, 2021–2023.
Taguchi, A., Kanisawa, K., 2006. Stable reconstruction and adsorbates of InAs(111)A surface. Appl. Surf. Sci. 252, 5263–5266.
Tian, B., et al., 2007. Coaxial silicon nanowires as solar cells and nanoelectronic power sources. Nature 449, 885–888.
Tomioka, K., et al., 2007. Growth of highly uniform InAs nanowire arrays by selective-area MOVPE. J. Cryst. Growth 298, 644–647.
Tomioka, K., et al., 2008. Control of InAs nanowire growth directions on Si. Nano Lett. 8, 3475–3480.
Tomioka, K., et al., 2012. A III–V nanowire channel on silicon for high-performance vertical transistors. Nature 488, 189–192.
Tongay, S., et al., 2012. Rectification at graphene-semiconductor interfaces: zero-gap semiconductor-based diodes. Phys. Rev. X 2, 011002.
Trushin, O.S., et al., 1997. Step roughening effect on adatom diffusion. Phys. Rev. B 56, 12135–12138.
Ueno, K., et al., 1990. Heteroepitaxial growth of layered transition metal dichalcogenides on sulfur-terminated GaAs{111} surfaces. Appl. Phys. Lett. 56, 327–329.
Utama, M.I.B., et al., 2011. Vertically aligned cadmium chalcogenide nanowire arrays on muscovite mica: a demonstration of epitaxial growth strategy. Nano Lett. 11, 3051–3057.
Utama, M.I.B., et al., 2012a. Incommensurate van der Waals epitaxy of nanowire arrays: a case study with ZnO on muscovite mica substrates. Nano Lett. 12, 2146–2152.
Utama, M.I.B., et al., 2012b. Epitaxial II–VI tripod nanocrystals: a generalization of van der Waals epitaxy for nonplanar polytypic nanoarchitectures. ACS Nano 6, 2281–2288.
Wallentin, J., et al., 2013. InP nanowire array solar cells achieving 13.8% efficiency by exceeding the ray optics limit. Science 339, 1057–1060.
Wang, Y., 2010. Electronic properties of two-dimensional hydrogenated and semi-hydrogenated hexagonal boron nitride sheets. Phys. Status Solidi RRL 4, 34–36.

Wang, S., et al., 2009. Wettability and surface free energy of graphene films. Langmuir 25, 11078–11081.
Wang, H., et al., 2011. BN/Graphene/BN transistors for RF applications. IEEE Electron Device Lett. 32, 1209–1211.
Wang, Q.H., et al., 2012. Electronics and optoelectronics of two-dimensional transition metal dichalcogenides. Nat. Nanotechnol. 7, 699–712.
Watanabe, K., et al., 2004. Direct-bandgap properties and evidence for ultraviolet lasing of hexagonal boron nitride single crystal. Nat. Mater. 3, 404–409.
Wisotzki, E., et al., 2000. Quasi van der Waals epitaxy of ZnSe on the layered chalcogenides InSe and GaSe. Thin Solid Films 380, 263–265.
Xiang, J., et al., 2006. Ge/Si nanowire heterostructures as high-performance field-effect transistors. Nature 441, 489–493.
Yoo, H., et al., 2012. Microstructures of GaN thin films grown on graphene layers. Adv. Mater. 24, 515–518.
Yu, Y.-J., et al., 2009. Tuning the graphene work function by electric field effect. Nano Lett. 9, 3430–3434.
Zhang, H., et al., 2009. Topological insulators in $Bi_2Se_3$, $Bi_2Te_3$ and $Sb_2Te_3$ with a single dirac cone on the surface. Nat. Phys. 5, 438–442.
Zhu, Y., et al., 2013. Solution phase van der Waals epitaxy of ZnO wire arrays. Nanoscale 5, 7242–7249.
Zunger, A., et al., 1976. Optical properties of hexagonal boron nitride. Phys. Rev. B 13, 5560–5573.

CHAPTER FOUR

# Position-Controlled Selective Growth of ZnO Nanostructures and Their Heterostructures

**Hosang Yoon, Gyu-Chul Yi**[1]
Department of Physics and Astronomy, Seoul National University, Seoul, South Korea
[1]Corresponding author: e-mail address: gcyi@snu.ac.kr

## Contents



## 1. INTRODUCTION

Modern society and industry are based largely on critical breakthroughs in semiconductor science and technology, such as in transistors, integrated circuits, and light-emitting devices (LEDs), where the fabrication and integration of very small, nanoscale devices allowed exponential growth of their functionality and complexity. The current semiconductor technology for fabricating nanoscale devices is based mainly on a "top-down"

*Semiconductors and Semimetals*, Volume 93
ISSN 0080-8784
http://dx.doi.org/10.1016/bs.semsem.2015.07.004

method using inorganic semiconductor thin-film growth, lithography, and etching techniques. Precise size and position control during nanomaterial preparation by the top-down method facilitate nanodevice integration.

Recently, new "bottom-up" approaches have attracted tremendous interest, due to their simplicity and multifunctionality (Lu and Lieber, 2007). In the bottom-up method, nanoscale devices and systems are fabricated by hierarchical assembly or selective growth of nanoscale materials; among the various nanostructures that can be fabricated, one-dimensional (1D) nanostructures (e.g., nanowires, nanorods, and nanotubes) are particularly attractive due to their unique semiconducting behavior and high aspect ratio (Duan et al., 2001; Pauzauskie and Yang, 2006; Yatsui et al., 2007). In particular, 1D heterostructures exhibiting a quantum confinement effect have been produced by carefully controlling both the composition and thicknesses of the layers grown on the nanostructures (Jang et al., 2006; Park et al., 2003c, 2004). The ability to embed quantum structures in a single nanostructure would enable the exploration of new physical properties, such as quantum confinement, in the form of continuous tuning of spectral wavelengths by varying the well thickness. Additional advantages of quantum nanostructures include the miniaturization of optoelectronic devices and the availability of a platform to investigate unique functional operations inherent to nanoscale electronics, optoelectronics, and photonics (Qian et al., 2005; Yatsui et al., 2005).

One of the most important goals in nanoscience and nanotechnology is the creation of high-performance devices. For many applications, the nanoscale devices must be integrated monolithically on a substrate; hence, position-controlled preparation of the nanomaterials is a critical prerequisite for integrated device technology. This chapter reviews recent research activities related to the selective growth of semiconductor (specifically, ZnO) nanostructures and their heterostructures, including quantum structures for visible LED applications. The main text of this chapter is organized into three sections. Section 2 briefly introduces selective growth methods of ZnO nanostructures and describes the underlying growth mechanisms. In Section 3, the position-controlled selective growth of ZnO nanostructures on various substrates and their resulting growth characteristics are reviewed. Section 4 focuses on the fabrication of LEDs based on ZnO nanostructures and their heterostructures. Finally, conclusions and perspectives on ZnO nanostructures for various device applications are presented.

# 2. SELECTIVE GROWTH METHODS AND MECHANISMS

Position-controlled selective growth of semiconductor nanostructures can be performed by numerous methods. In this section, we review the various methods used to synthesize the semiconductor nanostructures, along with the associated approaches for growing the nanostructures at specific position-controlled sites with controlled dimensions, a prerequisite for realizing bottom-up integration of semiconductor nanostructures in practical electronic and optical devices.

The growth method for semiconductor nanostructures can be classified into two distinct categories: catalyst assisted and catalyst free. In catalyst-assisted methods, the growth of nanostructures can be regulated by the presence or absence of the catalyst. Thus, position-controlled selective growth is usually achieved by placing the catalyst at the desired sites with controlled dimensions. In contrast, catalyst-free methods need to rely on different mechanisms to selectively regulate the nanostructure growth. Typically, this is achieved by selectively blocking the nucleation where the nanostructure growth needs to be prohibited using a growth mask, or by selectively creating artificial nucleation sites at desired locations where the growth can take place exclusively, with little to no growth on the undesired areas.

In the following, we will review various selective growth methods, with the catalyst-assisted vapor–liquid–solid (VLS) method, catalyst-free metalorganic vapor-phase epitaxy (MOVPE), and the catalyst-free hydrothermal method featured as representative methods for position-controlled selective growth of semiconductor nanostructures. For each method, we will briefly review the growth mechanism and then discuss the process used to achieve position-controlled selective growth.

## 2.1 VLS Method

The most commonly used method of growing semiconductor nanostructures in the catalyst-assisted approach is the VLS method. In this method, vapor-phase reactants are first dissolved into a liquid-phase catalyst metal, which is then deposited onto the substrate in the solid phase, thus giving it its name (Björk et al., 2002; Fan et al., 2006b; Gudiksen et al., 2002; Hu et al., 1999a,b; Huang et al., 2002; Kang et al., 2012; Ng et al., 2003; Solanki et al., 2002; Wagner and Ellis, 1964; Wu and Yang, 2000; Zhong et al., 2003).

The VLS process begins with the deposition of catalyst metal (typically Au) clusters (diameter: tens of nanometers) onto the substrate, as illustrated in Fig. 1A. The substrate is then placed in a reaction tube or chamber (furnace, chemical vapor deposition (CVD), molecular beam epitaxy, and pulsed laser deposition (PLD) setups are often used). The substrate is then heated until the catalyst metal melts and forms nanometer-sized droplets. The temperature, which depends on the specific metal and substrate used, typically ranges from 300 to 1100 °C (Fan et al., 2006b). In the next step, the growth material is delivered to the substrate in vapor form. The vapor dissolves into the liquid-phase metal to form a supersaturated liquid alloy. As a consequence, precipitation of the reactant occurs at the liquid–solid interface, which gradually moves the liquid droplet upward, resulting in the growth of a nanowire structure (Fig. 1B). The diameter of the nanowire is determined by the size of the liquid metal droplet, while the growth rate of the nanowire is determined by the reactant concentration in the gas precursor, as well as the temperature (Djurišić et al., 2007; Wu and Yang, 2000).

Based on this method, various semiconductor materials, such as Si, Ge, ZnO, InP, and GaN, can be grown as vertically aligned nanostructures on different semiconducting or insulating substrates, with the resulting nanostructure having an epitaxial relationship with the substrate (Fan et al., 2006b; Yoo et al., 2006; Zhong et al., 2003). It is also possible to form heterostructures in either the axial or radial direction of the nanowire structures, highly desirable for device applications, such as those involving LEDs (Mieszawska et al., 2007; Qian et al., 2004, 2005). However, for use in practical devices, it is necessary to grow these nanostructures in a controlled

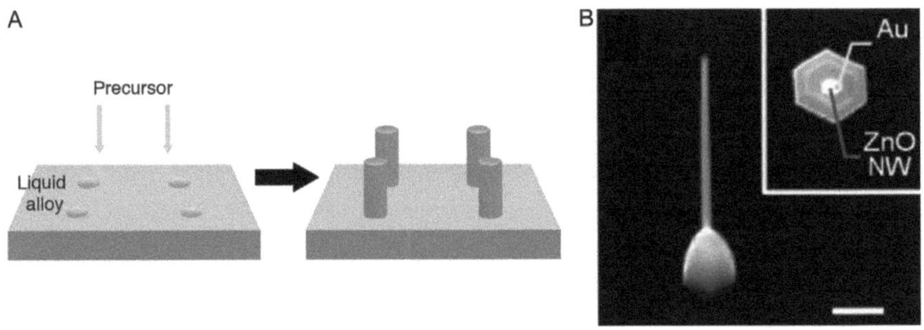


**Figure 1** (A) Schematic description of the VLS process and (B) SEM image of a ZnO nanowire vertically grown on a SiC substrate by the VLS method (scale bar: 200 nm). Inset is its top-view image. A hexagonal dome-like base was also formed at the bottom, which relaxes the lattice mismatch between ZnO and SiC. *Reprinted from Ng et al. (2004) with permission, © 2004 American Chemical Society.*

manner, at precisely known locations with low uncertainty in their dimensions.

A number of different techniques have been used to achieve position-controlled selective growth with the VLS method. In a more conventional approach, photolithography or e-beam lithography is used to pattern and deposit catalyst metal nanodots by a lift-off process (Greyson et al., 2004; Jensen et al., 2004; Mårtensson et al., 2003; Ng et al., 2004), as illustrated in Fig. 2A. This allows nanostructure growth at well-defined positions with uniform dimensions using the VLS method. While highly convenient, the downside of VLS growth with photolithography is the large resulting size of the Au pads, which can result in reduced vertical alignment of the nanowires (Greyson et al., 2004). VLS growth with e-beam lithography in general is limited by a low throughput, although acceptable at the research stage.

To create the nanometer-sized catalyst metal dots with a high throughput, various methods based on self-organization have been investigated, including nanosphere lithography (Fan et al., 2006a; Fuhrmann et al., 2005; Kempa et al., 2002; Kim et al., 2012b; Kosiorek et al., 2004; Pyun et al., 2010) and gold deposition with masks based on porous alumina (Fan et al., 2005; Masuda and Fukuda, 1995; Wu et al., 2002). Figure 2B shows an overview of the former approach, where a layer of polystyrene spheres held together by a metal coating was used as a mask for depositing

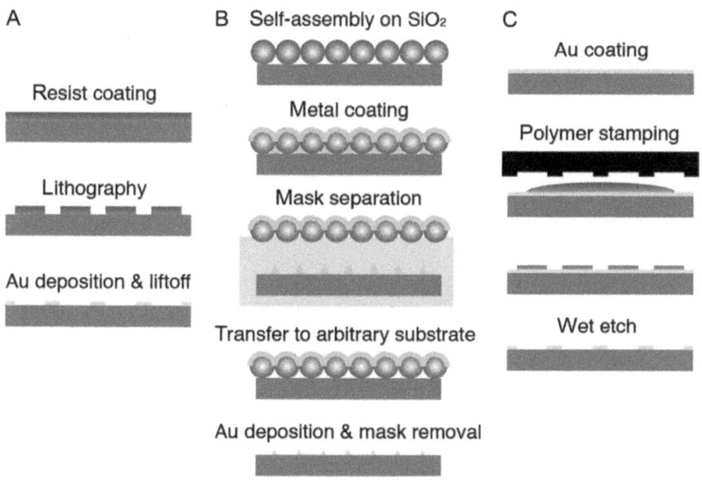


**Figure 2** Various methods of patterning catalyst metal spots for position-controlled selective growth of nanowires using the vapor–liquid–solid (VLS) method: (A) photo- or e-beam lithography, (B) nanosphere lithography (Fan et al., 2006a), and (C) nanoimprint lithography (Nikoobakht et al., 2004).

Au nanoislands, after which the VLS mechanism can be used to grow ZnO nanowires (Fan et al., 2006a). The latter approach similarly deposits nanoislands of Au, but instead using a porous alumina membrane with self-organized pores or a metal mask derived from it as the deposition mask (Fan et al., 2005; Wu et al., 2002). Another promising method for nanopatterning Au islands with a high throughput is nanoimprint lithography (Chou et al., 1996; Hochbaum et al., 2005; Mårtensson et al., 2004; Nikoobakht et al., 2004). Figure 2C shows an example of the nanoimprint method used for VLS growth, where a polydimethylsiloxane mold and a polymer mask is used to selectively pattern and etch away Au coating on a substrate (Nikoobakht et al., 2004). An in-depth summary of these methods, along with other methods for position-controlled selective growth using the VLS method, can be found in a review by Fan et al. (2006b).

Despite the advantages of the VLS method, a critical issue inherent to the resulting nanostructure is the potential unintentional incorporation of metal catalyst atoms during the growth process (Kang et al., 2012; Yoo et al., 2006). A droplet of the metal catalyst sits on top of the nanostructure during its growth at high temperatures (Fig. 1B); however, as the growth proceeds, the metal catalyst atoms diffuse into the nanostructure and act as impurity sites. This is a major disadvantage with regard to device applications; even trace amounts of impurities in the semiconductor nanostructure have a marked effect on the quality and performance of electrical and optical devices. For instance, in LED applications, the diffusion of metal atoms can promote nonradiative recombination in the active region of the device and degrade the internal quantum efficiency of the LED. Thus, accurate, systematic control over the impurity concentration is essential. Accordingly, it is highly desirable to grow semiconductor nanostructures using catalyst-free methods, which we will review in the following sections.

## 2.2 Catalyst-Free MOVPE

MOVPE is a CVD method used to produce crystalline materials. It is driven by the chemical reaction of metalorganic compounds in the vapor phase at the surface of the substrate, whose products can grow epitaxially on the substrate under suitable conditions. MOVPE is used to synthesize a variety of III–V, II–VI, and elemental semiconductor materials, such as ZnO, GaN, and InP, among others (An et al., 2004; Hong et al., 2007, 2009b, 2011; Jang et al., 2006; Kim et al., 2003, 2009b; Lee et al., 2004; Noborisaka et al., 2005; Park et al., 2002a,b, 2003a,c, 2007; Wang et al., 2006).

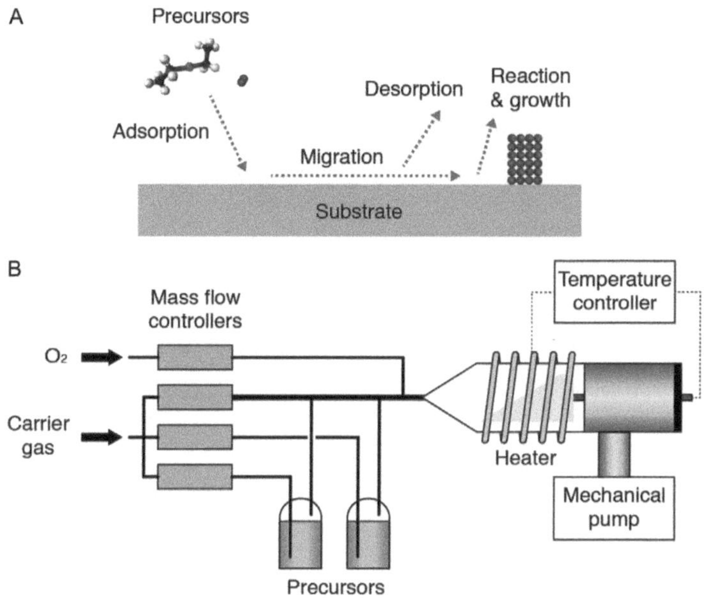


**Figure 3** Schematic illustrations of the (A) growth process and (B) experimental setup for catalyst-free metalorganic vapor-phase epitaxy (MOVPE) growth of ZnO nanostructures (Lee and Yi, 2012).

Figure 3 shows the growth process and experimental setup for MOVPE, using catalyst-free ZnO nanostructure growth as an example (Lee and Yi, 2012). The reactants (metalorganic molecules, dopants, $O_2$, etc.) are transported into the reactor by a carrier gas through a gas-handling manifold. The reactants then diffuse and adsorb to the heated surface of the substrate, where the metalorganic molecules thermally decompose and react to form ZnO nanostructures on the substrate. By-products of this reaction are desorbed and carried away from the surface. The reactants for ZnO nanostructure growth are usually diethyl-zinc ($Zn(C_2H_5)_2$; DEZn) and $O_2$, where the chemical reaction $Zn(C_2H_5)_2 + 7O_2 \rightarrow ZnO + 5H_2O + 4CO_2$ is responsible for the growth of ZnO. The typical growth temperature is in the range of 400–800 °C. However, the morphology of the grown nanostructure depends on the substrate, temperature, reactant substance, partial pressures, and the carrier gas type; thus, a specific parameter window must be identified to grow nanostructures having the desired shape. Detailed information on the required growth conditions for various nanostructures (e.g., nanorods, nanoneedles, and nanowalls) can be found in the literature (Kim et al., 2009b; Park et al., 2002a,b).

Under suitable growth conditions, ZnO grows into nanorod shapes, even on unpatterned flat substrate surfaces. This growth mechanism can be understood by investigating the surface formation energies of ZnO along the different crystal directions. First-principles calculations (Kim et al., 2008), performed using the Vienna *ab initio* simulation package (VASP), reveal the anisotropic surface formation energies $\gamma$ of the wurtzite ZnO crystal, summarized in Table 1. From these values, one can make a number of observations regarding the growth mechanism of ZnO. During deposition of ZnO, the growth proceeds in a direction that makes the $\{10\bar{1}0\}$ surface larger due to its lower surface formation energy while keeping the other surfaces minimal, given that the growth temperate is high enough with sufficiently mobile adatoms to ensure growth near thermal equilibrium. Thus, during the initial growth stage, the nanorod morphology having a small diameter and $\{10\bar{1}0\}$ side surfaces is energetically favored over a thin-film morphology with a large (0001) surface. This also means that once a nanorod is formed, the vertical growth rate will be larger than the lateral growth rate. This can be seen by a simple geometric argument that considers a hexagonal cylinder of height $h$ and a $\langle 10\bar{1}0\rangle$ semidiameter $r$, where the increase in surface energy for a given $\Delta h$ and $\Delta r$ is expressed as

$$\frac{\Delta E}{4\sqrt{3}} = \left(h\gamma_{10\bar{1}0} + r\gamma_{0001}\right)\Delta r + r\gamma_{10\bar{1}0}\Delta h$$

Here, the coefficient for $\Delta r$ is always larger than that for $\Delta h$, with their difference becoming even larger as $h$ grows, thus explaining the vertical-dominated growth. Similarly, when growth is terminated, $\{10\bar{1}1\}$ surfaces tend to form on top of the nanorod, as opposed to the (0001) surface, due to the lower surface formation energy of the former. These observations are consistent with Fig. 4, which shows an array of tall ZnO nanorods grown on $Al_2O_3$(0001), with hexagonal cross-sections and pyramidal shapes at the top (Park et al., 2002a). Also, the ratio of the vertical and lateral growth rate of ZnO nanostructures is measured to be as large as 100 (Hong et al., 2009b), allowing one to grow these nanorods with a high aspect ratio.

While the above discussion explains the formation of ZnO nanorods under ideal growth conditions for the [0001] direction, the morphology

**Table 1** Surface Formation Energies of Wurtzite ZnO (Kim et al., 2008)

| **ZnO Crystal Planes** | $\{10\bar{1}0\}$ | $\{11\bar{2}0\}$ | $\{10\bar{1}1\}$ | **(0001)** |
|---|---|---|---|---|
| Surface formation energy, $\gamma$ (J/m$^2$) | 0.91 | 1.64 | 1.58 | 1.74 |

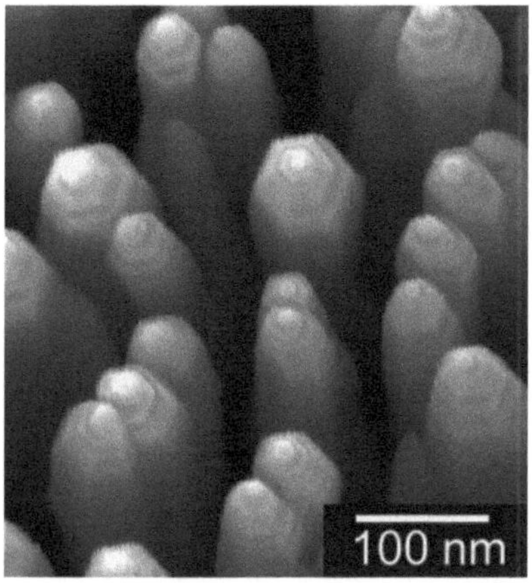


**Figure 4** SEM image of a vertically aligned ZnO nanorod array on an $Al_2O_3$(0001) substrate. *Reprinted from Park et al. (2002a) with permission, © 2002 AIP Publishing LLC.*

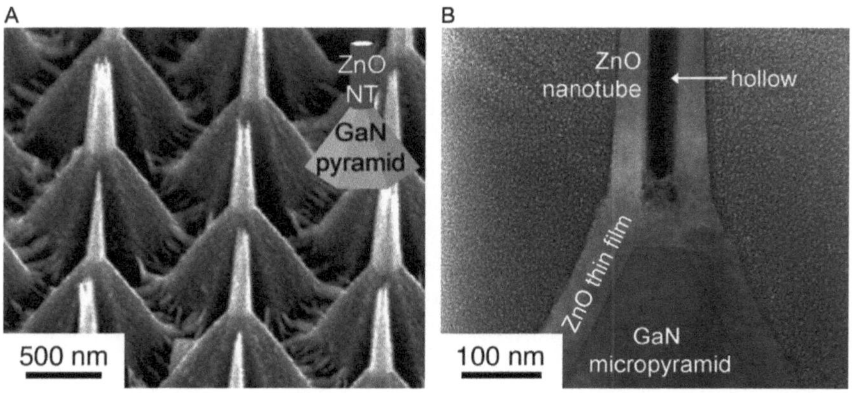


**Figure 5** Growth of ZnO on different crystal planes of GaN. (A) SEM image of ZnO nanotubes grown on top of GaN micropyramids. Each GaN micropyramid has six $\{10\bar{1}1\}$ side surfaces and a (0001) top surface. (B) Cross-sectional TEM image at the top of a micropyramid. *Reprinted from Hong et al. (2009c) with permission, © 2009 The Royal Society of Chemistry.*

of the actual nanostructures depends strongly on the crystal plane direction of the substrate, as well as the temperature. Figure 5 shows an example of the important role that the substrate crystal plane direction plays on the growth morphology; here, ZnO was grown on top of GaN micropyramids

(Fig. 5A), consisting of $\{10\bar{1}1\}$ side surfaces and a (0001) top surface (Hong et al., 2009c). Under exactly the same growth conditions, ZnO nanotubes with $\{10\bar{1}0\}$ side surfaces formed on GaN(0001) top surfaces, while thin films of ZnO$\{10\bar{1}1\}$ formed on GaN$\{10\bar{1}1\}$ side surfaces. This can be understood by considering the interface formation energy of ZnO on GaN, shown in Table 2 (Hong et al., 2009c), in conjunction with Table 1. The interface formation energy for $\{10\bar{1}1\}$ is much smaller than (0001); thus, growth is initiated by the formation of a ZnO film on the $\{10\bar{1}1\}$ surfaces. Once the film forms, adatoms attempt to minimize the added surface energy by moving to the top of the pyramids with subsequent ZnO$\{10\bar{1}0\}$ wall formation, resulting in a tube morphology. Hence, the growth rate in the $\langle 10\bar{1}1\rangle$ direction is expected to be considerably smaller than the vertical growth rate of the tubes, as shown in the cross-sectional transmission electron microscopy (TEM) image (Fig. 5B).

The effect of temperature on the growth of ZnO nanostructures is shown in Fig. 6A and B, where scanning electron microscopy (SEM) images of ZnO nanostructures grown at different temperatures are shown on ZnO/$Al_2O_3$(0001) and mechanically exfoliated few-layer graphene, respectively. As the temperature increased beyond that required for optimal nanorod growth on the ZnO/$Al_2O_3$(0001) substrate (Fig. 6A, 500 °C), neighboring nanorods merged into nanowalls (Fig. 6A, 600 °C). The nanowall morphology further reduces the surface energy by maximizing the ZnO$\{10\bar{1}0\}$ surface, if the nucleation density is sufficiently high to support lateral merging. As the temperature was further increased (Fig. 6A, 700 °C), the growth of the nanostructures was suppressed, and ultrafine nanorods formed. In contrast, for the graphene substrate, the growth was gradually suppressed as the temperature increased, without a significant change in morphology (Fig. 6B). This is due presumably to a much lower nucleation density on the graphene substrate, owing to its chemical inertness and the large lattice mismatch between the nanostructure growth material and graphene.

As a general tendency, at low temperatures, growth occurs on the surface with less degree of alignment due to the lack of mobility of the adatoms. As

**Table 2** Interface Formation Energies of ZnO Crystals on GaN Surfaces (Hong et al., 2009c)

| **Crystal Planes** | $\{10\bar{1}0\}$ | $\{11\bar{2}0\}$ | $\{10\bar{1}1\}$ | **(0001)** |
|---|---|---|---|---|
| Interface formation energy ($J/m^2$) | 1.13 | 1.24 | 1.37 | 2.69 |

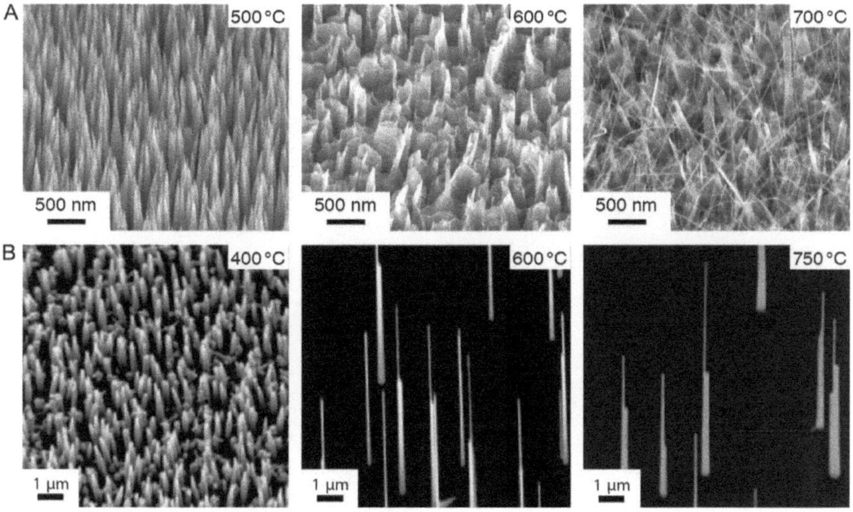


**Figure 6** SEM images of ZnO nanostructures grown at various growth temperatures on $Al_2O_3$(0001) substrates (A) with a ZnO buffer layer (unpublished data) and on mechanically exfoliated few-layer graphene (B) without a ZnO buffer layer. *Reprinted from Kim et al. (2009b) with permission, © 2009 AIP Publishing LLC.*

the temperature becomes higher, growth with a well-defined morphology is established until the temperature becomes too high to support the growth of these structures because the desorption rate of reactants from the surface becomes too high. Numerous cases of similar tendencies in the changing morphology of nano- and microstructures with growth temperature are reported in the literature (Kim et al., 2010; Kitamura et al., 2008; Park et al., 2005; Zhang et al., 2004).

Based on these growth behaviors, position-controlled selective growth of semiconductor nanostructures can be achieved with catalyst-free MOVPE. As discussed earlier, catalyst-free MOVPE itself has no selectivity in growth. Hence, selective growth can be achieved either by selectively prohibiting or by promoting the nucleation of nanostructure growth on certain areas of the substrate.

Figure 7 illustrates an overview of a work demonstrating the former approach, where a growth mask that prohibits nanostructure growth was used to grow shape- and position-controlled ZnO nanowalls using catalyst-free MOVPE on substrates with a GaN intermediate layer (Hong et al., 2009b). To control the shape and position of the ZnO nanowalls, a layer of amorphous $SiO_2$ was deposited on top of the substrate, which was patterned and partially etched away using conventional lithography

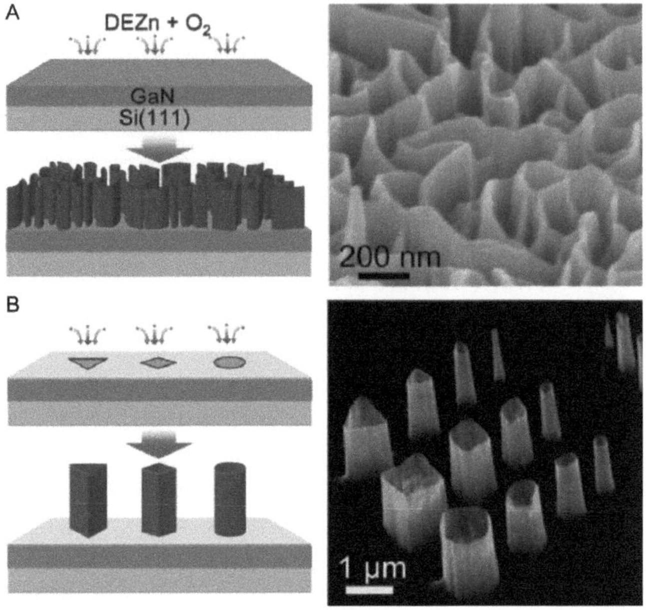


**Figure 7** Shape-controlled nanoarchitectures using a $SiO_2$ growth mask. (A) Schematics (left) and SEM image (right) of the growth of ZnO nanowalls with a random network morphology on bare GaN/Si(111) substrates. (B) Schematics (left) and SEM image (right) of shape-controlled ZnO nanotubes with triangular, square, and circular morphologies. *Reprinted from Hong et al. (2009b) with permission, © 2009 John Wiley and Sons.*

and chemical etching. ZnO nanowalls were then grown by MOVPE at 600 °C, using DEZn and $O_2$ as the reactant gases. A comparison of Fig. 7A and B reveals the role of the growth mask. Figure 7A shows that without the mask, vertical ZnO nanowalls grew everywhere on the substrate, creating a random network morphology, while in Fig. 7B, the nanowalls grew selectively along the edges of the patterns with triangular-, rectangular-, and circular-tube morphologies. Therefore, by changing the geometric parameters of the growth mask and the growth time, full control over the nanotube dimensions and position can be achieved.

The mechanism underlying the growth prohibition on the amorphous $SiO_2$ growth mask and the preferred nucleation and growth of ZnO at the pattern edges can be argued as follows. Figure 7B shows that almost no growth occurred on the $SiO_2$-covered parts of the substrate. This is due to the difference in the surface polarity and crystallinity of the GaN (0001) and amorphous $SiO_2$ surfaces, which results in different growth condition windows for ZnO on the respective surfaces. As an example, for the sample shown in Fig. 7B, the temperature was too high for ZnO to initiate nucleation on the $SiO_2$ surface, while a form-to-no-form transition occurs at a higher temperature on the GaN(0001) surface. Nonetheless, the reactant molecules can still adsorb and desorb on the $SiO_2$ surface, migrating on the surface in-between. This migration is responsible for the preferred growth at the perimeter of the hole patterns. Once adsorbed, the average migration distance of the molecules can be in the order of microns on the growth mask (Wang et al., 2014); in contrast, the migration distance is presumably smaller on the GaN(0001) surface, due to its larger reactivity and polarity. Hence, as illustrated in Fig. 8A, the perimeter of the patterns will be subjected to a larger influx of migrating molecules from the neighboring areas, compared with the internal regions of the patterns, explaining the preferred growth at the perimeter. According to this picture, growth should happen primarily at the perimeter of the hole when the size of the hole is comparable to or smaller than the adsorbent migration distance on GaN(0001). Above a certain critical size, growth should be initiated inside the pattern. This behavior is indeed observed in Fig. 8B, where ZnO nanowalls inside the hole patterns begin to form as the hole size approaches 1 μm.

Another approach in achieving position-controlled selective growth of semiconductor nanostructures by catalyst-free MOVPE is to selectively create nucleation sites on substrates that otherwise exhibit low nucleation density during growth. Graphene is an example of such a substrate. The

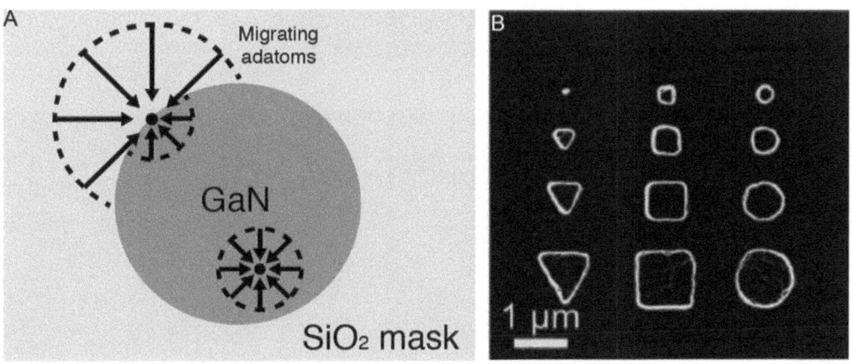


**Figure 8** (A) Top-view schematic diagram explaining the edge-dominant growth behavior in the masked growth approach and (B) top-view SEM image of the ZnO nanowall structures in Fig. 7B. *Reprinted from Hong et al. (2009b) with permission, © 2009 John Wiley and Sons.*

nucleation density of the semiconductor nanostructures on graphene is much lower than that on semiconductor buffer layers such as ZnO or GaN, due presumably to the chemical inertness of graphene and its large lattice mismatch between the nanostructure growth material and graphene (Kim et al., 2009b). However, studies have shown that nanorods and nanowalls grow preferentially along the step edges or boundaries that exist naturally on graphene layers.

Using this observation of enhanced nucleation and growth at the step edges, high-density ZnO nanowalls can be grown on a graphene surface by artificially creating the step edges using an oxygen plasma treatment (Chung et al., 2010; Kim et al., 2009b, 2012c; Park et al., 2011). Figure 9A shows vertically aligned ZnO nanowall growth on graphene, having a density similar to that shown in Fig. 7A (on bare GaN/Si(111) substrates). Position-controlled selective growth can be readily achieved by patterning a polymer mask layer on graphene and subsequently exposing it to oxygen plasma. Figure 9B and C shows the results of this selective

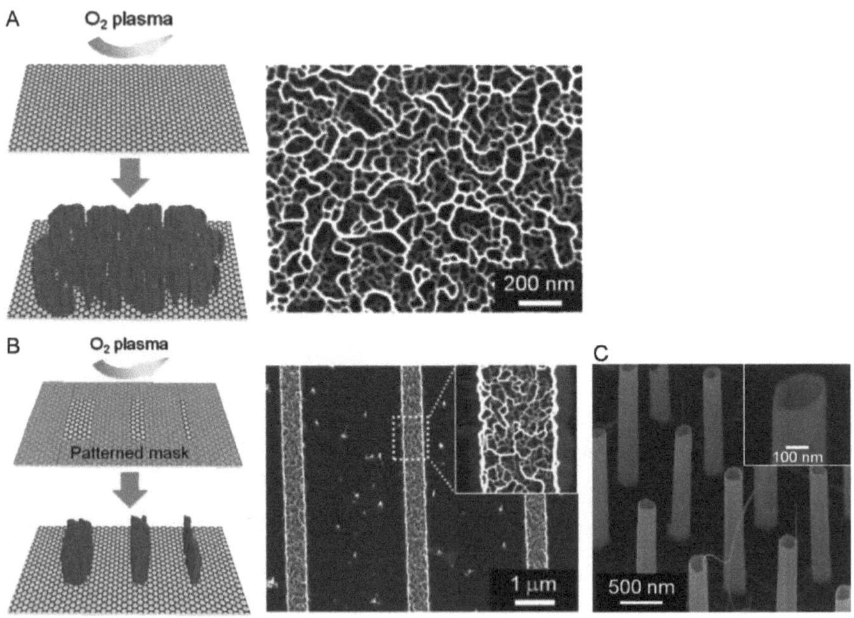


**Figure 9** Schematic illustrations and corresponding SEM images of (A) ZnO nanowalls grown on oxygen-plasma-treated graphene and (B) selectively grown ZnO nanowalls on graphene. (C) SEM image of selectively grown ZnO nanotubes on graphene. *Reprinted from Kim et al. (2012c) with permission, © 2012 John Wiley and Sons.*

growth. A pattern with a wide exposed area led to high-density ZnO nanowall formation (Fig. 9B), but a pattern with small holes led to individually separated ZnO nanotubes (Fig. 9C). The mechanism for this change in morphology with the pattern size is the same as that described in Fig. 8. This suggests that graphene layers with artificially formed step edges show great promise as a substrate for position-controlled selective growth of semiconductor nanostructures. Similarly, selective growth of nanostructures is possible on hexagonal boron nitride (hBN), by selectively treating the surface with oxygen plasma (Oh et al., 2014). This will be reviewed in Section 3.4, along with a more detailed characterization of the grown nanostructures.

Morphologies other than nanotubes and nanowalls can also be obtained with the selective growth methods introduced above. For instance, Fig. 10A shows that for very small hole diameters in the growth mask, a nanoneedle

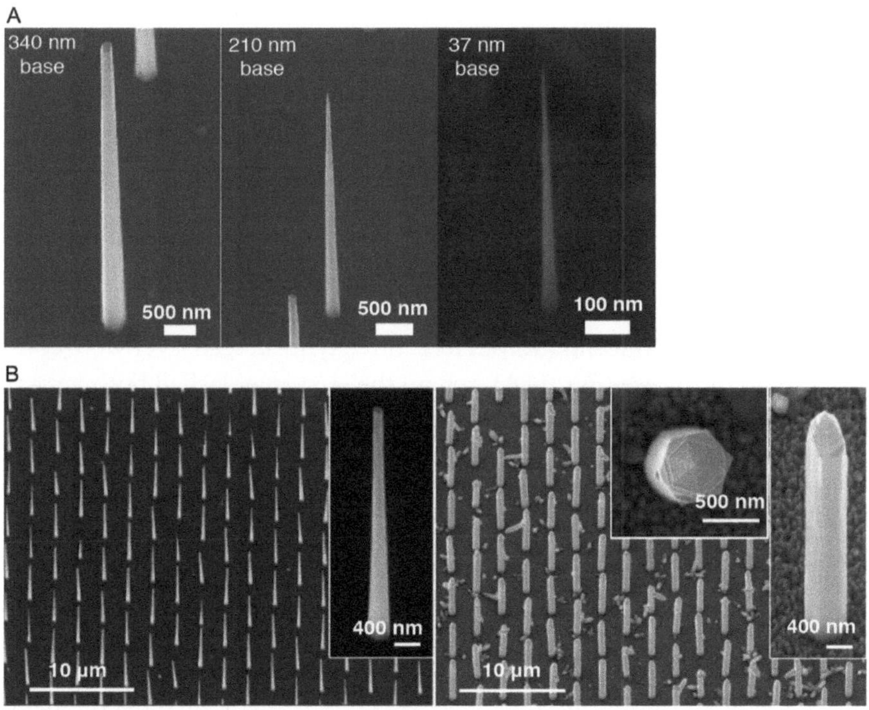


**Figure 10** (A) Transition from a nanotube to nanoneedle morphology as the hole pattern diameter decreases: (left, middle) ZnO growth with a $SiO_2$ mask on a graphene/$SiO_2$/Si substrate; (right) ZnO growth with a $SiO_2$ mask on a graphene/W/Si substrate (Park and Yi, unpublished data). (B) Lateral overgrowth on ZnO nanotubes (left) to form ZnO nanorods (right) on a graphene/Si(100) substrate (Baek and Yi, manuscript in preparation).

forms instead of a nanotube this is also evident in Fig. 8B. This is due to the tapering of the nanotube as it grows, which results in merging of the tube surfaces to form a hollow needle shape for the very small hole diameters. Figure 10B shows a different approach that includes an additional overgrowth step to create a nanorod shape with a nearly uniform diameter from its base to the top (Baek and Yi, manuscript in preparation); the outer shape of the nanotubes changed from a smooth polygon to a hexagonal structure, indicating single-crystalline heteroepitaxial growth of ZnO on the nanotube surface. The photoluminescent characteristics of this ZnO nanorod array were investigated; these characteristics will be reviewed in the Section 3.4.

## 2.3 Hydrothermal Growth

Hydrothermal growth involves the growth of nanostructures by thermal decomposition of metal complex molecules in an aqueous solution at elevated temperatures. In the case of ZnO growth, Vayssieres (2003) first demonstrated ZnO nanostructure growth in an aqueous nutrient solution of zinc nitrate hexahydrate [$Zn(NO_3)_2{\cdot}6H_2O$] and hexamethylenetetramine [$(CH_2)_6N_4$] at moderate temperatures (80–100 °C). Often, the hydrothermal growth is initiated with seed layer deposition to improve the nucleation density and alignment of the nanostructures to be grown.

The morphology of nanostructures grown by the hydrothermal method is strongly dependent on the crystallinity of the seed layer, as shown in Fig. 11 (Lee et al., 2011d), among other growth parameters, such as temperature and nutrient concentrations. Here, three types of seed layer were compared. In the first case (Fig. 11A), where the seed layer lacked both vertical and in-plane alignment, the resulting ZnO nanostructure exhibited a flower-like shape with nanorods pointing in directions likely related to the direction of ZnO[0001] of individual grains at the base. In the second case (Fig. 11B), the seed layer was vertically aligned but had random in-plane orientations. The resulting structure was a merged column of vertical ZnO nanorods with different in-plane orientations. In the third case (Fig. 11C), both vertical and in-plane alignment were preserved in the seed layer, which resulted in a single nanorod shape with hexagonal sidewalls.

Although the position-controlled selective growth of semiconductor nanostructures has been achieved by catalyst-assisted VLS techniques and catalyst-free MOVPE, these methods require high growth temperatures. In contrast, the growth temperature can be as low as 80–100 °C for wet chemical processes. As a result, growth can be performed on many types

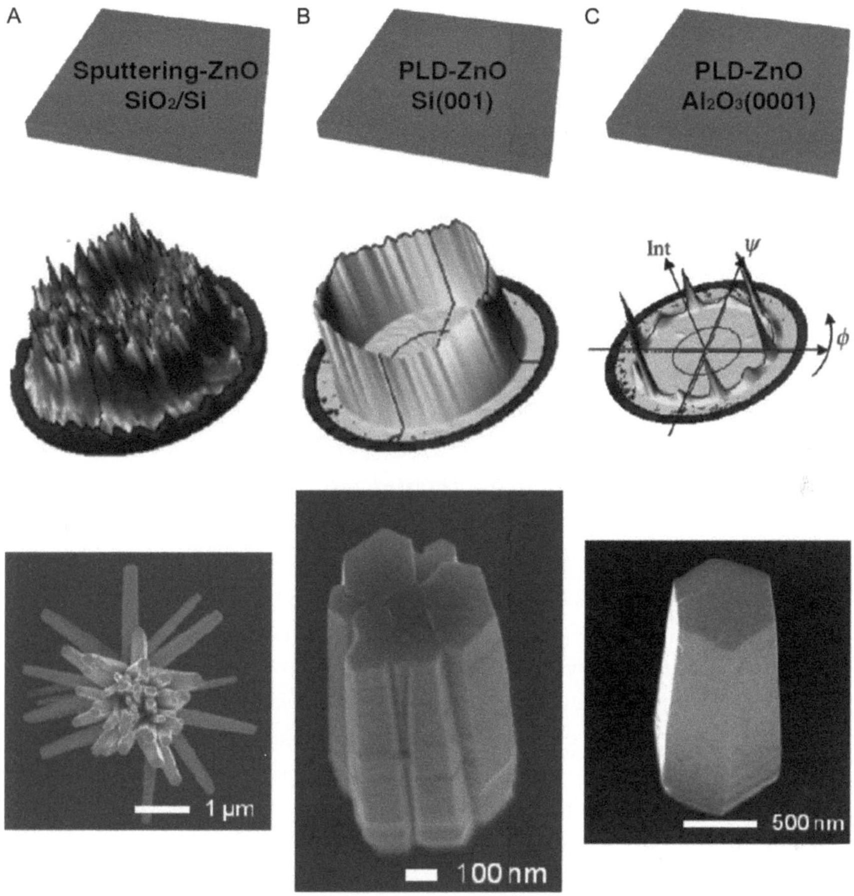


**Figure 11** Various morphology of ZnO nanostructures grown by the hydrothermal method depending on the seed layer crystallinity. For each panel: (top) seed layer type and substrate; (middle) X-ray pole figure analysis through the ZnO$(10\bar{1}1)$ plane; (bottom) SEM image of resulting ZnO nanostructures. (A) Sputtered ZnO seed layer on $SiO_2$/Si substrate, with random vertical and in-plane orientation. (B) ZnO seed layer on Si(001) substrate by PLD, with vertical alignment and random in-plane orientation. (C) PLD ZnO seed layer on $Al_2O_3$(0001) substrate, with both vertical and in-plane alignments. *Reprinted from Lee et al. (2011d) with permission, © 2011 American Chemical Society.*

of substrate, including Si (Kim et al., 2006; Xu et al., 2008), $Al_2O_3$ (Zhou et al., 2008), indium tin oxide (ITO) (Yang et al., 2014), various metals (Wen et al., 2012; Zheng et al., 2013), and even polymers (Weintraub et al., 2007), to name a few. Furthermore, due to the low growth temperature, polymer growth masks, such as those made of poly(methyl methacrylate) (PMMA), can be used to grow nanostructures selectively at controlled positions via a simple lithography process, unlike with other high-temperature growth methods. These polymer masks effectively prevent the nucleation of the growth material in the masked area (Kim et al., 2006; Weintraub et al., 2007; Xu et al., 2008; Yang et al., 2014; Zhou et al., 2008). Mask materials, such as $SiO_2$ and $Si_3N_4$, used in other methods, can also be applied to achieve selective growth with the hydrothermal method as well (Consonni et al., 2014).

Figure 12A shows a typical position-controlled selective growth process for ZnO nanorod formation using the hydrothermal method (Kim et al., 2006). Here, as a seed layer for growing ZnO nanostructures, Si substrates with a native oxide layer were coated with ZnO thin films using PLD or MOCVD (An et al., 2002; Park et al., 2001). After spin coating a layer of PMMA onto the ZnO-coated Si substrates, circular hole array patterns were created on the PMMA layer using e-beam lithography. The patterned substrates were then treated with oxygen plasma to remove the organic residue from the patterned holes. The plasma-treated patterned substrates were then immersed in an aqueous nutrient solution containing zinc nitrate hexahydrate and hexamethylenetetramine (0.1 M equimolar) for the growth of ZnO nanorods and microrods (Tak and Yong, 2005; Vayssieres, 2003). Presumably due to the lack of polarity and crystallinity of the PMMA surface, ZnO grew only on the patterned sites, where the ZnO seed layer was exposed. In addition to the position control, the diameter and length of the ZnO micro- and nanostructures were easily controlled by varying the pattern sizes and growth conditions (e.g., growth time and temperature). The average growth rate of the ZnO nanorods and microrods at 95 °C

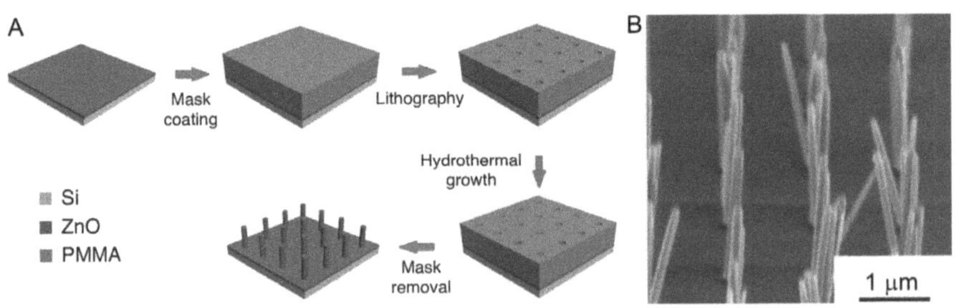


**Figure 12** (A) Schematic of the patterning and fabrication procedures for periodic ZnO nanorod arrays in the hydrothermal method. Adapted from Yang et al. (2014) with permission, © 2014 American Chemical Society. (B) SEM image of a ZnO nanorod array. *Reprinted from Kim et al. (2006) with permission, © 2006 AIP Publishing LLC.*

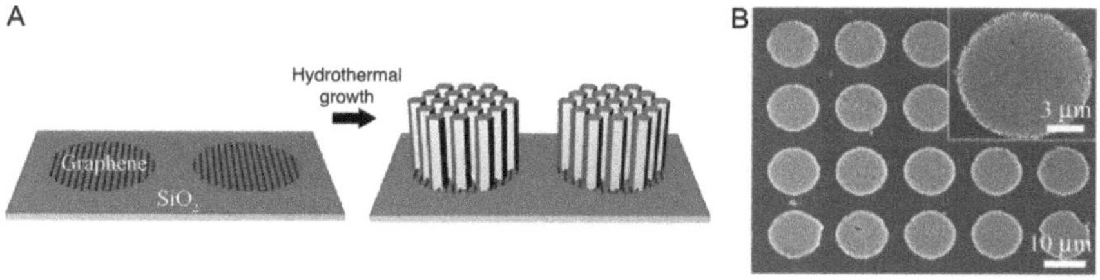


**Figure 13** Selective area growth of ZnO nanorods on patterned graphene. (A) Schematic of the growth procedure. (B) SEM image of ZnO nanorods selectively grown on the graphene pattern array. *Reprinted from Choi et al. (2011) with permission, © 2011 Springer.*

was in the range of 0.15–0.30 μm/h. After the growth process was complete, the ZnO nanostructures grown on substrates were immersed in acetone to dissolve the PMMA layers, and the residual material was removed. Figure 12B shows the resulting structures, consisting of individually separated ZnO nanorods, grown with good vertical alignment.

A different approach to achieving selective growth and position control is to selectively etch away the substrate material on which the nanostructures can grow. Figure 13 shows an example of this approach involving the selective growth of ZnO nanorods on graphene disks on a $SiO_2/Si$ substrate (Choi et al., 2011). After graphene was synthesized with a Ni film as the catalyst, it was transferred to a $SiO_2/Si$ substrate, and then photolithography and reactive ion etching were used to create patterned graphene disks (Fig. 13A). Unlike the method used by Kim et al. (2012c), in which graphene was exposed to an oxygen plasma to create nucleation sites in the form of artificial step edges, Choi et al. (2011) used a solution containing nanocrystalline ZnO seed particles that were spin-coated onto graphene/$SiO_2/Si$; ZnO nanorods were then grown using the hydrothermal method at 90 °C. The ZnO nanorods formed a dense array on the graphene disks and showed vertical alignment (Fig. 13B).

## 3. SELECTIVE GROWTH AND CHARACTERIZATION OF NANOSTRUCTURES ON VARIOUS SUBSTRATES

In Section 1, we discussed how the bottom-up approach of nanostructure growth can provide a simpler means of creating high-quality semiconductor nanomaterials, compared with the top-down approach. In particular, the top-down approach, based on thin-film deposition, lithography, and etching, can pose a significant challenge with regard to the fabrication of high-quality functional nanostructure devices. The heteroepitaxial films produced often include structural defects, due to the mismatch in lattice constants or thermal expansion coefficients of the materials (An et al., 2006; Etzkorn and Clarke, 2001); additionally, the films can easily become contaminated or damaged during the lithography and etching processes (Yagi et al., 2002). The bottom-up approach based on nanoepitaxy can circumvent these problems (Huang et al., 2001; Morales and Lieber, 1998; Park et al., 2002a; Xia et al., 2003) to produce high-quality nanomaterials (Huang et al., 2001; Park et al., 2003a) although precise control over the position and shape of the grown nanomaterial, critical for high-performance integrated nanodevice applications, is not trivial (Lu and Lieber, 2007; Thelander et al., 2006). Thus far, we have discussed the position-controlled selective growth of nanostructures using various bottom-up growth techniques. It is of interest to characterize in detail the nanostructures grown by these methods, in an attempt to better understand the key factors required to create nanostructures with high crystallinity and low defect density for high-quality electronic and optoelectronic devices.

The diverse choice of substrates applicable to position-controlled selective growth of nanostructures includes various semiconducting substrates (e.g., Si, GaN, ZnO, SiC, and ITO), metallic substrates (e.g., Au, Pt, Cu, and Ag), and insulating substrates (e.g., $Al_2O_3$), covering the vast majority of substrates used in the device industry. Meanwhile, layered two-dimensional (2D) materials, such as graphene and hBN, have recently been emerging as novel substrates for transferable and flexible optoelectronic and electronic devices. The individual layers of these materials are held together by van der Waals (vdW) interactions, rather than covalent bonds as in typical substrates, which results in a so-called vdW epitaxy (Hong et al., 2013; Koma, 1999; Mohseni et al., 2014; Utama et al., 2010); the vdW configuration has the ability to produce single-crystalline semiconductor nanostructures or films, with an abrupt clean heterointerface and suppressed threading dislocation density, even for highly lattice-mismatched heteroepitaxial systems (Utama et al., 2012). Furthermore, the excellent mechanical and chemical stability of these layered materials (Song et al., 2010) allows their transfer to any arbitrary substrate (Chung et al., 2010; Kobayashi et al., 2012), thus enabling the bottom-up growth of nanostructures on materials inaccessible to the top-down approach, such as flexible polymer substrates (Choi et al., 2010; Lee et al., 2011b; Peng et al., 2012).

In the following, we review the structural, optical, and electronic characteristics of semiconductor nanostructures obtained by position-controlled selective growth. We will first focus on nanostructures grown on more conventional semiconducting, metallic, and insulating substrates and then move to more recent developments in semiconductor nanostructure growth on graphene and hBN and characterizations of heterointerface between the layered 2D materials and the semiconductor nanostructures, to see how the bottom-up approach based on nanoepitaxy can produce high-quality semiconductor nanostructures.

## 3.1 Semiconducting Substrates

For semiconducting substrates, all of the growth methods reviewed in Section 2, namely VLS, MOVPE, and hydrothermal methods, can be used to grow nanostructures selectively at position-controlled sites. As an example of the VLS method, preparation and electronic device applications of ZnO nanowires selectively grown on a SiC(0001) substrate were demonstrated by Ng et al. (2004). In this work, ZnO nanowires were grown with ZnO and graphite powder as the reactant sources in an Ar carrier gas in a tube furnace at 925 °C. Figure 14A shows an SEM image of the resulting ZnO nanowires; the small lattice mismatch of 5.5% between ZnO(0001) and hexagonal SiC(0001) promoted vertical growth of ZnO nanowires. To obtain the individually separated ZnO nanowires shown in Fig. 14A, the diameter and thickness of the Au catalyst pads were controlled to 200 nm and 20 Å, respectively, with which ZnO nanowires with diameters of $40 \pm 5$ nm grew at an average vertical growth rate of 1 μm/h. Figure 14B shows a high-resolution TEM (HR-TEM) image, which revealed sharp lattice fringes, having an interplanar spacing of 5.45 Å corresponding to ZnO (0001). A highly ordered selected-area electron diffraction (SAED) pattern, shown in the inset of Fig. 14B, indicated high single crystallinity.

The performance of the ZnO nanowires as an electronic material was evaluated by fabricating a vertical surround-gate field effect transistor (FET), as illustrated in Fig. 14C. The drain-source current versus drain-source voltage ($I_{ds}$–$V_{ds}$) curve for various gate voltages ($V_{gs}$) is shown in Fig. 14D for an n-type FET; $I_{ds}$ increased with $V_{gs}$, characteristic of electron transport. The $I_{ds}$–$V_{ds}$ profile indicated a rapid increase in $I_{ds}$ without saturation; thus, the traditional channel pinch-off or the carrier velocity saturation was absent during device operation. Regarding the $V_{gs}$-dependent behavior, the FET was on at zero bias, which was attributed to unintentional

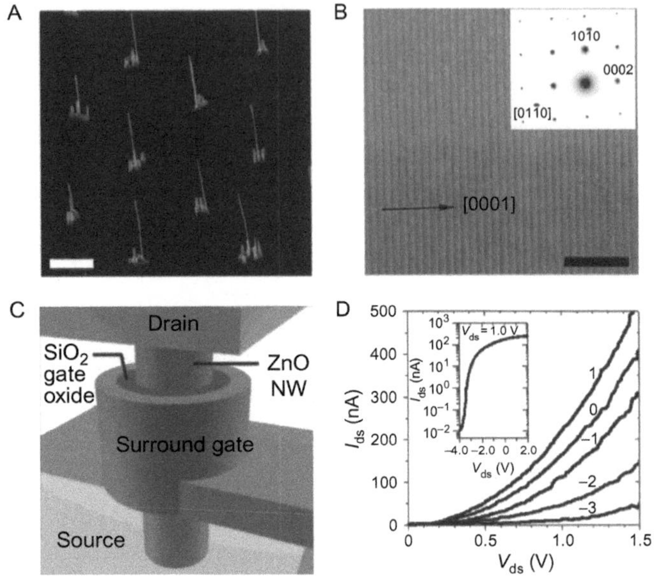


**Figure 14** (A) A 45° perspective SEM view of an array of ZnO nanowires grown from Au catalyst spots (scale bar: 1 μm). (B) A HR-TEM image showing sharp lattice fringes with an interplanar spacing of 5.45 Å, corresponding to ZnO(0001) planes (scale bar: 5 nm). The inset shows a SAED pattern taken perpendicular to the stem of the nanowire. (C) A schematic illustrating the vertical surround-gate field effect transistor (FET) based on the ZnO nanowire. (D) $I_{ds}$ versus $V_{ds}$ characteristics for different $V_{gs}$ (inset: $I_{ds}$ vs. $V_{gs}$ for $V_{ds} = 1.0$ V) of an n-type vertical surround-gate FET. *Reprinted from Ng et al. (2004) with permission, © 2004 American Chemical Society.*

n-type doping of the ZnO channel in air, and turned off at $V_{gs} = -3.5$ V, with a high on–off ratio of $>10^4$. However, the subthreshold gradient of the FET was 170 mV per decade, larger than that of Si-MOSFETs. The estimated mobility was quite low at 0.5 $cm^2$/V s, likely to be a lower bound of the channel mobility hindered by poor contact performance.

Structural and optical characterization of ZnO nanotube arrays grown by catalyst-free MOVPE were performed by Yoo et al. (2009). The crystal structure and orientation of the position-controlled nanostructures can be investigated using X-ray diffraction (XRD) measurements. Figure 15 shows a high-resolution synchrotron radiation XRD characterization of position-controlled ZnO nanotube arrays, grown on a GaN/AlN/Si(111) substrate by catalyst-free MOVPE. In Fig. 15C, the $\omega$–$2\theta$ scan of ZnO nanotube arrays shows diffraction peaks at 20.54°, 20.65°, and 21.40°, resulting from *c*-plane Bragg reflection from the ZnO nanotubes, GaN, and AlN layers,

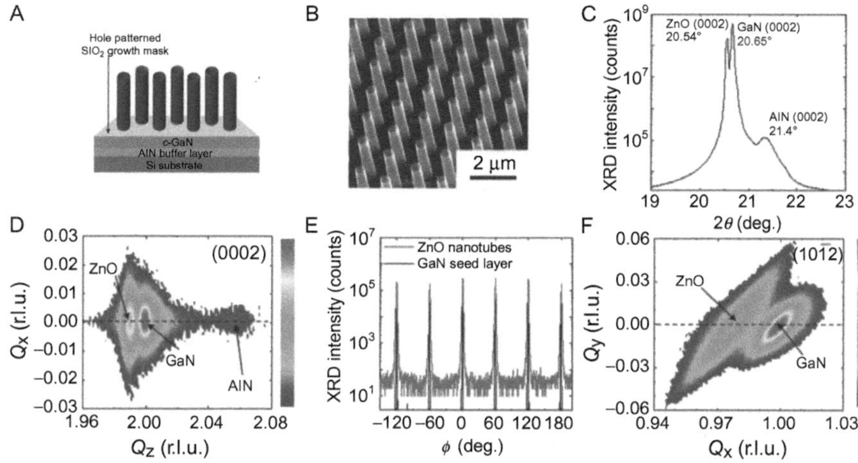


**Figure 15** Selective growth schematic, SEM image, and XRD characteristics of ZnO nanotube arrays grown on patterned GaN/AlN/Si substrates. (A) Schematic diagram of the sample structure and (B) tilted-view SEM image of the sample. (C) $\omega$–$2\theta$ scan measured in the range of 19–23° with a 1.10701 Å wavelength X-ray beam. (D) *HL*-mesh contour reciprocal space map (RSM) around the (0002) diffraction. The reciprocal lattice unit is abbreviated as r.l.u., while $Q_x$, $Q_y$, and $Q_z$ correspond to *H*, *K*, and *L*, respectively. (E) Azimuthal ($\phi$)-scan of ZnO$\{10\bar{1}2\}$ diffractions. (F) *HK*-mesh contour RSM around the $(10\bar{1}2)$ diffraction. *Reprinted from Yoo et al. (2009) with permission, © 2009 John Wiley and Sons.*

respectively. In Fig. 15D, the reciprocal space mapping (RSM) also exhibited XRD peaks at $Q_z$ values of 1.99, 2.00, and 2.06 reciprocal lattice units (r.l.u.), corresponding to ZnO(0002), GaN(0002), and AlN(0002) diffraction peaks, respectively. Furthermore, these diffraction peaks were located at the same $H$ value ($H=0$), indicating that the ZnO nanotubes and the intermediate GaN and AlN layers exhibited identical *c*-axis crystallographic orientations.

The in-plane epitaxial relationship between the ZnO nanotubes and the GaN and AlN intermediate epilayers on the Si(111) substrate was also investigated. Figure 15E shows the $\phi$-scans of ZnO (red (gray in the print version) line) and GaN (black line) $\{10\bar{1}2\}$ diffractions, which exhibited sixfold rotational symmetry with identical $(10\bar{1}2)$ diffraction peak positions for the ZnO nanotubes and the GaN intermediate layer. From the *HK*-mesh contour maps of RSM around the $(10\bar{1}2)$ diffraction shown in Fig. 15F, two well-resolved peaks corresponding to ZnO$(10\bar{1}2)$ and GaN$(10\bar{1}2)$ diffraction are clearly visible; these diffraction peaks lie along the same $K$

value ($K=0$) value. These results clearly indicate that the *c*-axis oriented, single-crystal ZnO nanotubes were heteroepitaxially grown on GaN, with homogeneous in-plane alignment between the nanotubes.

Photoluminescence (PL) and cathodoluminescence (CL) spectroscopy can be used to investigate the purity and crystalline quality of nanostructures. Figure 16 shows a study of the PL and CL characteristics of position-controlled ZnO nanotube arrays (Yoo et al., 2009). In Fig. 16A, the room-temperature PL spectrum of ZnO nanotubes shows a dominant near-band-edge (NBE) emission peak at 3.29 eV. It is noteworthy that no mid-gap emission, typically a feature of the PL spectra of bulk materials, was observed. The PL spectrum measured at 10 K in Fig. 16B shows NBE emission peaks at 3.358 and 3.451 eV, due to the ZnO nanotubes and the GaN intermediate layer, respectively. The full width at half-maximum

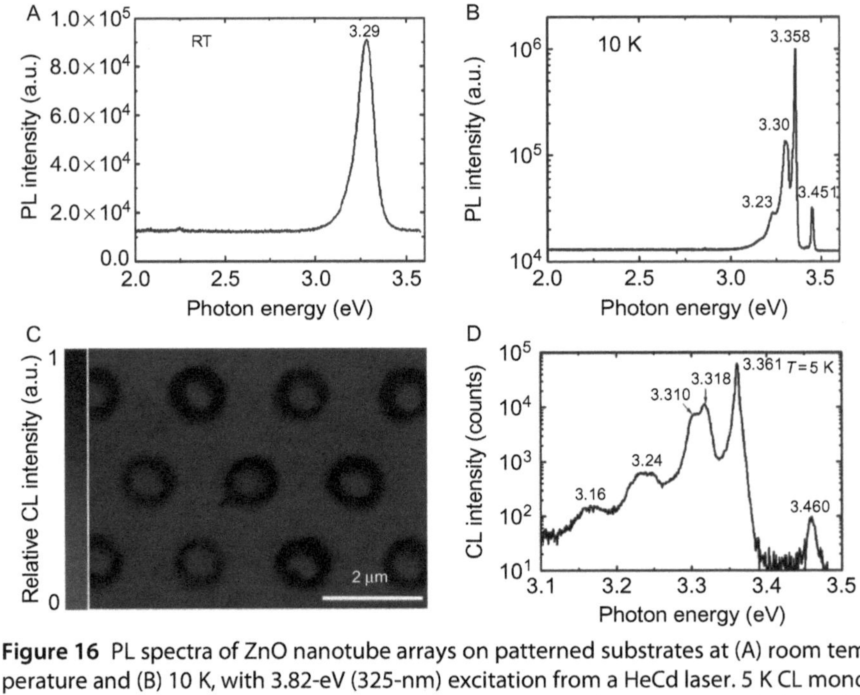


**Figure 16** PL spectra of ZnO nanotube arrays on patterned substrates at (A) room temperature and (B) 10 K, with 3.82-eV (325-nm) excitation from a HeCd laser. 5 K CL monochromatic (photon energy 3.36±0.02 eV) image (C) and CL spectrum (D) of ZnO nanotube arrays on patterned GaN/AlN/Si substrates. CL response, excited with a 100 pA electron beam at 30 kV acceleration and collected by a parabolic mirror, was detected using a grating monochromator equipped with a liquid-$N_2$-cooled CCD camera. *Reprinted from Yoo et al. (2009) with permission, © 2009 John Wiley and Sons.*

(FWHM) of the PL peaks was as small as 7 meV. This strong NBE emission with a small FWHM from the ZnO nanotubes suggests that the nanotube arrays were of high optical quality, resulting presumably from their high crystallinity and purity. Regarding the origins of the PL peaks, the dominant PL peak at 3.358 eV in Fig. 16B was attributed to neutral donor-bound excitons in the ZnO nanotubes (Meyer et al., 2004). Other PL peaks at 3.30 and 3.23 eV, with an energy separation of about 70 meV, were assigned to donor–acceptor pair recombination and the first-order longitudinal optical phonon replica, respectively (Meyer et al., 2004; Sun et al., 2002).

CL spectroscopy of the nanotube array showed results similar to the PL results. Figure 16C shows a CL monochromatic image of the ZnO nanotube array, with strong CL emission from the position-controlled ZnO nanotubes. In Fig. 16D, the CL spectrum of the nanotube array, which were measured at 5 K, shows strong NBE emission peaks at 3.460, 3.361, 3.318, 3.310, 3.240, and 3.160 eV. Most of the CL peaks were attributed to emission from the ZnO nanotubes, with the exception of the peak at 3.460 eV from the GaN intermediate layer (Meyer et al., 2004; Viswanath et al., 1998).

Similar characterizations have also been performed on nanostructures selectively grown by the hydrothermal method (Consonni et al., 2014; Yang et al., 2014). Using the hydrothermal method, Consonni et al. (2014) recently demonstrated that the polarity of selectively grown ZnO nanostructures (that is, growth in either the [0001] or $\left[000\overline{1}\right]$ direction of the wurtzite ZnO crystal) can be controlled to be both Zn- and O-polar, as opposed to other methods that produced mostly Zn-polar ZnO crystals only. In this work, bulk *c*-plane ZnO single crystals, with either O- or Zn-polarity, were used as the substrate. The growth mask was created from a $SiO_x/Si_xN_y$ thin layer, patterned by e-beam lithography. Hydrothermal growth was performed in an equimolar solution of zinc nitrate and hexamethylenetetramine. Figure 17B, C, H, and I shows SEM and TEM images of the nanowires grown on O- and Zn-polar ZnO substrates, where, in either cases, regular arrays of nanowires with uniform dimensions have been obtained. In order to confirm the polarity of the nanowires grown on the respective substrates, a convergent beam electron diffraction (CBED) technique was utilized. Comparing the experimental and simulated $\left(000\overline{2}\right)$ and (0002) diffracted discs along the $\left\langle 10\overline{1}0\right\rangle$ zone axis with different crystal thicknesses confirmed that the as-grown ZnO crystals were O- and Zn-polar, respectively, on the O- and Zn-polar substrates (Fig. 17D, E, J, and K).

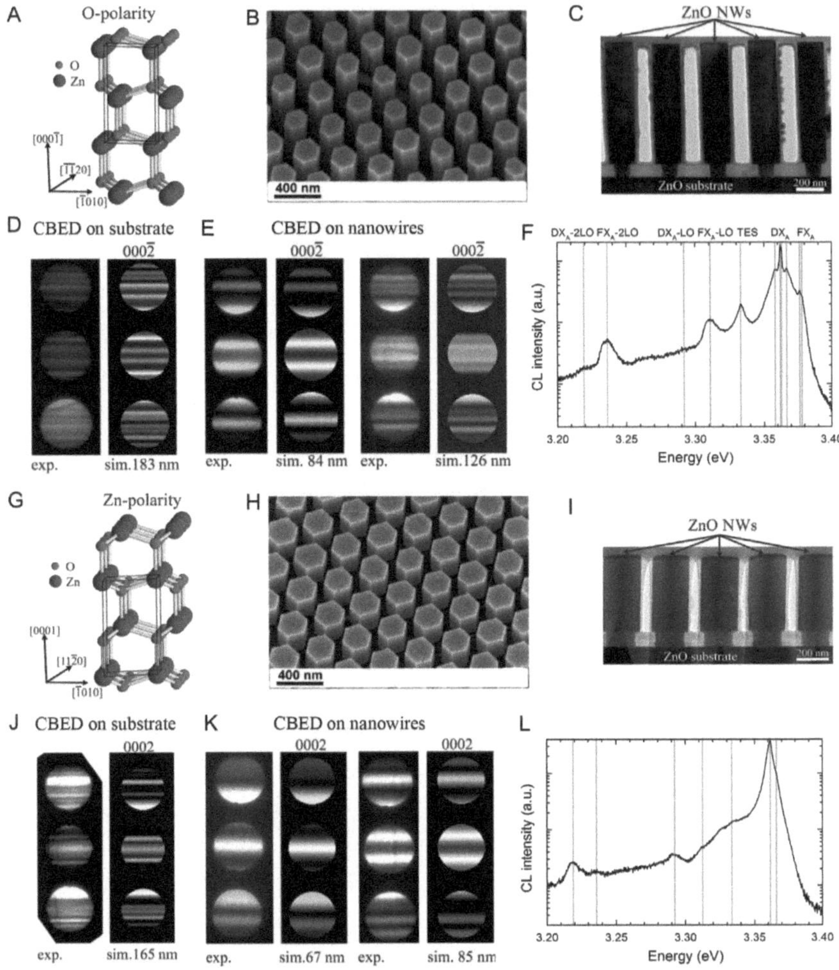


**Figure 17** SEM, TEM, and CL characterization of O- and Zn-polar ZnO nanowires grown by the hydrothermal method. (A) Schematic of the O-polar ZnO structure. (B) Tilted-view SEM image of ZnO nanowires grown on O-polar *c*-plane ZnO single-crystal substrate. (C) Low-magnification cross-sectional TEM image along the $\langle 10\bar{1}0 \rangle$ zone axis of O-polar ZnO nanowires. (D and E) Experimental and simulated CBED along the $\langle 10\bar{1}0 \rangle$ zone axis on (D) the substrate and (E) the nanowires, revealing the O-polarity of the substrate and the nanowires. (F) The 5 K CL spectrum of a single O-polar ZnO nanowire. (G–L) Same analysis as (A–F) but with Zn-polar substrate and nanowire growth. *Reprinted from Consonni et al. (2014) with permission, © 2014 American Chemical Society.*

Further characterization of the nanowires was performed using low-temperature CL spectroscopy, as shown in Fig. 17F and L. For both O- and Zn-polar nanowires, the intensities of NBE emission peaks were two to three orders of magnitude larger than that of the broad defect emission band centered at 2.0 eV, showing good optical quality of the nanowires. Interestingly, the NBE emission peaks of O-polar nanowires were much more structured than Zn-polar nanowires, indicating that impurities of a more various nature may be present in O-polar ZnO than the other.

## 3.2 Metallic Substrates

Direct growth of nanostructures on metals is desired in applications where the metals can also function as individually addressable electrodes. Due to the relatively high growth temperatures of the VLS and MOVPE methods, the position-controlled selective growth of nanostructures on top of metallic surfaces or substrates (while the metal stays in solid phase as opposed to the VLS method) is mainly performed by the hydrothermal method. In addition to employing a polymer mask, as introduced in Section 2.3, a number of other methods can be used to achieve selective growth on metallic substrates.

Wen et al. (2012) demonstrated selective growth of ZnO nanowires on patterned Au, Cu, Ag, and Sn electrodes on a $SiO_2$/Si substrate using a hydrothermal method based on zinc nitrate hexahydrate and ammonium hydroxide (Fig. 18A–D). They observed that the nucleation of nanowires was suppressed on the $SiO_2$ substrate, while densely packed, vertically aligned nanowires (especially in the case of Au) were obtained on patterned metal electrodes. Although the mechanism underlying this selectivity was not discussed in this work, it is likely similar to the prohibition of nucleation with $SiO_2$/$Si_3N_4$ or polymer masks. The TEM and SAED patterns shown in

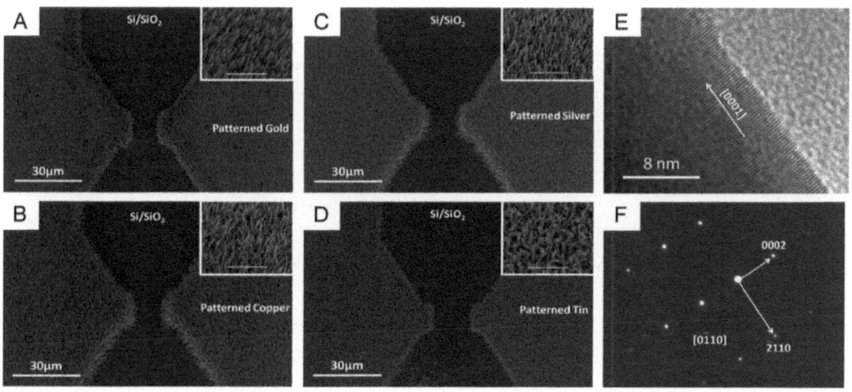


**Figure 18** SEM images of ZnO nanowires selectively grown on patterned (A) gold, (B) copper, (C) silver, and (D) tin electrodes. Scale bars in the insets are 6 μm. (E) HR-TEM image and (F) SAED pattern of ZnO nanowires on Au electrodes. *Reprinted from Wen et al. (2012) with permission, © 2012 Royal Society of Chemistry.*

Fig. 18E and F, respectively, reveal that the ZnO nanowires on Au grew in the [0001] direction with high crystallinity. The combination of this method with an additional polymer mask is promising for easy integration of ZnO nanowires into electronic or optoelectronic device configurations.

More recently, a novel approach specific to metal substrates (and highly conducting semiconductor substrates) was demonstrated by Zheng et al. (2013), where the position-controlled selective growth of ZnO nanowires was obtained by a galvanic cell-based approach, as shown in Fig. 19A. In this technique, a sacrificing anode made from Al was placed in contact with a relatively inert substrate material, such as Pt, Au, Cu, indium tin oxide (ITO), or fluorine-doped tin oxide (FTO). The work function difference between two metals created a bias that drove the reactions in the nutrient solution, as shown in Fig. 19A. As a result, ZnO nanorods formed on the cathode, and hydroxides of Al and Zn were deposited on the anode. This process is similar to electrochemical deposition, except that a galvanic cell is used instead of an external power source. In Fig. 19B and C, position-controlled selective growth using this approach was demonstrated by the

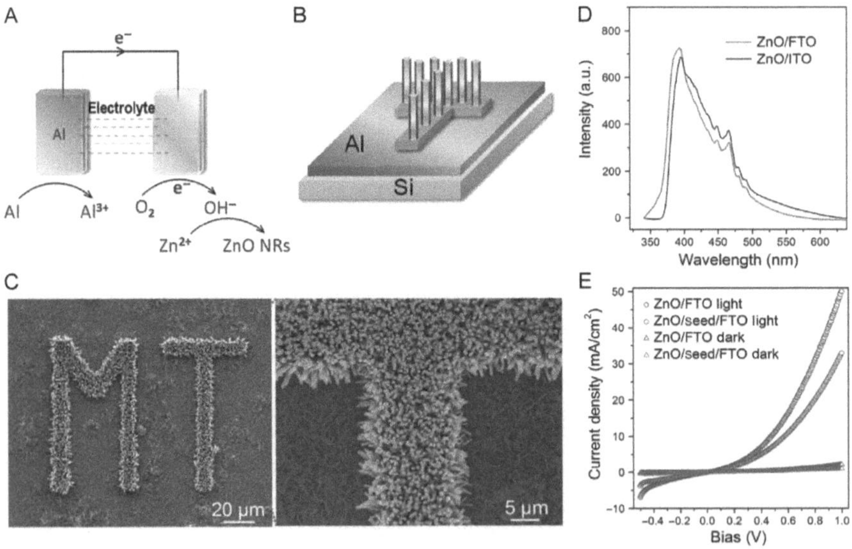


**Figure 19** (A) Schematic illustration of the galvanic-cell-based growth of ZnO nanorod arrays, (B) schematic representation and (C) SEM image of position-controlled selective growth of ZnO nanorods on Pt patterns on Al/Si substrate, (D) PL spectra of ZnO nanorods grown on fluorine-doped tin oxide (FTO) and ITO glasses, and (E) *I–V* curves of ZnO nanorods grown on FTO by the galvanic cell-based approach and a conventional approach employing a polycrystalline ZnO seed layer prepared by zinc acetate deposition. *Reprinted from Zheng et al. (2013) with permission, © 2013 Macmillan Publishers Ltd.*

deposition of a Pt layer defined by standard lithography. Densely packed and reasonably vertically aligned ZnO nanorods grew only on Pt, while considerably shorter nanowall-like structures formed on the Al surface.

Figure 19D shows PL spectra of ZnO nanorods grown on FTO and ITO glasses using the galvanic cell-based approach. Clear NBE emission at 393 nm (3.16 eV) was observed with negligible deep-level emission, confirming the high optical quality and low defect density of the ZnO nanorods. In Fig. 19E, photoconductive *I*–*V* measurements were performed with ZnO nanorods grown directly on FTO substrates. To demonstrate the advantages of the direct growth, the authors compared the photoconductivity results to those obtained from ZnO nanorods grown using a polycrystalline ZnO seed layer prepared by zinc acetate deposition under similar growth conditions. A significantly enhanced photoelectric response was obtained for the ZnO nanorods grown using the galvanic cell-based method, in which the nanorods made direct contact with the metallic substrate. The improved response was attributed to enhanced charge transport from the ZnO nanorods to the substrate and reduced recombination centers for photo-generated carriers associated with the seed layer.

## 3.3 Insulating Substrates

The direct growth of position-controlled semiconductor nanostructures on insulating substrates is desired in device applications, in which minimization of leakage current is critical. Although selective growth on these substrates should be possible using all of the growth methods reviewed in Section 2, most existing work has been based on the VLS and MOVPE methods.

Fan et al. (2004) demonstrated the selective VLS growth of ZnO nanowires on *a*-plane $Al_2O_3$. The *a*-plane $Al_2O_3$ is particularly suited for the growth of ZnO nanowires, due to the small mismatch in the lattice constants (0.08%) of ZnO(0001) and $Al_2O_3(11\bar{2}0)$. Figure 20A shows an array of square regions of selectively grown ZnO nanowires, which was obtained by VLS growth on a square array of Au, deposited using a TEM grid as the deposition mask. Cross-sectional TEM and SAED patterns were used to verify the epitaxial relationship between the ZnO nanowires and *a*-plane $Al_2O_3$, as shown in Fig. 20B–E, in which the clear epitaxial relationship of $ZnO(0001)[11\bar{2}0]||Al_2O_3(11\bar{2}0)[0001]$ was confirmed.

Nikoobakht et al. (2004) demonstrated that the growth mode of the ZnO nanowires on *a*-plane $Al_2O_3$ using the VLS method depends strongly on the size of the Au catalyst droplets; a Au droplet size smaller than 20 nm resulted in horizontal $([1\bar{1}00])$ growth as opposed to vertical ([0001])

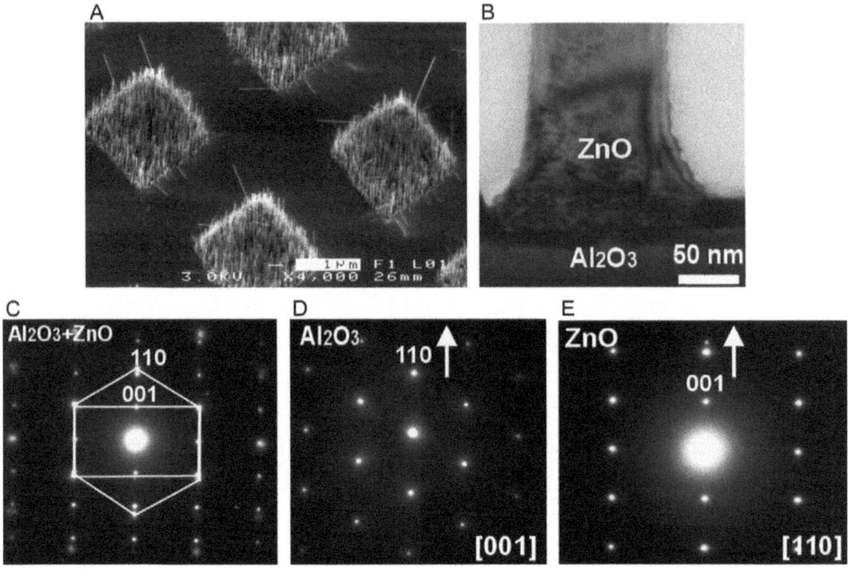


**Figure 20** (A) SEM image of ZnO nanowires selectively grown on patterned *a*-plane $Al_2O_3$, (B) cross-sectional TEM image at the ZnO/$Al_2O_3$ interface, and (C–E) SAED patterns at different regions of the TEM image revealing the epitaxial relationship between ZnO nanowires and $Al_2O_3$. *Reprinted from Fan et al. (2004) with permission, © 2004 Elsevier.*

growth. Nanoimprint lithography was used to create narrow lines of thin Au by placing nanometer-sized Au droplets at position-controlled sites (Fig. 2C). Upon heating for the VLS process, the lines of Au were transformed into nanodroplets, from which the horizontal growth of ZnO occurred, as shown in Fig. 21A. Electron backscattering diffraction was used to confirm that the ZnO(0001) planes grew parallel to $Al_2O_3(11\bar{2}0)$, and that the axis of the nanowires was ZnO$[1\bar{1}00]$, which was aligned with $Al_2O_3[1\bar{1}00]$, as illustrated in Fig. 21B. The difference in the growth behavior of the smaller nanodroplets was attributed to the difference in strain on the ZnO nanocrystal as it was being formed inside the Au nanodroplet; the smaller strain inside the smaller nanodroplets promoted better interaction with the substrate, which resulted in horizontal growth.

Selective area growth of ZnO nanowalls on an insulating AlN/Si(111) substrate using catalyst-free MOVPE was performed by Lee et al. (2011c). In this work, an insulating 200-nm-thick AlN film was heteroepitaxially grown on a Si(111) substrate using MOVPE with trimethylaluminum and ammonia as reactants. A $SiO_2$ mask layer was

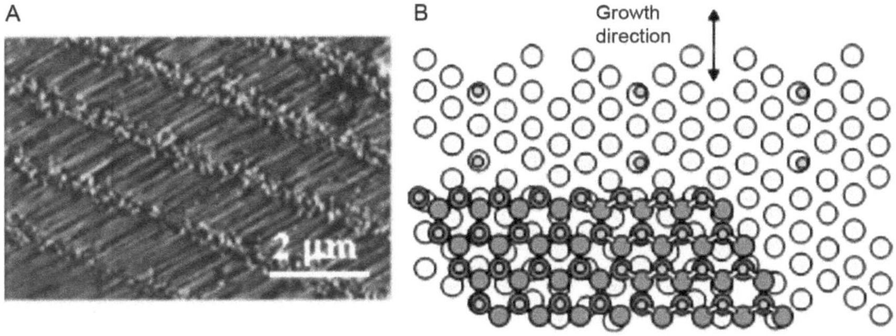


**Figure 21** Horizontal growth of ZnO nanowires on *a*-plane $Al_2O_3$. (A) SEM image of horizontally grown ZnO nanowires. (B) Schematic illustration of the growth mode with oxygen atoms in *a*-plane $Al_2O_3$ (big white circles) and Zn/O atoms in *c*-plane ZnO (small/large gray circles). *Reprinted from Nikoobakht et al. (2004) with permission, © 2004 AIP Publishing LLC.*

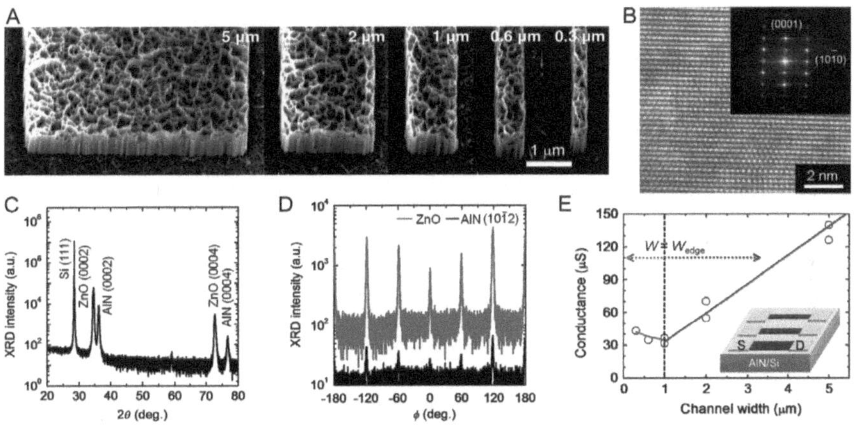


**Figure 22** Selective area MOVPE growth of ZnO nanowalls on AlN. (A) SEM image of ZnO nanowall channels of various widths. (B) Cross-sectional HR-TEM (inset: fast Fourier transform (FFT) pattern) of ZnO nanowalls. (C) $\omega$–$2\theta$ scan and (D) azimuthal ($\phi$) scan for the ZnO and AlN$\{10\bar{1}2\}$ diffractions determined by XRD. (E) Conductance of 3-µm-long ZnO nanowall networks of varying widths. Inset shows an illustration of the measured structure. *Reprinted from Lee et al. (2011c) with permission, © 2011 IOP Publishing.*

deposited and patterned via conventional lithography and wet etching, to promote growth of nanowalls of the desired shape, similar to Fig. 7B. The resulting ZnO nanowall network structures are shown in Fig. 22A. The cross-sectional HR-TEM image in Fig. 22B reveals a well-ordered crystal lattice without structural defects; the lattice spacing between adjacent planes was 0.26 nm, corresponding to the *d*-spacing of the ZnO(0002) planes. The electron diffraction pattern in the inset confirmed the single-crystalline nature of the nanowalls.

The crystal structure and orientation of the ZnO nanowall networks were investigated using XRD (Lee et al., 2011c). Figure 22C shows clear diffraction peaks in the $\omega$–$2\theta$ scan at 34.51° and 72.63°, resulting from *c*-plane Bragg reflections of the ZnO nanowalls; the two peaks at 36.14° and 76.63° were associated with AlN layers. This indicates that the ZnO nanowalls and the AlN layer exhibited identical *c*-axis crystallographic orientations; i.e., both the *c*-axis orientation and the growth direction were normal with respect to the substrate surface. The in-plane epitaxial relationship between the ZnO nanowalls and AlN layer was also studied by azimuthal ($\phi$) scans, as shown in Fig. 22D. The $\phi$-scans of ZnO and AlN$\{10\overline{1}2\}$ exhibited six distinct XRD peaks separated from one another by 60° and identical $(10\overline{1}2)$ peaks due to the ZnO nanowalls and the AlN layer. This indicates that the *c*-axis-oriented single-crystalline ZnO nanowalls were heteroepitaxially grown on AlN, with homogeneous in-plane alignment between the nanowalls.

The insulating nature of the AlN layer allowed the authors to investigate the intrinsic electrical properties of the nanowall networks (Lee et al., 2011c). The inset in Fig. 22E shows the structures used to measure the electrical conductivity of the ZnO nanowall networks, consisting of rectangular ZnO nanowall network channels, with a fixed 3-μm length and varying widths. All devices exhibited linear and symmetric *I*–*V* curves, indicating good ohmic contact between the ZnO nanowalls and the metal electrodes. The measured conductance as a function of the channel width shows a general trend of increasing conductivity for wider channels, which is expected as the greater number of nanowalls in the wider channels provide more current paths. However, as the channel width is reduced below a critical value, the conductance shows an upturn. This is explained by the improved alignment of the nanowalls along the channel direction when the channel is narrow, which results in shorter effective current paths.

## 3.4 Layered Two-Dimensional Substrates

Here, we will discuss the characterization of selectively grown nanostructures on layered 2D materials. In particular, hybrid heterostructures of semiconductor nanostructures with 2D graphene layers have been emerging as new materials for fabrication of transferable and flexible optoelectronic and electronic devices (Chung et al., 2010; Kim et al.,

2009b; Lee et al., 2010; Lin et al., 2010b; Yoo et al., 2012). Graphene layers, which have excellent electrical and thermal conductivity, high mechanical strength and elasticity, and optical transparency, provide a transferable, flexible, conducting substrate (Balandin et al., 2008; Castro et al., 2007; Kim et al., 2009a; Novoselov et al., 2004, 2005; Zhang et al., 2005, 2009c). The semiconductor nanostructures that are grown on top of graphene have functions as both efficient channels for carrier transport and electrical pumping sources for radiative recombination, thereby significantly improving device performance, compared with traditional devices. In Section 2, we reviewed how the position-controlled selective growth of semiconductor nanostructures could be achieved on graphene, a prerequisite for practical semiconductor nanostructure/graphene heterostructure applications. Here, we are interested in the characterization of these hybrid nanostructures.

The microstructural characteristics of ZnO nanostructures grown on graphene layers, especially at the heteroepitaxial interface between them, were investigated using cross-sectional TEM analysis by Kim et al. (2012c). The specimen was an array of line patterns of selectively grown ZnO nanowalls obtained by catalyst-free MOVPE on graphene/$SiO_2$/Si substrate, as shown in Fig. 23A. The width and spacing of the line patterns were 600 nm and 5.4 μm, respectively. The edge regions of the individual ZnO nanostructures, grown on patterned graphene layers, are usually taller

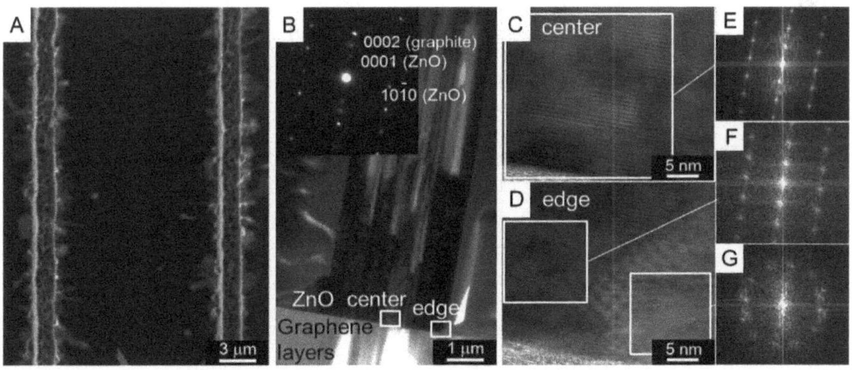


**Figure 23** Microstructural characteristics of ZnO nanostructures. (A) Top-view SEM image of line-patterned ZnO nanostructures on graphene layers. (B) Cross-sectional TEM bright-field image of the ZnO nanostructures on graphene layers. The inset represents the corresponding SAED pattern. (C and D) HR images of the ZnO nanostructures taken from the regions marked "center" and "edge" in (B), respectively. (E–G) FFTs obtained from HR images in the regions indicated by white boxes in (C and D). *Reprinted from Kim et al. (2012c) with permission, © 2012 John Wiley and Sons.*

than the inner wall structure; in this case, the edge regions and inner walls were 3.5 and 2.0–2.3 μm in height, respectively. In Section 2.2, we discussed the associated underlying mechanism with respect to the results shown in Fig. 8.

Figure 23B shows a bright-field TEM image of ZnO nanostructures on graphene layers; the heteroepitaxial relationship between ZnO and graphene layers was investigated using the SAED. The SAED pattern (inset of Fig. 23B) shows that the ZnO nanostructures have a wurtzite crystal structure, with their *c*-axis parallel with respect to the *c*-axis of the honeycomb structure of the graphene layers. The heteroepitaxy between the ZnO nanostructures and graphene layers resulted in the orientation relationships of $(0002)_{ZnO} \parallel (0002)_{graphene}$ and $\left[2\overline{1}\overline{1}0\right]_{ZnO} \parallel \left[2\overline{1}\overline{1}0\right]_{graphene}$, which is in agreement with previous reports (Yoo et al., 2012). Few microstructural defects (e.g., dislocations) were observed; no distinct streaks associated with the presence of stacking faults were evident in the SAED pattern. Thus, the nanoepitaxy between ZnO and the graphene layers resulted in ZnO nanostructures with a high crystallinity.

The atomic arrangement of the ZnO nanostructures was investigated in more detail using HR-TEM and its fast Fourier transform (FFT). Figure 23C shows a HR-TEM image of the center region, indicated in Fig. 23B; a highly ordered atomic arrangement of the ZnO nanostructures on the graphene layers was observed and confirmed by the corresponding FFT (Fig. 23E). However, the edge of the ZnO nanostructure was not sharply defined at the base; lateral tailing of ZnO was observed, extending from the submicron scale to several microns, as shown in Fig. 23B. The HR image obtained from the lateral tail of ZnO and the corresponding FFT indicated that the crystal orientation was rotated by about 30°; i.e., heteroepitaxy was not achieved in this region. Additionally, the crystallinity of the ZnO tail was significantly less than that of ZnO epitaxially grown on graphene layers, as shown in the HR image in Fig. 23D and the corresponding FFT in Fig. 23G.

The optical characteristics of the selectively grown ZnO nanostructures on graphene were examined using CL and PL spectroscopy. The specimen for these experiments was an array of position-controlled ZnO nanotubes grown by catalyst-free MOVPE on a graphene/$SiO_2$/Si substrate, which was prepared by selectively exposing the graphene surface to oxygen plasma (Fig. 9) (Kim et al., 2012c). Figure 24A shows that the room-temperature CL spectrum of the selectively grown ZnO nanostructures had the NBE

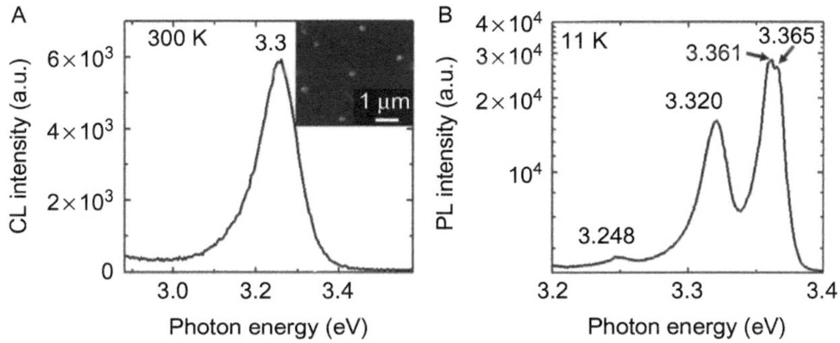


**Figure 24** (A) CL spectra of ZnO nanotubes selectively grown on graphene layers. Inset represents the corresponding monochromatic image at 3.3 eV. (B) Low-temperature (11 K) PL spectrum of ZnO nanotubes on graphene layers. *Reprinted from Kim et al. (2012c) with permission, © 2012 John Wiley and Sons.*

emission peak at 3.3 eV. The monochromatic CL image corresponding to the NBE emission at 3.3 eV indicated high growth selectivity of the ZnO nanostructures on the graphene layers (inset of Fig. 24A). CL was observed only for the ZnO nanostructures on the oxygen plasma-treated graphene layers; CL was not detected from the remaining graphene layers.

The possible incorporation of impurities from graphene into the ZnO nanostructures was investigated using PL spectroscopy. As shown in Fig. 24B, the PL spectrum of selectively grown ZnO nanostructures, which was measured at 11 K, had three distinct NBE emission peaks at 3.320, 3.361, and 3.365 eV. The dominant PL peaks at 3.361 and 3.365 eV were tentatively assigned to neutral-donor-bound exciton peaks (Park et al., 2003a). No additional peaks associated with carbon impurities were observed (Tan et al., 2007). This suggests that neither degradation nor impurity incorporation by carbon atoms occurred during the growth process, despite the fact that the chemical reactivity of the carbon atoms was enhanced by the oxygen plasma treatment.

Baek and Yi (manuscript in preparation) studied the lasing characteristics of position-controlled ZnO nanorod arrays on a graphene-coated Si(100) substrate, shown in Fig. 10B. The ZnO nanorods, irradiated by 355-nm radiation from a Nd:YAG laser, host multiple Fabry–Pérot modes along the vertical direction of the nanorods that can be amplified via stimulated emission of the NBE transitions of ZnO. The PL spectra as a function of the excitation intensity (Fig. 25A) exhibit this lasing behavior; sharp peaks (FWHM: 0.3 nm) emerge from the broad spontaneous emission peak at intensity levels above $I_{th} = 70$ kW/cm$^2$. By plotting the nonlinear intensity

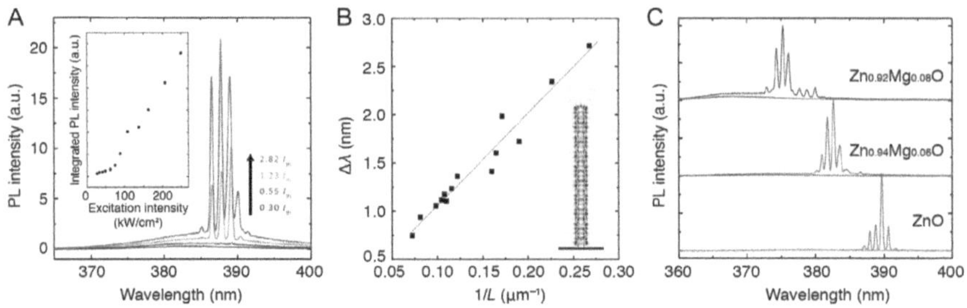


**Figure 25** Lasing characteristics of ZnO nanorods (Baek and Yi, manuscript in preparation). (A) Room-temperature PL spectra of position-controlled ZnO nanorods (Fig. 10B) on a graphene/Si(100) substrate with optical pumping by 355-nm radiation from a Nd:YAG laser. The colors (different gray shades in the print version) indicate different excitation intensity levels with respect to the threshold intensity, $I_{th} = 70$ kW/cm$^2$. The inset shows the integrated PL intensity as a function of the excitation intensity. (B) Mode spacing, $\Delta\lambda$, as a function of the inverse length of the nanorods. The inset shows a simulation of the field distribution ($E_x$) for a resonant mode at 393 nm. (C) PL spectra at different levels of Mg contents.

curve in log scale, three different regions corresponding to spontaneous emission, amplified spontaneous emission, and laser oscillation were identified; a spontaneous emission factor $\beta$ of 0.05 was revealed by fitting the data. The nanorod length dependence of the mode spacing is shown in Fig. 25B. The inverse proportionality of the mode spacing on the nanorod length confirmed the origin of these sharp peaks as Fabry–Pérot modes along the vertical direction of the nanorods. Figure 25C shows the effect of alloying ZnO with MgO. $Zn_{1-x}Mg_xO$ has a larger bandgap than ZnO, which increases as a function of $x$ (Park et al., 2003c). Hence, the lasing wavelength is expected to shift toward higher energy with larger $x$, as shown in Fig. 25C.

Another 2D layered material that shows great promise as a transferrable (Chung et al., 2010; Kobayashi et al., 2012) and flexible (Choi et al., 2010; Lee et al., 2011b; Peng et al., 2012) substrate is hBN with a wide direct bandgap (5.2–5.9 eV). Among various 2D-layered nanomaterials (Butler et al., 2013; Osada and Sasaki, 2012), hBN possesses significant advantages as a substrate for many advanced electronics and optoelectronics applications (Dean et al., 2010; Wang et al., 2011), due to its excellent thermal conductivity, optical properties (Watanabe et al., 2004), chemical/mechanical stability (Song et al., 2010), and low dielectric constant (Kim et al., 2012a). The diatomic honeycomb crystal lattices of layered hBN sheets are structurally compatible with many semiconductors having wurtzite, zinc blende, and diamond structures (Kobayashi et al., 2012). Moreover, the ability of hBN to form alloy (Gong et al., 2014) or lateral heterostructures (Levendorf et al., 2012) with graphene is noteworthy for fabricating sophisticated on-demand devices. The recent progress in large-scale hBN synthesis with precisely controlled thickness via CVD techniques (Song et al., 2010) may allow the integration of semiconductors with wafers based on conventional microfabrication processes. Accordingly, the heteroepitaxial integration of semiconductors on hBN is expected to generate significant advances in design, fabrication, and performance for diverse semiconductor optoelectronic and electronic circuitry.

The position-controlled selective growth and characterization of ZnO nanostructures on an hBN surface were demonstrated by Oh et al. (2014), as shown in Fig. 26. As discussed in Section 2.2, the preferential formation of high-density nanowalls on a rough hBN surface prepared by selective exposure to $O_2$ (or Ar) plasma could be exploited for creating shape- and position-controlled ZnO nanostructures (Fig. 9). The SEM images in Fig. 26A and B show the shape- and position-controlled ZnO nanostructures filled with vertical nanowall networks. Shape-controlled ZnO nanowall architectures grown on hBN exhibited the epitaxial relationship $(000\bar{1})[10\bar{1}0]_{ZnO}||(0001)[10\bar{1}0]_{hBN}$ and exclusive growth

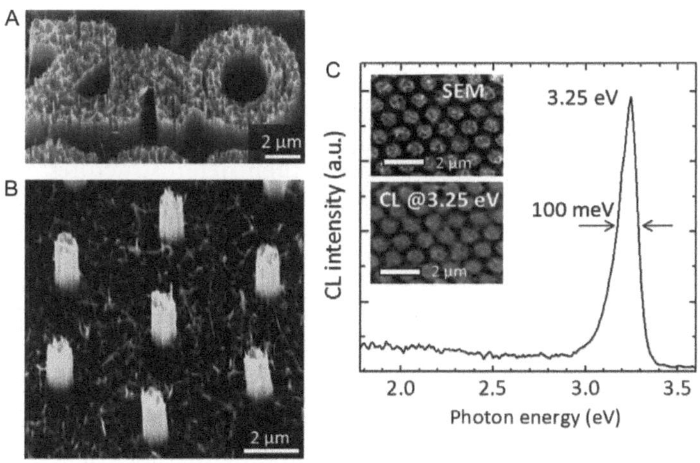


**Figure 26** (A and B) Tilt-view SEM image of ZnO nanoarchitectures on hBN films with designed shape and spatial arrangements and (C) room-temperature CL spectra of ZnO nanostructures. The insets are top-view scanning electron microscopy images (upper) and the corresponding monochromatic CL image obtained at a photon energy of 3.25 eV (bottom). *Reprinted from Oh et al. (2014) with permission.*

orientation along the oxygen-terminated $[000\overline{1}]$ direction, as confirmed by SAED and CBED analysis (Oh et al., 2014).

In Fig. 26C, the optical properties of the ZnO nanoarchitectures selectively grown on hBN were investigated using CL spectroscopy. The room-temperature CL spectrum of the ZnO nanostructure arrays shows a dominant NBE emission peak at 3.25 eV, which was attributed to free exciton emission (Djurišić and Leung, 2006). No deep-level emission, typically observed from bulk ZnO materials, was observed. The FWHM of the NBE peak was as narrow as 100 meV; this value is comparable to previous reports (Park et al., 2002b) and indicates the high quality and purity of the ZnO nanostructures grown on hBN by vdW epitaxy. The insets to Fig. 26C show a SEM image and the corresponding CL monochromatic image of the ZnO nanostructure array; the latter was measured at a photon energy of $3.25 \pm 0.01$ eV using a monochromator. The monochromatic CL image shows homogeneous, strong CL emission from the ZnO nanostructure arrays. The SEM image and CL image are similar, suggesting that significant non-radiative or other radiative transitions from defects or impurities were not observed. The NBE emissions of the nanostructure arrays were homogeneous in intensity. Additionally, nitrogen- or boron-associated luminescence was not detected by PL spectroscopic analyses (Kim et al., 2013; Wang et al., 2001). This suggests that the atoms in hBN were not incorporated into the ZnO nanostructures during the vdW–MOVPE process (Oh et al., 2014).

## 4. LED APPLICATIONS OF NANOSTRUCTURES AND THEIR HETEROSTRUCTURES

Semiconductor nanostructures and their heterostructures offer functional components for diverse nanostructure-based LEDs (Guo et al., 2010; Kikuchi et al., 2004; Kim et al., 2004; Lin et al., 2010a; Pearton et al., 2008; Qian et al., 2005; Sun et al., 2005; Willander et al., 2009). Recently, vertical ZnO nanostructures have widely been used as active elements for light emission or lasing in nanophotonic devices and as passive elements for improving the light extraction of conventional thin-film LEDs (An et al., 2008; Huang et al., 2001; Park and Yi, 2004). Although ZnO p–n junction LEDs prepared using *in situ* doping methods have rarely been achieved, due to the difficulty in preparing p-type ZnO, a few works reported fabrication of ZnO p–n junction nanorod LEDs by ion implantation (Sun et al., 2009b; Yang et al., 2008; Zhang et al., 2008). In particular, Sun et al. (2009b) demonstrated UV emission from ZnO p–n junction nanorods; p-type doping was achieved by phosphorous or arsenic ion implantation. Despite this successful demonstration, ion implantation for p-type doping may not offer sophisticated device configurations with thickness- and composition-controlled structures, such as multiple quantum wells (MQWs), required for high-efficiency LEDs (Lee and Yi, 2012).

In overcoming the challenges associated with these ZnO nanostructure p–n junction devices, a heterostructure formed at either the axial or radial surfaces of ZnO nanostructures is a promising device configuration (Hong et al., 2009a; Lee et al., 2009a; Park and Yi, 2004). This nano-heterostructure device could have several potential advantages over traditional thin-film heterojunction devices. First, nanoepitaxy enables the growth of high-quality semiconductor nanostructures with a clean interface on a thin film, leading to high-efficiency LEDs by reducing the nonradiative recombination at the junction. Second, nanoscale junctions promote enhanced carrier injection efficiency. Park et al. (2003b) and Smit et al. (2002) reported a significant increase in the carrier injection rate for nanocontacts in Schottky diodes. Third, the light extraction efficiency can be enhanced considerably due to the reduced internal reflection on a rough surface composed of nanostructures (An et al., 2008).

The basic strategy for the fabrication of nano-heterostructure LEDs is to use GaN as the p-type material on the axial or radial surfaces of ZnO nanostructures, as well-developed growth techniques for p-GaN are available unlike p-type ZnO. Among the various p-type semiconductors available in industry, GaN thin film is the most adequate choice for this structure because it has a fundamental bandgap energy (3.4 eV) similar to ZnO, the same wurtzite crystal structure, and low mismatch (1.9%) between the lattice constants (ZnO: $a=b=3.249$ Å, $c=5.206$ Å for ZnO; GaN: $a=b=3.186$ Å, $c=5.178$ Å) (Park and Yi, 2004), allowing high-quality GaN/ZnO heteroepitaxy with few interfacial defects (Hong et al., 2009a). Other p-type thin films, including SiC, Si, NiO, and $CuAlO_2$, have also been used (Hsieh et al., 2009; Ling et al., 2009; Mofor et al., 2007; Sun et al., 2006, 2009a; Wang et al., 2009; Willander et al., 2009; Zhang et al., 2009a); however, the associated heterojunction devices exhibited less than optimal performance, due to a large energy barrier and/or many interface states at the junction. Furthermore, the ability to control the electrical conductivity and modulate the band gap (from the infrared to far ultraviolet (UV) regime) by alloying of In and Al in III nitrides enables the fabrication of high-efficiency, full-color optoelectronic devices (Ponce and Bour, 1997; Strite and Morkoç, 1992).

Below, we review the advances in fabricating LEDs based on ZnO nanostructures. We first introduce the initial works that used ZnO nanostructures grown without selective mechanisms, and then move onto more advanced nanoarchitectures enabled by position-controlled selective growth.

## 4.1 LEDs Without Selective Growth

Figure 27A shows a schematic diagram of a generic n-ZnO nanorod/p-type thin-film LED, where nanoscale p–n junctions are formed at the bases of the

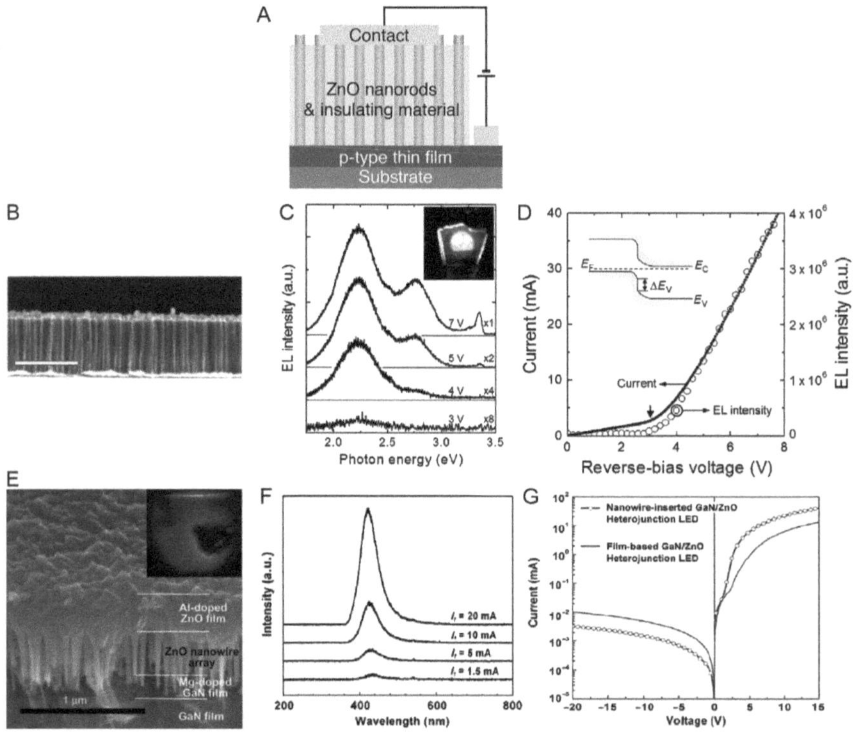


**Figure 27** (A) Schematic illustration of a generic n-ZnO nanostructure/p-type thin-film heterojunction LED. (B) Cross-sectional SEM image of a heterojunction LED device based on $Al_2O_3$(0001) substrate, p-GaN thin film, n-ZnO nanorods, and Ti/Au contacts (scale bar: 1 μm). (C) EL spectra (inset: photograph at a reverse bias of 5 V) and (D) *I–V* characteristics superposed with EL intensity curve (inset: band diagram of the heterojunction) of the device in (B). Reprinted from Park and Yi (2004) with permission, © 2004 John Wiley and Sons. (E) Tilted-view SEM image of a heterojunction LED device based on $Al_2O_3$(0001) substrate, p-GaN thin film, n-ZnO nanorods, and n-ZnO film. (F) EL spectra and (G) *I–V* characteristics of the device in (E). *Reprinted from Jeong et al. (2007) with permission, © 2007 John Wiley and Sons.*

nanorods. To fabricate this structure, vertically aligned n-ZnO nanorods on a p-type substrate are first grown by the numerous methods reviewed in Section 2. The empty gaps between the individual nanorods are filled with an insulating supporting material, such as a polymer or spin-on-glass (SOG) for electrical isolation, after which ohmic contacts are formed separately on the n- and p-type material by depositing thin metal layers or transparent ITO.

Figure 27B shows a first demonstration of this structure introduced by Park and Yi (2004), in which n-ZnO nanorods were grown on p-GaN (0001) by catalyst-free MOVPE; Ti/Au ohmic contacts were created on the nanorod tips after filling the gaps with photoresist and exposing the nanorod tips by plasma etching. The electroluminescence (EL) spectra shown in Fig. 27C exhibit a broad yellow emission band centered at 2.2 eV (560 nm), only under reverse-bias conditions. Weak blue emission at 2.8 eV (450 nm) was observed as the reverse bias increased. Based on PL data, the yellow and blue emissions in the EL spectra were assigned to deep-level emission in ZnO and to the Mg-acceptor level in GaN, respectively. This unusual carrier transport under reverse bias may be explained in terms of type II band alignment of the p-GaN/n-ZnO heterojunction, with a large valence band offset ($\Delta E_v$), as shown in the inset to Fig. 27D. The large band offset at the heterojunction makes the tunneling barrier very thin, by lowering the n-ZnO conduction band or raising the p-GaN valence band. Under a small reverse-bias voltage, the unoccupied conduction band minimum of n-ZnO is lower than the occupied valance band maximum of p-GaN. Thus, carrier transport by tunneling can occur even under small reverse-bias voltages, and the tunneling probability increases with the reverse-bias voltage.

Many types of n-ZnO nanorod/p-type thin-film heterojunction LED with similar concepts and geometries have been reported by other groups (Chen et al., 2009, 2010; Guo et al., 2008; Lai et al., 2008; Lee et al., 2009b; Yan et al., 2010; Zhang et al., 2009b). For example, a hybrid heterojunction LED which exhibited blue light emission under forward-bias conditions was demonstrated (Zhang et al., 2009b). The device geometry is almost identical to that shown in Fig. 27A, except that a transparent ITO electrode was used. As a result of using ITO contacts, the device exhibited resistive characteristics and emitted light under high forward-bias voltages. As a modified geometry, Jeong et al. reported an Al-doped ZnO film/n-ZnO nanowire array/p-GaN film structure by fabricating continuous ZnO thin films on top of the ZnO nanowire array (Jeong et al.,

2006, 2007), as shown in Fig. 27E; blue light emission was observed under a forward bias (Fig. 27F). This geometry has the advantage of simple device fabrication by depositing metal contacts on thin films without gap-filling or selective-etching processes. Furthermore, compared with film-based GaN/ZnO heterojunction LEDs, nanowire-inserted heterojunction LEDs showed stronger EL emission and enhanced electrical characteristics with a smaller leakage current, a lower turn-on voltage, and a larger forward current, as shown in Fig. 27G. Such superior characteristics were explained by a low density of interfacial defects at the lattice-matched heterojunction interface and enhanced electron injection through nanoscale contacts.

A different approach to fabricate the LEDs is to deposit GaN films on the radial surfaces of ZnO nanostructures. The p–n junctions are then formed by GaN films at the bases of the nanorods with the substrate or at the radial surfaces of the nanostructures by depositing additional GaN films with alternating dopants. This approach has the advantage of forming a homojunction between n-GaN and p-GaN, instead of a heterojunction between n-ZnO and p-GaN. In general, p–n heterojunction devices are less efficient than homojunction devices, due to the energy barrier at the junction interface, which decreases carrier injection efficiency for heterojunction devices with a large band offset (Wang et al., 1995).

Figure 28A shows an illustration of the former approach demonstrated by An and Yi (2007), in which n-GaN films deposited on the surfaces of ZnO nanorods formed p–n junctions with the p-GaN substrate. The GaN/ZnO coaxial structure is shown in Fig. 28B; the heteroepitaxial growth of the GaN film on the ZnO nanorod surface was confirmed. The EL spectra in Fig. 28C show dominant emission peaks at 2.96 eV (430 nm) and 3.24 eV (382 nm),

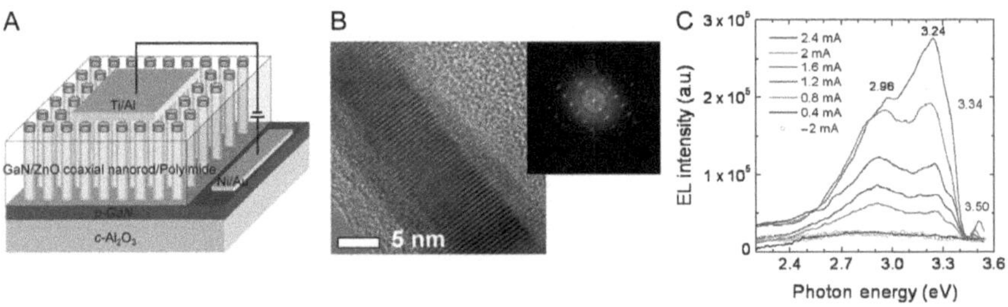


**Figure 28** (A) Schematic illustration of a GaN/ZnO coaxial nanorod LED device based on the homojunction between the n-GaN film and p-GaN substrate. (B) Cross-sectional HR-TEM image of the GaN/ZnO coaxial nanorod heterostructure (inset: FFT pattern). (C) EL spectra at various current levels. *Reprinted from An and Yi (2007) with permission, © 2007 AIP Publishing LLC.*

associated with deep acceptors in Mg-doped p-GaN and shallow donors in the n-GaN/ZnO nanostructures, respectively. The suppression of the EL emission at 2.2 eV by ZnO and enhancement of the blue emission by GaN indicates that the carrier injection occurred predominantly between the n-GaN films and p-GaN substrate, rather than with ZnO nanorods.

Figure 29A shows an illustration of an n-GaN/MQW/p-GaN coaxial LED on the latter approach demonstrated by Lee et al. (2011b), that is, the p–n junction formation at the radial surface of the nanorods. In this device structure, n-GaN, $In_xGa_{1-x}N$/GaN MQWs, and p-GaN were coated heteroepitaxially onto the surfaces of ZnO nanorods grown on graphene. MQWs were included to enhance the radiative recombination efficiency of the p–n junction by confining the carriers in the wells. Because these ZnO nanorods and the subsequent GaN films were grown directly on a graphene/$SiO_2$/Si substrate, the LED nanostructures on graphene could be released from the substrate using a buffered oxide etchant and transferred onto an arbitrary substrate. In particular, the authors transferred the LEDs

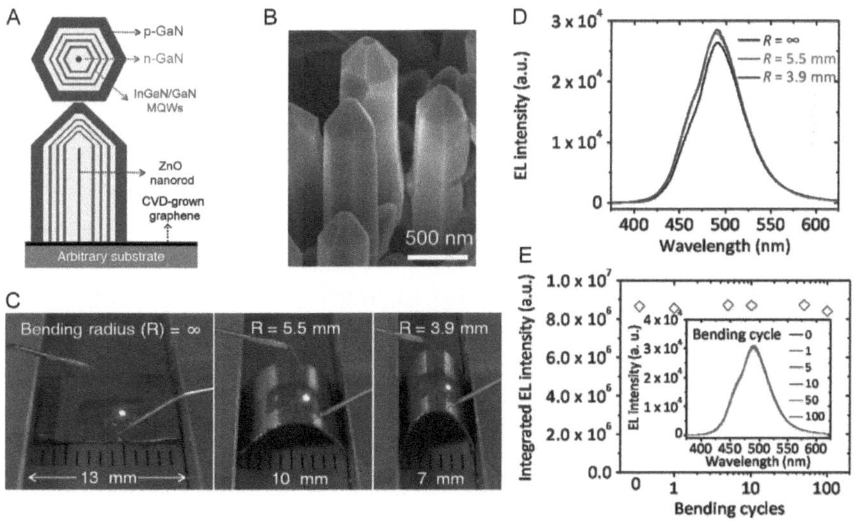


**Figure 29** (A) Cross-sectional schematic of an n-GaN/MQW/p-GaN coaxial LED on the surface of a ZnO nanorod. (B) Tilted-view SEM image of the LED nanostructures. (C) Light emission photographs at bending radii of ∞, 5.5, and 3.9 mm. (D) EL spectra at different bending radii at a fixed current of 10 mA. (E) Integrated EL intensity as a function of bending cycles at a fixed current of 10 mA. Inset shows the corresponding EL spectra. *Reprinted from Lee et al. (2011b) with permission, © 2011 John Wiley and Sons.*

onto a Cu-coated polyethylene terephthalate substrate to demonstrate a flexible LED (Fig. 29C).

The EL spectra of the flexible LED are shown in Fig. 29D. Prior to bending ($R=\infty$), the dominant EL peak was observed at 490–493 nm from the $In_xGa_{1-x}N$/GaN MQWs. In addition to the dominant peak, a broad shoulder also appeared around 465 nm, presumably resulting from the non-uniform thicknesses and compositions of the quantum-well layers formed on the multifaceted nanorod surfaces. Upon bending, the EL spectra showed no significant changes in the EL peak position and intensity, confirming that the flexible nanostructure LEDs maintained their optical characteristics as the bending radii approached 3.9 mm. In particular, the absence of any shift in the EL peak on bending indicated that the strain applied to the active region in the individual nanostructures was negligibly small, because strain generally induces changes in the luminescent spectra. In addition to flexibility, the reliability of the flexible LEDs was investigated by measuring the EL spectra with repeated bending cycles. The nanostructure LEDs exhibited almost identical EL spectra after repeated bending, and the integrated emission intensity remained nearly constant up to 100 bending cycles (Fig. 29E), which demonstrated the reliable operation of nanostructure LEDs fabricated on graphene films in a flexible form.

## 4.2 Position- and Dimension-Controlled LEDs

As discussed previously, position-controlled selective growth is essential in creating nanostructure devices suitable for practical applications, where the nanostructures are located at precisely known positions with designed dimensions. Selective growth is needed not only to control the position of the grown nanomaterial but also to improve the performance of the device. For instance, the MQW structures, essential to enhancing the quantum efficiency of LEDs, require that the well and barrier layers have precisely controlled thicknesses and composition. The reliability of this process would be improved if the nanostructures were fabricated into a regular array, as opposed to randomly nucleated nanostructures, so that the lateral growth rate and composition could be better controlled.

Figure 30A shows an example of a GaN/$In_xGa_{1-x}N$/GaN/ZnO nanoarchitecture LED fabricated on the surface of position-controlled ZnO nanotubes. Epitaxial p- and n-GaN layers, with $In_xGa_{1-x}N$/GaN MQW layers in-between, were grown on the radial surfaces of position-controlled ZnO nanotube arrays on an n-GaN/$c$-$Al_2O_3$ substrate with a $SiO_2$ growth mask (Hong et al., 2009a; Lee et al., 2009a). Figure 30B shows a horizontal cross-sectional scanning TEM image of the $GaN/In_{1-x}Ga_xN/GaN/ZnO$ coaxial nanotube. Three bright lines corresponding to $In_xGa_{1-x}N$ quantum-well layers were revealed, which alternated with darker regions corresponding to the GaN quantum barrier layers; thus, three-period MQW layers coaxially coated the GaN/ZnO nanotube heterostructures. The corresponding FFT image of the coaxial nanotube quantum structure revealed single crystallinity, with a clear sixfold rotational symmetry.

For the fabrication of vertical nanoarchitecture LED arrays, ohmic contacts were made on the outermost p-GaN surface of the nanoarchitecture and the n-GaN seed layer on the substrate, as shown in Fig. 31A (Lee et al., 2009a). For electrical isolation between the two metal electrodes, gaps

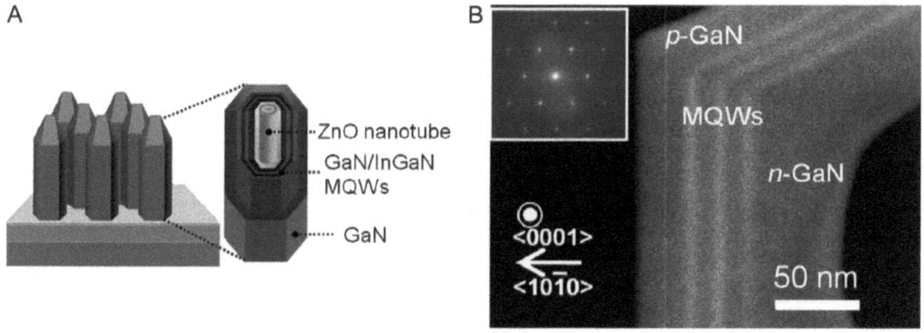


**Figure 30** Position-controlled GaN-based coaxial nanotube quantum structure arrays. (A) Schematics of position-controlled GaN/InGaN/GaN/ZnO coaxial nanotube quantum structure arrays and their cross-sectional structure. (B) Horizontal cross-sectional scanning TEM image of the $GaN/In_{1-x}Ga_xN/GaN/ZnO$ coaxial nanotube. Inset is the corresponding diffraction pattern obtained from a FFT process. *Reprinted from Hong et al. (2009a) with permission, © 2009 IOP Publishing.*

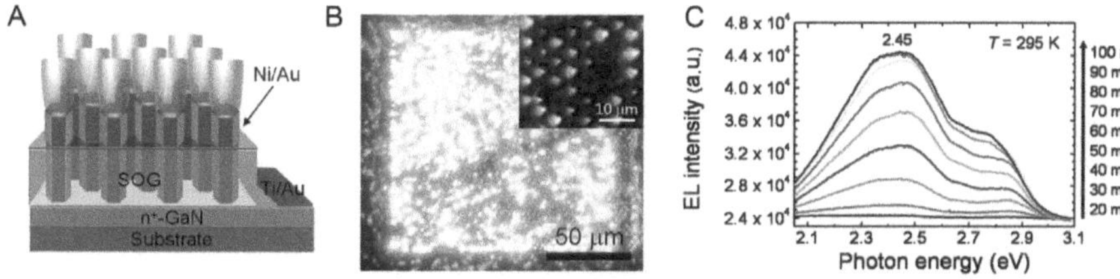


**Figure 31** $GaN/In_{1-x}Ga_xN/GaN/ZnO$ nanoarchitecture LED microarrays. (A) Schematic representation showing the device structure of position-controlled nanoarchitecture LED arrays. (B) Optical microscopic images of light emission from nanoarchitecture LED microarrays. The inset is a magnified image. (C) EL spectra of nanoarchitecture LED microarrays at various applied current levels of 20–100 mA. *Reprinted from Lee et al. (2009a) with permission, © 2009 AIP Publishing LLC.*

between nanoarchitectures can be filled with an insulating SOG layer. Figure 31B shows the hexagonal arrays of turquoise-colored light emission from the nanoarchitecture LED microarrays under a forward current of 100 mA; each spot corresponds to light emission from individual nanoarchitecture LEDs. The EL spectra of nanoarchitecture LED microarrays (Fig. 31C) show a dominant peak centered at 2.45 eV and a shoulder near 2.85 eV, which were attributed to $In_xGa_{1-x}N/GaN$ MQWs. In addition to the EL characteristics, the electrical characteristics were consistent with typical rectifying behavior, with a turn-on voltage of 3 V and a leakage current of $5 \times 10^{-4}$ A at −4 V. Above the turn-on voltage, the current and light emission intensity increased rapidly with the bias voltage.

The above approach using position-controlled GaN/ZnO coaxial nanotube heterostructures enabled the fabrication of nanostructure LEDs on Si substrates as well (Kang et al., 2012; Lee et al., 2011a). Assembling LEDs on Si substrates is of particular interest, because it allows the integration of various novel optoelectronic technologies with well-established Si-based electronics. However, such integration remains challenging, due to the unfavorable heteroepitaxy between Si and optical materials. This issue could be resolved using a crack-free thin GaN buffer layer (1 μm thickness) on a Si substrate and fabricating LEDs on top of the buffer layer based on nanoepitaxy, as shown in Fig. 32A and B (Lee et al., 2011a). The resulting LEDs, consisting of position-controlled arrays of n-GaN/MQW/p-GaN/ZnO coaxial nanotube heterostructures, emitted green light (2.35 eV), as shown in Fig. 32C.

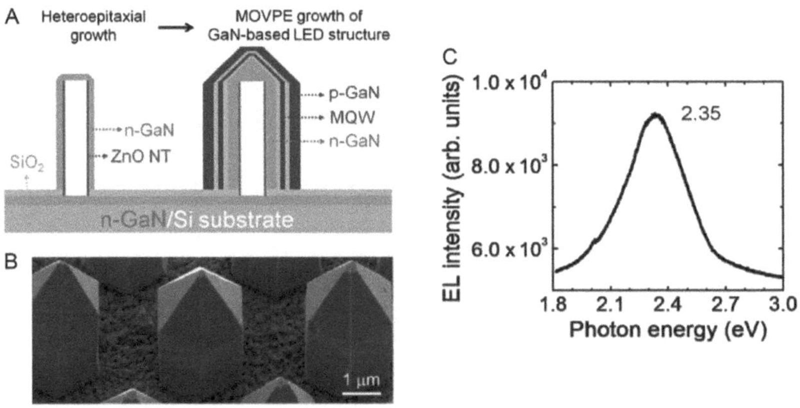


**Figure 32** (A) Schematic representation of the fabrication of GaN-based micro-LED structures. (B) SEM image and (C) EL spectra of the micro-LED arrays fabricated on Si substrates. *Reprinted from Lee et al. (2011a) with permission, © 2011 IEEE.*

Similarly, GaN/ZnO coaxial nanotube heterostructure LEDs have also been demonstrated with graphene as a substrate (Kim et al., 2012c). Figure 33A shows a thin Si-doped n-GaN layer, a three-period GaN/$In_xGa_{1-x}N$ MQW, and a Mg-doped p-GaN layer that were heteroepitaxially and coaxially grown on ZnO nanotubes (Lee et al., 2009a). Here, the graphene layer was transferred onto a heavily doped Si substrate that was used as the bottom electrode; however, this is not a necessity as the graphene layer can be transferred onto any arbitrary substrate. As shown in Fig. 33B and C, the light-emitting properties of the nanostructure LEDs on graphene layers investigated by EL measurements exhibited bright violet light emission at 3.0 eV from the GaN/$In_xGa_{1-x}N$ MQW. In Fig. 33D, the electrical properties of the nanostructure LEDs were also investigated by measuring their $I$–$V$ characteristic curves, which indicated a rectifying behavior with a turn-on voltage of 5 V and good ohmic contact between the as-grown ZnO nanostructure and graphene layers. This is because

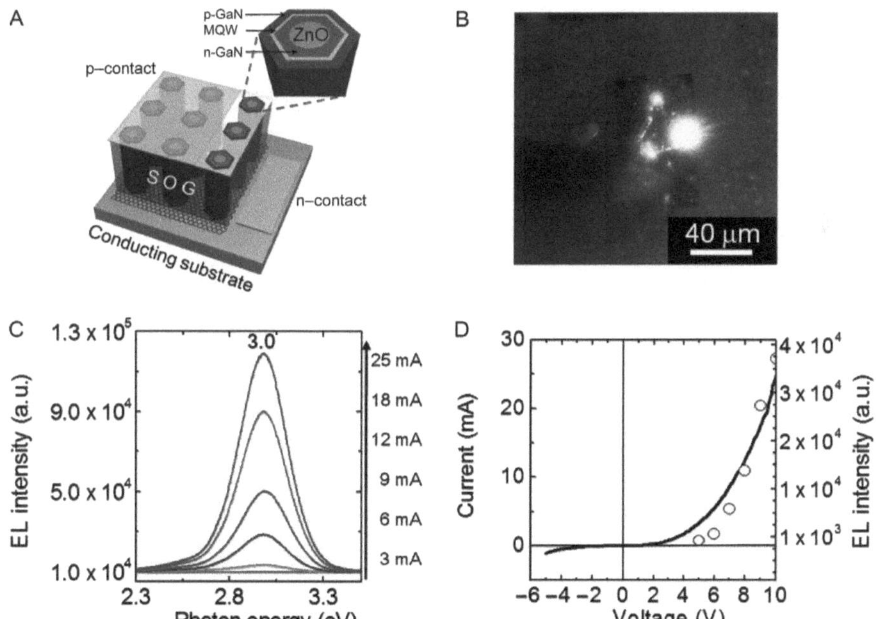


**Figure 33** (A) Schematic image of GaN/$In_xGa_{1-x}N$/GaN/ZnO coaxial quantum nanostructure LEDs with GaN/$In_xGa_{1-x}N$ MQW layers. (B) Optical image of light emission from the LEDs fabricated on the graphene layers. (C) EL spectra from the nanostructure LEDs at various applied voltage biases. (D) $I$–$V$ characteristic curve measured from nanostructure LEDs. *Reprinted from Kim et al. (2012c) with permission, © 2012 John Wiley and Sons.*

current can be injected into n-GaN without encountering a substantial energy barrier. Such behavior may be explained in terms of the negligible energy barrier at the junction between the two material systems, due to the small difference between the electron affinity of ZnO (4.50 eV) and the work function of graphene (4.48 eV) (Giovannetti et al., 2008).

## 5. CONCLUSIONS

This chapter provided an overview of position-controlled selective growth of ZnO nanostructures and their device applications. We have reviewed the recent research progress on catalyst-assisted and catalyst-free growth methods of ZnO nanostructures, including metalorganic CVD. The nanostructures can have different device applications depending on the substrates employed. Thus, we investigated the characteristics of the nanostructures grown on various substrates, categorized into four distinct subgroups: (i) semiconducting, (ii) metallic, (iii) insulating, and (iv) 2D materials, along with their device applications. The ZnO nanostructures provide significant opportunities to fabricate various electronic and optoelectronic devices. In particular, vertically aligned high-density ZnO nanostructure arrays enable the fabrication of diverse ZnO-nanostructure-based LEDs, including nanostructure/film hybrid heterojunction LEDs and GaN homojunction LEDs with coaxial GaN p–n junction layers coated on ZnO nanorods. In contrast to high-density ZnO nanostructures, position-controlled growth of ZnO nanostructure arrays can be applied to 3D nanoarchitectures based on coaxial heterostructures, including InGaN/GaN/ZnO quantum structures, which are highly desirable for high-efficiency LED applications. In our opinion, position-controlled 3D nanoarchitectures can be expanded to many other material systems for the fabrication of diverse electronic devices, as well as optoelectronic devices, including LEDs, laser diodes, and solar cells.

We would like to conclude this review with discussion of the current challenges and future research directions toward the synthesis of position-controlled ZnO nanostructures for diverse applications. First, site-specific growth of ZnO nanostructures with better position and size control is required for integrated device fabrication. For example, nanostructures may be used for many sophisticated electronic nanodevices, including ultrahigh-density memory devices. For ultrahigh-density dynamic random access memory, position-controlled and vertically well-aligned ultrathin 1D nanostructures (diameter of $<50$ nm) and a very high aspect ratio above 100 are required. Moreover, the nanostructures should be grown on a metal layer such as W or TiN deposited on Si. Second, the technology used to fabricate artificially designed nanomaterial heterostructures should be developed further to create devices with greatly enhanced performance and functionality. Finally, the catalyst-free growth mechanism of ZnO nanostructures on many types of substrate should be investigated for better control of their selective growth. In particular, an improved understanding of the growth mechanism of the nanostructures on 2D substrates, including graphene layers, would enable uniform growth of nanostructures on large-size flexible substrates, which is crucial for the practical use of devices based on hybrid nanostructures.

## ACKNOWLEDGMENTS

This work was financially supported by Seoul National University (SNU)-Yonsei Research Cooperation Program through SNU in 2014. This work was also supported by the Future-based Technology Development Program (Nano Fields) through the National Research Foundation (NRF) of Korea funded by the Ministry of Education, Science and Technology (MEST) (Nos. 2014M3A7B4051589 and 2010-0029325).

## REFERENCES

An, S.J., Yi, G.-C., 2007. Near ultraviolet light emitting diode composed of n-GaN/ZnO coaxial nanorod heterostructures on a p-GaN layer. Appl. Phys. Lett. 91, 123109.

An, S.J., Park, W.I., Yi, G.-C., Cho, S., 2002. Laser–MBE growth of high-quality ZnO thin films on $Al_2O_3$(0001) and $SiO_2$/Si(100) using the third harmonics of a Nd:YAG laser. Appl. Phys. A 74, 509–512.

An, S.J., Park, W.I., Yi, G.-C., Kim, Y.-J., Kang, H.-B., Kim, M., 2004. Heteroepitaxal fabrication and structural characterizations of ultrafine GaN/ZnO coaxial nanorod heterostructures. Appl. Phys. Lett. 84, 3612–3614.

An, S.J., Hong, Y.J., Yi, G.-C., Kim, Y.J., Lee, D.K., 2006. Heteroepitaxial growth of high-quality GaN thin films on Si substrates coated with self-assembled sub-micrometer-sized silica balls. Adv. Mater. 18, 2833–2836.

An, S.J., Chae, J.H., Yi, G.-C., Park, G.H., 2008. Enhanced light output of GaN-based light-emitting diodes with ZnO nanorod arrays. Appl. Phys. Lett. 92, 121108.

Balandin, A.A., Ghosh, S., Bao, W., Calizo, I., Teweldebrhan, D., Miao, F., Lau, C.N., 2008. Superior thermal conductivity of single-layer graphene. Nano Lett. 8, 902–907.

Björk, M.T., Ohlsson, B.J., Sass, T., Persson, A.I., Thelander, C., Magnusson, M.H., Deppert, K., Wallenberg, L.R., Samuelson, L., 2002. One-dimensional heterostructures in semiconductor nanowhiskers. Appl. Phys. Lett. 80, 1058–1060.

Butler, S.Z., Hollen, S.M., Cao, L., Cui, Y., Gupta, J.A., Gutiérrez, H.R., Heinz, T.F., Hong, S.S., Huang, J., Ismach, A.F., Johnston-Halperin, E., Kuno, M., Plashnitsa, V.V., Robinson, R.D., Ruoff, R.S., Salahuddin, S., Shan, J., Shi, L., Spencer, M.G., Terrones, M., Windl, W., Goldberger, J.E., 2013. Progress, challenges, and opportunities in two-dimensional materials beyond graphene. ACS Nano 7, 2898–2926.

Castro, E.V., Novoselov, K.S., Morozov, S.V., Peres, N.M.R., dos Santos, J.M.B.L., Nilsson, J., Guinea, F., Geim, A.K., Neto, A.H.C., 2007. Biased bilayer graphene: semiconductor with a gap tunable by the electric field effect. Phys. Rev. Lett. 99, 216802.

Chen, C.-H., Chang, S.-J., Chang, S.-P., Li, M.-J., Chen, I.C., Hsueh, T.-J., Hsu, C.-L., 2009. Electroluminescence from n-ZnO nanowires/p-GaN heterostructure light-emitting diodes. Appl. Phys. Lett. 95, 223101.

Chen, X.Y., Ng, A.M.C., Fang, F., Djurišić, A.B., Chan, W.K., Tam, H.L., Cheah, K.W., Fong, P.W.K., Lui, H.F., Surya, C., 2010. The influence of the ZnO seed layer on the ZnO nanorod/GaN LEDs. J. Electrochem. Soc. 157, H308–H311.

Choi, D., Choi, M.-Y., Choi, W.M., Shin, H.-J., Park, H.-K., Seo, J.-S., Park, J., Yoon, S.-M., Chae, S.J., Lee, Y.H., Kim, S.-W., Choi, J.-Y., Lee, S.Y., Kim, J.M., 2010. Fully rollable transparent nanogenerators based on graphene electrodes. Adv. Mater. 22, 2187–2192.

Choi, W.M., Shin, K.-S., Lee, H.S., Choi, D., Kim, K., Shin, H.-J., Yoon, S.-M., Choi, J.-Y., Kim, S.-W., 2011. Selective growth of ZnO nanorods on $SiO_2$/Si substrates using a graphene buffer layer. Nano Res. 4, 440–447.

Chou, S.Y., Krauss, P.R., Renstrom, P.J., 1996. Imprint lithography with 25-nanometer resolution. Science 272, 85–87.

Chung, K., Lee, C.-H., Yi, G.-C., 2010. Transferable GaN layers grown on ZnO-coated graphene layers for optoelectronic devices. Science 330, 655–657.

Consonni, V., Sarigiannidou, E., Appert, E., Bocheux, A., Guillemin, S., Donatini, F., Robin, I.-C., Kioseoglou, J., Robaut, F., 2014. Selective area growth of well-ordered ZnO nanowire arrays with controllable polarity. ACS Nano 8, 4761–4770.

Dean, C.R., Young, A.F., Meric, I., Lee, C., Wang, L., Sorgenfrei, S., Watanabe, K., Taniguchi, T., Kim, P., Shepard, K.L., Hone, J., 2010. Boron nitride substrates for high-quality graphene electronics. Nat. Nanotechnol. 5, 722–726.

Djurišić, A.B., Leung, Y.H., 2006. Optical properties of ZnO nanostructures. Small 2, 944–961.

Djurišić, A.B., Tam, K.H., Hsu, Y.F., Zhang, S.L., Xie, M.H., Chan, W.K., 2007. GaN nanowires—influence of the starting material on nanowire growth. Thin Solid Films 516, 238–242.

Duan, X., Huang, Y., Cui, Y., Wang, J., Lieber, C.M., 2001. Indium phosphide nanowires as building blocks for nanoscale electronic and optoelectronic devices. Nature 409, 66–69.

Etzkorn, E.V., Clarke, D.R., 2001. Cracking of GaN films. J. Appl. Phys. 89, 1025–1034.

Fan, H.J., Fleischer, F., Lee, W., Nielsch, K., Scholz, R., Zacharias, M., Gösele, U., Dadgar, A., Krost, A., 2004. Patterned growth of aligned ZnO nanowire arrays on sapphire and GaN layers. Superlattice. Microst. 36, 95–105.

Fan, H.J., Lee, W., Scholz, R., Dadgar, A., Krost, A., Nielsch, K., Zacharias, M., 2005. Arrays of vertically aligned and hexagonally arranged ZnO nanowires: a new template-directed approach. Nanotechnology 16, 913.

Fan, H.J., Werner, P., Zacharias, M., 2006b. Semiconductor nanowires: from self-organization to patterned growth. Small 2, 700–717.

Fan, H.J., Fuhrmann, B., Scholz, R., Syrowatka, F., Dadgar, A., Krost, A., Zacharias, M., 2006a. Well-ordered ZnO nanowire arrays on GaN substrate fabricated via nanosphere lithography. J. Cryst. Growth 287, 34–38.

Fuhrmann, B., Leipner, H.S., Höche, H.-R., Schubert, L., Werner, P., Gösele, U., 2005. Ordered arrays of silicon nanowires produced by nanosphere lithography and molecular beam epitaxy. Nano Lett. 5, 2524–2527.

Giovannetti, G., Khomyakov, P.A., Brocks, G., Karpan, V.M., van den Brink, J., Kelly, P.J., 2008. Doping graphene with metal contacts. Phys. Rev. Lett. 101, 026803.

Gong, Y., Shi, G., Zhang, Z., Zhou, W., Jung, J., Gao, W., Ma, L., Yang, Y., Yang, S., You, G., Vajtai, R., Xu, Q., MacDonald, A.H., Yakobson, B.I., Lou, J., Liu, Z., Ajayan, P.M., 2014. Direct chemical conversion of graphene to boron- and nitrogen- and carbon-containing atomic layers. Nat. Commun. 5, 3193.

Greyson, E.C., Babayan, Y., Odom, T.W., 2004. Directed growth of ordered arrays of small-diameter ZnO nanowires. Adv. Mater. 16, 1348–1352.
Gudiksen, M.S., Lauhon, L.J., Wang, J., Smith, D.C., Lieber, C.M., 2002. Growth of nanowire superlattice structures for nanoscale photonics and electronics. Nature 415, 617–620.
Guo, R., Nishimura, J., Matsumoto, M., Higashihata, M., Nakamura, D., Okada, T., 2008. Electroluminescence from ZnO nanowire-based p-GaN/n-ZnO heterojunction light-emitting diodes. Appl. Phys. B 94, 33–38.
Guo, W., Zhang, M., Banerjee, A., Bhattacharya, P., 2010. Catalyst-free InGaN/GaN nanowire light emitting diodes grown on (001) silicon by molecular beam epitaxy. Nano Lett. 10, 3355–3359.
Hochbaum, A.I., Fan, R., He, R., Yang, P., 2005. Controlled growth of Si nanowire arrays for device integration. Nano Lett. 5, 457–460.
Hong, Y.J., An, S.J., Jung, H.S., Lee, C.H., Yi, G.-C., 2007. Position-controlled selective growth of ZnO nanorods on Si substrates using facet-controlled GaN micropatterns. Adv. Mater. 19, 4416–4419.
Hong, Y.J., Jeon, J.-M., Kim, M., Jeon, S.-R., Park, K.H., Yi, G.-C., 2009a. Structural and optical characteristics of GaN/ZnO coaxial nanotube heterostructure arrays for light-emitting device applications. New J. Phys. 11, 125021.
Hong, Y.J., Jung, H.S., Yoo, J., Kim, Y.-J., Lee, C.-H., Kim, M., Yi, G.-C., 2009b. Shape-controlled nanoarchitectures using nanowalls. Adv. Mater. 21, 222–226.
Hong, Y.J., Yoo, J., Doh, Y.-J., Kang, S.H., Kong, K.-j., Kim, M., Lee, D.R., Oh, K.H., Yi, G.-C., 2009c. Controlled epitaxial growth modes of ZnO nanostructures using different substrate crystal planes. J. Mater. Chem. 19, 941–947.
Hong, Y.J., Lee, C.-H., Yoon, A., Kim, M., Seong, H.-K., Chung, H.J., Sone, C., Park, Y.J., Yi, G.-C., 2011. Visible-color-tunable light-emitting diodes. Adv. Mater. 23, 3284–3288.
Hong, Y.J., Yang, J.W., Lee, W.H., Ruoff, R.S., Kim, K.S., Fukui, T., 2013. Van der Waals epitaxial double heterostructure: InAs/single-layer graphene/InAs. Adv. Mater. 25, 6847–6853.
Hsieh, Y.-P., Chen, H.-Y., Lin, M.-Z., Shiu, S.-C., Hofmann, M., Chern, M.-Y., Jia, X., Yang, Y.-J., Chang, H.-J., Huang, H.-M., Tseng, S.-C., Chen, L.-C., Chen, K.-H., Lin, C.-F., Liang, C.-T., Chen, Y.-F., 2009. Electroluminescence from ZnO/Si-nanotips light-emitting diodes. Nano Lett. 9, 1839–1843.
Hu, J., Odom, T.W., Lieber, C.M., 1999a. Chemistry and physics in one dimension: synthesis and properties of nanowires and nanotubes. Acc. Chem. Res. 32, 435–445.
Hu, J., Ouyang, M., Yang, P., Lieber, C.M., 1999b. Controlled growth and electrical properties of heterojunctions of carbon nanotubes and silicon nanowires. Nature 399, 48–51.
Huang, M.H., Mao, S., Feick, H., Yan, H., Wu, Y., Kind, H., Weber, E., Russo, R., Yang, P., 2001. Room-temperature ultraviolet nanowire nanolasers. Science 292, 1897–1899.
Huang, Y., Duan, X., Cui, Y., Lieber, C.M., 2002. Gallium nitride nanowire nanodevices. Nano Lett. 2, 101–104.
Jang, E.-S., Bae, J.Y., Yoo, J., Park, W.I., Kim, D.-W., Yi, G.-C., Yatsui, T., Ohtsu, M., 2006. Quantum confinement effect in $ZnO/Mg_{0.2}Zn_{0.8}O$ multishell nanorod heterostructures. Appl. Phys. Lett. 88, 023102.
Jensen, L.E., Björk, M.T., Jeppesen, S., Persson, A.I., Ohlsson, B.J., Samuelson, L., 2004. Role of surface diffusion in chemical beam epitaxy of InAs nanowires. Nano Lett. 4, 1961–1964.
Jeong, M.-C., Oh, B.-Y., Ham, M.-H., Myoung, J.-M., 2006. Electroluminescence from ZnO nanowires in n-ZnO film/ZnO nanowire array/p-GaN film heterojunction light-emitting diodes. Appl. Phys. Lett. 88, 202105.
Jeong, M.-C., Oh, B.-Y., Ham, M.-H., Lee, S.-W., Myoung, J.-M., 2007. ZnO-nanowire-inserted GaN/ZnO heterojunction light-emitting diodes. Small 3, 568–572.

Kang, M.S., Lee, C.-H., Park, J.B., Yoo, H., Yi, G.-C., 2012. Gallium nitride nanostructures for light-emitting diode applications. Nano Energy 1, 391–400.

Kempa, K., Kimball, B., Rybczynski, J., Huang, Z.P., Wu, P.F., Steeves, D., Sennett, M., Giersig, M., Rao, D.V.G.L.N., Carnahan, D.L., Wang, D.Z., Lao, J.Y., Li, W.Z., Ren, Z.F., 2002. Photonic crystals based on periodic arrays of aligned carbon nanotubes. Nano Lett. 3, 13–18.

Kikuchi, A., Kawai, M., Tada, M., Kishino, K., 2004. InGaN/GaN multiple quantum disk nanocolumn light-emitting diodes grown on (111) Si substrate. Jpn. J. Appl. Phys. 43, L1524.

Kim, H.M., Kang, T.W., Chung, K.S., 2003. Nanoscale ultraviolet-light-emitting diodes using wide-bandgap gallium nitride nanorods. Adv. Mater. 15, 567–569.

Kim, H.-M., Cho, Y.-H., Lee, H., Kim, S.I., Ryu, S.R., Kim, D.Y., Kang, T.W., Chung, K.S., 2004. High-brightness light emitting diodes using dislocation-free indium gallium nitride/gallium nitride multiquantum-well nanorod arrays. Nano Lett. 4, 1059–1062.

Kim, Y.-J., Lee, C.-H., Hong, Y.J., Yi, G.-C., Kim, S.S., Cheong, H., 2006. Controlled selective growth of ZnO nanorod and microrod arrays on Si substrates by a wet chemical method. Appl. Phys. Lett. 89, 163128.

Kim, M., Hong, Y.J., Yoo, J., Yi, G.-C., Park, G.-S., Kong, K.-j., Chang, H., 2008. Surface morphology and growth mechanism of catalyst-free ZnO and $Mg_xZn_{1-x}O$ nanorods. Phys. Status Solidi (RRL) 2, 197–199.

Kim, K.S., Zhao, Y., Jang, H., Lee, S.Y., Kim, J.M., Kim, K.S., Ahn, J.-H., Kim, P., Choi, J.-Y., Hong, B.H., 2009a. Large-scale pattern growth of graphene films for stretchable transparent electrodes. Nature 457, 706–710.

Kim, Y.-J., Lee, J.-H., Yi, G.-C., 2009b. Vertically aligned ZnO nanostructures grown on graphene layers. Appl. Phys. Lett. 95, 213101.

Kim, Y.-J., Jeon, J.-M., Choi, J.H., Park, S.S., Kim, S.I., Baik, C.W., Kim, M., Kim, J.M., Yi, G.-C., 2010. ZnO nanostructures with controlled morphologies on a glass substrate. Nanotechnology 21, 265603.

Kim, K.K., Hsu, A., Jia, X., Kim, S.M., Shi, Y., Dresselhaus, M., Palacios, T., Kong, J., 2012a. Synthesis and characterization of hexagonal boron nitride film as a dielectric layer for graphene devices. ACS Nano 6, 8583–8590.

Kim, S.B., Lee, W.W., Yi, J., Park, W.I., Kim, J.-S., Nichols, W.T., 2012b. Simple, large-scale patterning of hydrophobic ZnO nanorod arrays. ACS Appl. Mater. Interfaces 4, 3910–3915.

Kim, Y.-J., Yoo, H., Lee, C.-H., Park, J.B., Baek, H., Kim, M., Yi, G.-C., 2012c. Position- and morphology-controlled ZnO nanostructures grown on graphene layers. Adv. Mater. 24, 5565–5569.

Kim, S., Park, H., Nam, G., Yoon, H., Kim, J.S., Kim, J.S., Son, J.-S., Lee, S.-H., Leem, J.-Y., 2013. Temperature-dependent photoluminescence of boron-doped ZnO nanorods. Bull. Kor. Chem. Soc. 34, 3335–3339.

Kitamura, K., Yatsui, T., Ohtsu, M., Yi, G.-C., 2008. Fabrication of vertically aligned ultrafine ZnO nanorods using metal–organic vapor phase epitaxy with a two-temperature growth method. Nanotechnology 19, 175305.

Kobayashi, Y., Kumakura, K., Akasaka, T., Makimoto, T., 2012. Layered boron nitride as a release layer for mechanical transfer of GaN-based devices. Nature 484, 223–227.

Koma, A., 1999. Van der Waals epitaxy for highly lattice-mismatched systems. J. Cryst. Growth 201–202, 236–241.

Kosiorek, A., Kandulski, W., Chudzinski, P., Kempa, K., Giersig, M., 2004. Shadow nanosphere lithography: simulation and experiment. Nano Lett. 4, 1359–1363.

Lai, E., Kim, W., Yang, P., 2008. Vertical nanowire array-based light emitting diodes. Nano Res. 1, 123–128.

Lee, C.-H., Yi, G.-C., 2012. Catalyst-free metal-organic vapor-phase epitaxy of ZnO and GaN nanostructures for visible light-emitting devices. In: Yi, G.-C. (Ed.), Semiconductor Nanostructures for Optoelectronic Devices. Springer-Verlag, Berlin, Heidelberg, pp. 37–66.

Lee, W., Jeong, M.-C., Myoung, J.-M., 2004. Catalyst-free growth of ZnO nanowires by metal-organic chemical vapour deposition (MOCVD) and thermal evaporation. Acta Mater. 52, 3949–3957.

Lee, C.-H., Yoo, J., Hong, Y.J., Cho, J., Kim, Y.-J., Jeon, S.-R., Baek, J.H., Yi, G.-C., 2009a. GaN/$In_{1-x}Ga_xN$/GaN/ZnO nanoarchitecture light emitting diode microarrays. Appl. Phys. Lett. 94, 213101.

Lee, S.-D., Kim, Y.-S., Yi, M.-S., Choi, J.-Y., Kim, S.-W., 2009b. Morphology control and electroluminescence of ZnO nanorod/GaN heterojunctions prepared using aqueous solution. J. Phys. Chem. C 113, 8954–8958.

Lee, J.M., Choung, J.W., Yi, J., Lee, D.H., Samal, M., Yi, D.K., Lee, C.-H., Yi, G.-C., Paik, U., Rogers, J.A., Park, W.I., 2010. Vertical pillar-superlattice array and graphene hybrid light emitting diodes. Nano Lett. 10, 2783–2788.

Lee, C.-H., Hong, Y.J., Kim, Y.-J., Yoo, J., Baek, H., Jeon, S.-R., Lee, S.-J., Yi, G.-C., 2011a. GaN/ZnO nanotube heterostructure light-emitting diodes fabricated on Si. IEEE J. Sel. Top. Quantum Electron. 17, 966–970.

Lee, C.-H., Kim, Y.-J., Hong, Y.J., Jeon, S.-R., Bae, S., Hong, B.H., Yi, G.-C., 2011b. Flexible inorganic nanostructure light-emitting diodes fabricated on graphene films. Adv. Mater. 23, 4614–4619.

Lee, C.-H., Kim, Y.-J., Lee, J., Hong, Y.J., Jeon, J.-M., Kim, M., Hong, S., Yi, G.-C., 2011c. Scalable network electrical devices using ZnO nanowalls. Nanotechnology 22, 055205.

Lee, W.W., Yi, J., Kim, S.B., Kim, Y.-H., Park, H.-G., Park, W.I., 2011d. Morphology-controlled three-dimensional nanoarchitectures produced by exploiting vertical and in-plane crystallographic orientations in hydrothermal ZnO crystals. Cryst. Growth Des. 11, 4927–4932.

Levendorf, M.P., Kim, C.-J., Brown, L., Huang, P.Y., Havener, R.W., Muller, D.A., Park, J., 2012. Graphene and boron nitride lateral heterostructures for atomically thin circuitry. Nature 488, 627–632.

Lin, H.-W., Lu, Y.-J., Chen, H.-Y., Lee, H.-M., Gwo, S., 2010a. InGaN/GaN nanorod array white light-emitting diode. Appl. Phys. Lett. 97, 073101.

Lin, J., Penchev, M., Wang, G., Paul, R.K., Zhong, J., Jing, X., Ozkan, M., Ozkan, C.S., 2010b. Heterogeneous graphene nanostructures: ZnO nanostructures grown on large-area graphene layers. Small 6, 2448–2452.

Ling, B., Sun, X.W., Zhao, J.L., Tan, S.T., Dong, Z.L., Yang, Y., Yu, H.Y., Qi, K.C., 2009. Electroluminescence from a n-ZnO nanorod/p-$CuAlO_2$ heterojunction light-emitting diode. Phys. E 41, 635–639.

Lu, W., Lieber, C.M., 2007. Nanoelectronics from the bottom up. Nat. Mater. 6, 841–850.

Mårtensson, T., Borgström, M., Seifert, W., Ohlsson, B.J., Samuelson, L., 2003. Fabrication of individually seeded nanowire arrays by vapour–liquid–solid growth. Nanotechnology 14, 1255.

Mårtensson, T., Carlberg, P., Borgström, M., Montelius, L., Seifert, W., Samuelson, L., 2004. Nanowire arrays defined by nanoimprint lithography. Nano Lett. 4, 699–702.

Masuda, H., Fukuda, K., 1995. Ordered metal nanohole arrays made by a two-step replication of honeycomb structures of anodic alumina. Science 268, 1466–1468.

Meyer, B.K., Alves, H., Hofmann, D.M., Kriegseis, W., Forster, D., Bertram, F., Christen, J., Hoffmann, A., Straßburg, M., Dworzak, M., Haboeck, U., Rodina, A.V., 2004. Bound exciton and donor–acceptor pair recombinations in ZnO. Phys. Status Solidi B 241, 231–260.

Mieszawska, A.J., Jalilian, R., Sumanasekera, G.U., Zamborini, F.P., 2007. The synthesis and fabrication of one-dimensional nanoscale heterojunctions. Small 3, 722–756.
Mofor, A.C., Bakin, A., Chejarla, U., Schlenker, E., El-Shaer, A., Wagner, G., Boukos, N., Travlos, A., Waag, A., 2007. Fabrication of ZnO nanorod-based p–n heterojunction on SiC substrate. Superlattice. Microst. 42, 415–420.
Mohseni, P.K., Behnam, A., Wood, J.D., Zhao, X., Yu, K.J., Wang, N.C., Rockett, A., Rogers, J.A., Lyding, J.W., Pop, E., Li, X., 2014. Monolithic III-V nanowire solar cells on graphene via direct van der Waals epitaxy. Adv. Mater. 26, 3755–3760.
Morales, A.M., Lieber, C.M., 1998. A laser ablation method for the synthesis of crystalline semiconductor nanowires. Science 279, 208–211.
Ng, H.T., Li, J., Smith, M.K., Nguyen, P., Cassell, A., Han, J., Meyyappan, M., 2003. Growth of epitaxial nanowires at the junctions of nanowalls. Science 300, 1249.
Ng, H.T., Han, J., Yamada, T., Nguyen, P., Chen, Y.P., Meyyappan, M., 2004. Single crystal nanowire vertical surround-gate field-effect transistor. Nano Lett. 4, 1247–1252.
Nikoobakht, B., Michaels, C.A., Stranick, S.J., Vaudin, M.D., 2004. Horizontal growth and in situ assembly of oriented zinc oxide nanowires. Appl. Phys. Lett. 85, 3244–3246.
Noborisaka, J., Motohisa, J., Fukui, T., 2005. Catalyst-free growth of GaAs nanowires by selective-area metalorganic vapor-phase epitaxy. Appl. Phys. Lett. 86, 213102.
Novoselov, K.S., Geim, A.K., Morozov, S.V., Jiang, D., Zhang, Y., Dubonos, S.V., Grigorieva, I.V., Firsov, A.A., 2004. Electric field effect in atomically thin carbon films. Science 306, 666–669.
Novoselov, K.S., Geim, A.K., Morozov, S.V., Jiang, D., Katsnelson, M.I., Grigorieva, I.V., Dubonos, S.V., Firsov, A.A., 2005. Two-dimensional gas of massless Dirac fermions in graphene. Nature 438, 197–200.
Oh, H., Hong, Y.J., Kim, K.-S., Yoon, S., Baek, H., Kang, S.-H., Kwon, Y.-K., Kim, M., Yi, G.-C., 2014. Architectured van der Waals epitaxy of ZnO nanostructures on hexagonal BN. NPG Asia Mater. 6, e145.
Osada, M., Sasaki, T., 2012. Two-dimensional dielectric nanosheets: novel nanoelectronics from nanocrystal building blocks. Adv. Mater. 24, 210–228.
Park, W.I., Yi, G.-C., 2004. Electroluminescence in n-ZnO Nanorod arrays vertically grown on p-GaN. Adv. Mater. 16, 87–90.
Park, W.I., Yi, G.-C., Jang, H.M., 2001. Metalorganic vapor-phase epitaxial growth and photoluminescent properties of $Zn_{1-x}Mg_xO(0 \leq x \leq 0.49)$ thin films. Appl. Phys. Lett. 79, 2022–2024.
Park, W.I., Kim, D.H., Jung, S.W., Yi, G.-C., 2002a. Metalorganic vapor-phase epitaxial growth of vertically well-aligned ZnO nanorods. Appl. Phys. Lett. 80, 4232–4234.
Park, W.I., Yi, G.-C., Kim, M., Pennycook, S.J., 2002b. ZnO nanoneedles grown vertically on Si substrates by non-catalytic vapor-phase epitaxy. Adv. Mater. 14, 1841–1843.
Park, W.I., Jun, Y.H., Jung, S.W., Yi, G.-C., 2003a. Excitonic emissions observed in ZnO single crystal nanorods. Appl. Phys. Lett. 82, 964–966.
Park, W.I., Yi, G.-C., Kim, J.W., Park, S.M., 2003b. Schottky nanocontacts on ZnO nanorod arrays. Appl. Phys. Lett. 82, 4358–4360.
Park, W.I., Yi, G.-C., Kim, M.Y., Pennycook, S.J., 2003c. Quantum confinement observed in ZnO/ZnMgO nanorod heterostructures. Adv. Mater. 15, 526–529.
Park, W.I., An, S.J., Yang, J.L., Yi, G.-C., Hong, S., Joo, T., Kim, M., 2004. Photoluminescent properties of $ZnO/Zn_{0.8}Mg_{0.2}O$ nanorod single-quantum-well structures. J. Phys. Chem. B 108, 15457–15460.
Park, W.I., Yoo, J., Yi, G.-C., 2005. Catalyst-free metalorganic chemical-vapor deposition of ultrafine ZnO nanorods. J. Korean Phys. Soc. 46, L1067.

Park, C.J., Choi, D.-K., Yoo, J., Yi, G.-C., Lee, C.J., 2007. Enhanced field emission properties from well-aligned zinc oxide nanoneedles grown on the Au/Ti/n-Si substrate. Appl. Phys. Lett. 90, 083107.

Park, W.I., Lee, C.-H., Lee, J.M., Kim, N.-J., Yi, G.-C., 2011. Inorganic nanostructures grown on graphene layers. Nanoscale 3, 3522–3533.

Pauzauskie, P.J., Yang, P., 2006. Nanowire photonics. Mater. Today 9, 36–45.

Pearton, S.J., Lim, W.T., Wright, J.S., Tien, L.C., Kim, H.S., Norton, D.P., Wang, H.T., Kang, B.S., Ren, F., Jun, J., Lin, J., Osinsky, A., 2008. Zno and related materials for sensors and light-emitting diodes. J. Electron. Mater. 37, 1426–1432.

Peng, H., Dang, W., Cao, J., Chen, Y., Wu, D., Zheng, W., Li, H., Shen, Z.-X., Liu, Z., 2012. Topological insulator nanostructures for near-infrared transparent flexible electrodes. Nat. Chem. 4, 281–286.

Ponce, F.A., Bour, D.P., 1997. Nitride-based semiconductors for blue and green light-emitting devices. Nature 386, 351–359.

Pyun, Y.B., Yi, J., Lee, D.H., Son, K.S., Liu, G., Yi, D.K., Paik, U., Park, W.I., 2010. Synthesis of ZnO nanotubes and nanotube-nanorod hybrid hexagonal networks using a hexagonally close-packed colloidal monolayer template. J. Mater. Chem. 20, 5136–5140.

Qian, F., Li, Y., Gradečak, S., Wang, D., Barrelet, C.J., Lieber, C.M., 2004. Gallium nitride-based nanowire radial heterostructures for nanophotonics. Nano Lett. 4, 1975–1979.

Qian, F., Gradečak, S., Li, Y., Wen, C.-Y., Lieber, C.M., 2005. Core/multishell nanowire heterostructures as multicolor, high-efficiency light-emitting diodes. Nano Lett. 5, 2287–2291.

Smit, G.D.J., Rogge, S., Klapwijk, T.M., 2002. Scaling of nano-Schottky-diodes. Appl. Phys. Lett. 81, 3852–3854.

Solanki, R., Huo, J., Freeouf, J.L., Miner, B., 2002. Atomic layer deposition of ZnSe/CdSe superlattice nanowires. Appl. Phys. Lett. 81, 3864–3866.

Song, L., Ci, L., Lu, H., Sorokin, P.B., Jin, C., Ni, J., Kvashnin, A.G., Kvashnin, D.G., Lou, J., Yakobson, B.I., Ajayan, P.M., 2010. Large scale growth and characterization of atomic hexagonal boron nitride layers. Nano Lett. 10, 3209–3215.

Strite, S., Morkoç, H., 1992. GaN, AlN, and InN: a review. J. Vac. Sci. Technol. B 10, 1237–1266.

Sun, H.D., Segawa, Y., Kawasaki, M., Ohtomo, A., Tamura, K., Koinuma, H., 2002. Phonon replicas in ZnO/ZnMgO multiquantum wells. J. Appl. Phys. 91, 6457–6460.

Sun, Y., Cho, Y.-H., Kim, H.-M., Kang, T.W., 2005. High efficiency and brightness of blue light emission from dislocation-free InGaN/GaN quantum well nanorod arrays. Appl. Phys. Lett. 87, 093115.

Sun, H., Zhang, Q.-F., Wu, J.-L., 2006. Electroluminescence from ZnO nanorods with an n-ZnO/p-Si heterojunction structure. Nanotechnology 17, 2271.

Sun, M., Zhang, Q.-F., Sun, H., Zhang, J., Wu, J.-L., 2009a. Enhanced ultraviolet electroluminescence from p-Si/n-ZnO nanorod array heterojunction. J. Vac. Sci. Technol. B 27, 618–621.

Sun, X.W., Ling, B., Zhao, J.L., Tan, S.T., Yang, Y., Shen, Y.Q., Dong, Z.L., Li, X.C., 2009b. Ultraviolet emission from a ZnO rod homojunction light-emitting diode. Appl. Phys. Lett. 95, 133124.

Tak, Y., Yong, K., 2005. Controlled growth of well-aligned ZnO nanorod array using a novel solution method. J. Phys. Chem. B 109, 19263–19269.

Tan, S.T., Sun, X.W., Yu, Z.G., Wu, P., Lo, G.Q., Kwong, D.L., 2007. p-type conduction in unintentional carbon-doped ZnO thin films. Appl. Phys. Lett. 91, 072101.

Thelander, C., Agarwal, P., Brongersma, S., Eymery, J., Feiner, L.F., Forchel, A., Scheffler, M., Riess, W., Ohlsson, B.J., Gösele, U., Samuelson, L., 2006. Nanowire-based one-dimensional electronics. Mater. Today 9, 28–35.

Utama, M.I.B., Peng, Z., Chen, R., Peng, B., Xu, X., Dong, Y., Wong, L.M., Wang, S., Sun, H., Xiong, Q., 2010. Vertically aligned cadmium chalcogenide nanowire arrays on muscovite mica: a demonstration of epitaxial growth strategy. Nano Lett. 11, 3051–3057.

Utama, M.I.B., Belarre, F.J., Magen, C., Peng, B., Arbiol, J., Xiong, Q., 2012. Incommensurate van der Waals epitaxy of nanowire arrays: a case study with ZnO on muscovite mica substrates. Nano Lett. 12, 2146–2152.

Vayssieres, L., 2003. Growth of arrayed nanorods and nanowires of ZnO from aqueous solutions. Adv. Mater. 15, 464–466.

Viswanath, A.K., Lee, J.I., Yu, S., Kim, D., Choi, Y., Hong, C.-H., 1998. Photoluminescence studies of excitonic transitions in GaN epitaxial layers. J. Appl. Phys. 84, 3848–3859.

Wagner, R.S., Ellis, W.C., 1964. Vapor-liquid-solid mechanism of single crystal growth. Appl. Phys. Lett. 4, 89–90.

Wang, M.W., McCaldin, J.O., Swenberg, J.F., McGill, T.C., Hauenstein, R.J., 1995. Schottky-based band lineups for refractory semiconductors. Appl. Phys. Lett. 66, 1974–1976.

Wang, X., Yang, S., Wang, J., Li, M., Jiang, X., Du, G., Liu, X., Chang, R.P.H., 2001. Nitrogen doped ZnO film grown by the plasma-assisted metal-organic chemical vapor deposition. J. Cryst. Growth 226, 123–129.

Wang, X., Sun, X., Fairchild, M., Hersee, S.D., 2006. Fabrication of GaN nanowire arrays by confined epitaxy. Appl. Phys. Lett. 89, 233115.

Wang, J.-Y., Lee, C.-Y., Chen, Y.-T., Chen, C.-T., Chen, Y.-L., Lin, C.-F., Chen, Y.-F., 2009. Double side electroluminescence from p-NiO/n-ZnO nanowire heterojunctions. Appl. Phys. Lett. 95, 131117.

Wang, H., Taychatanapat, T., Hsu, A., Watanabe, K., Taniguchi, T., Jarillo-Herrero, P., Palacios, T., 2011. BN/graphene/BN transistors for RF applications. IEEE Electron Device Lett. 32, 1209–1211.

Wang, X., Hartmann, J., Mandl, M., Mohajerani, M.S., Wehmann, H.-H., Strassburg, M., Waag, A., 2014. Growth kinetics and mass transport mechanisms of GaN columns by selective area metal organic vapor phase epitaxy. J. Appl. Phys. 115, 163104.

Watanabe, K., Taniguchi, T., Kanda, H., 2004. Direct-bandgap properties and evidence for ultraviolet lasing of hexagonal boron nitride single crystal. Nat. Mater. 3, 404–409.

Weintraub, B., Deng, Y., Wang, Z.L., 2007. Position-controlled seedless growth of ZnO nanorod arrays on a polymer substrate via wet chemical synthesis. J. Phys. Chem. C 111, 10162–10165.

Wen, X., Wu, W., Ding, Y., Wang, Z.L., 2012. Seedless synthesis of patterned ZnO nanowire arrays on metal thin films (Au, Ag, Cu, Sn) and their application for flexible electromechanical sensing. J. Mater. Chem. 22, 9469–9476.

Willander, M., Nur, O., Zhao, Q.X., Yang, L.L., Lorenz, M., Cao, B.Q., Pérez, J.Z., Czekalla, C., Zimmermann, G., Grundmann, M., Bakin, A., Behrends, A., Al-Suleiman, M., El-Shaer, A., Mofor, A.C., Postels, B., Waag, A., Boukos, N., Travlos, A., Kwack, H.S., Guinard, J., Dang, D.L.S., 2009. Zinc oxide nanorod based photonic devices: recent progress in growth, light-emitting diodes and lasers. Nanotechnology 20, 125020.

Wu, Y., Yang, P., 2000. Germanium nanowire growth via simple vapor transport. Chem. Mater. 12, 605–607.

Wu, Z.H., Mei, X.Y., Kim, D., Blumin, M., Ruda, H.E., 2002. Growth of Au-catalyzed ordered GaAs nanowire arrays by molecular-beam epitaxy. Appl. Phys. Lett. 81, 5177–5179.

Xia, Y., Yang, P., Sun, Y., Wu, Y., Mayers, B., Gates, B., Yin, Y., Kim, F., Yan, H., 2003. One-dimensional nanostructures: synthesis, characterization, and applications. Adv. Mater. 15, 353–389.

Xu, S., Wei, Y., Kirkham, M., Liu, J., Mai, W., Davidovic, D., Snyder, R.L., Wang, Z.L., 2008. Patterned growth of vertically aligned ZnO nanowire arrays on inorganic substrates at low temperature without catalyst. J. Am. Chem. Soc. 130, 14958–14959.

Yagi, H., Muranushi, K., Nunoya, N., Sano, T., Tamura, S., Arai, S., 2002. Low-damage etched/regrown interface of strain-compensated GaInAsP/InP quantum-wire laser fabricated by CH4/H2 dry etching and regrowth. Appl. Phys. Lett. 81, 966–968.

Yan, J.-T., Chen, C.-H., Yen, S.-F., Lee, C.-T., 2010. Ultraviolet ZnO nanorod/p-GaN-heterostructured light-emitting diodes. IEEE Photon. Technol. Lett. 22, 146–148.

Yang, Y., Sun, X.W., Tay, B.K., You, G.F., Tan, S.T., Teo, K.L., 2008. A p-n homojunction ZnO nanorod light-emitting diode formed by As ion implantation. Appl. Phys. Lett. 93, 253107.

Yang, T., Cheng, K., Cheng, G., Hu, B., Wang, S., Du, Z., 2014. Position-controlled hydrothermal growth of periodic individual ZnO nanorod arrays on indium tin oxide substrate. J. Phys. Chem. C 118, 20613–20619.

Yatsui, T., Ohtsu, M., Yoo, J., An, S.J., Yi, G.-C., 2005. Near-field measurement of spectral anisotropy and optical absorption of isolated ZnO nanorod single-quantum-well structures. Appl. Phys. Lett. 87, 033101.

Yatsui, T., Sangu, S., Kawazoe, T., Ohtsu, M., An, S.J., Yoo, J., Yi, G.-C., 2007. Nanophotonic switch using ZnO nanorod double-quantum-well structures. Appl. Phys. Lett. 90, 223110.

Yoo, J., Hong, Y.-J., An, S.J., Yi, G.-C., Chon, B., Joo, T., Kim, J.-W., Lee, J.-S., 2006. Photoluminescent characteristics of Ni-catalyzed GaN nanowires. Appl. Phys. Lett. 89, 043124.

Yoo, J., Hong, Y.J., Jung, H.S., Kim, Y.-J., Lee, C.-H., Cho, J., Doh, Y.-J., Dang, L.S., Park, K.H., Yi, G.-C., 2009. Fabrication and optical characteristics of position-controlled ZnO nanotubes and ZnO/$Zn_{0.8}$ $Mg_{0.2}O$ coaxial nanotube quantum structure arrays. Adv. Funct. Mater. 19, 1601–1608.

Yoo, H., Chung, K., Choi, Y.S., Kang, C.S., Oh, K.H., Kim, M., Yi, G.-C., 2012. Microstructures of GaN thin films grown on graphene layers. Adv. Mater. 24, 515–518.

Zhang, B.P., Binh, N.T., Wakatsuki, K., Segawa, Y., Yamada, Y., Usami, N., Kawasaki, M., Koinuma, H., 2004. Formation of highly aligned ZnO tubes on sapphire (0001) substrates. Appl. Phys. Lett. 84, 4098–4100.

Zhang, Y., Tan, Y.-W., Stormer, H.L., Kim, P., 2005. Experimental observation of the quantum Hall effect and Berry's phase in graphene. Nature 438, 201–204.

Zhang, J.-Y., Li, P.-J., Sun, H., Shen, X., Deng, T.-S., Zhu, K.-T., Zhang, Q.-F., Wu, J.-L., 2008. Ultraviolet electroluminescence from controlled arsenic-doped ZnO nanowire homojunctions. Appl. Phys. Lett. 93, 021116.

Zhang, J.-Y., Zhang, Q.-F., Deng, T.-S., Wu, J.-L., 2009a. Electrically driven ultraviolet lasing behavior from phosphorus-doped p-ZnO nanonail array/n-Si heterojunction. Appl. Phys. Lett. 95, 211107.

Zhang, X.-M., Lu, M.-Y., Zhang, Y., Chen, L.-J., Wang, Z.L., 2009b. Fabrication of a high-brightness blue-light-emitting diode using a ZnO-nanowire array grown on p-GaN thin film. Adv. Mater. 21, 2767–2770.

Zhang, Y., Tang, T.-T., Girit, C., Hao, Z., Martin, M.C., Zettl, A., Crommie, M.F., Shen, Y.R., Wang, F., 2009c. Direct observation of a widely tunable bandgap in bilayer graphene. Nature 459, 820–823.

Zheng, Z., Lim, Z.S., Peng, Y., You, L., Chen, L., Wang, J., 2013. General route to ZnO nanorod arrays on conducting substrates via galvanic-cell-based approach. Sci. Rep. 3, 2434.

Zhong, Z., Qian, F., Wang, D., Lieber, C.M., 2003. Synthesis of p-type gallium nitride nanowires for electronic and photonic nanodevices. Nano Lett. 3, 343–346.

Zhou, H.L., Shao, P.G., Chua, S.J., van Kan, J.A., Bettiol, A.A., Osipowicz, T., Ooi, K.F., Goh, G.K.L., Watt, F., 2008. Selective growth of ZnO nanorod arrays on a GaN/sapphire substrate using a proton beam written mask. Cryst. Growth Des. 8, 4445–4448.

CHAPTER FIVE

# Hybrid III–V/Silicon Nanowires

**Moïra Hocevar*, Sonia Conesa-Boj†, Erik Bakkers†,‡,1**
*Institut Néel CNRS/UJF, Grenoble, France
†Eindhoven University of Technology, Eindhoven, The Netherlands
‡Kavli Institute of Nanoscience, Delft University of Technology, Delft, The Netherlands
1Corresponding author: e-mail address: ebakkers@tue.nl

## Contents



## 1. INTRODUCTION

The success of the semiconductor industry originates from the ability to precisely tune the electronic properties of semiconductor materials. Different approaches are used to enhance the functionality of semiconductors, such as impurity doping, alloying, heterostructuring, and straining. All these methods are based on the addition of chemical elements to a pure semiconductor. An important example is the invention of the GaAs/AlGaAs heterostructure, which has enabled the development of laser diode and high-electron-mobility transistor (HEMT). In this chapter, we will discuss the combination of two different classes of semiconductors, group IV, like Si and Ge, and group III–V, like GaAs and InP. We will focus on a relatively new materials system, i.e., nanowires (NWs), in which strain can be effectively relieved at the surface due to the small dimensions, therefore relaxing the requirements on lattice matching. This system thus offers enhanced flexibility over the conventional layered structures in combining different semiconductor materials. These new combinations may boost the performance of already widely explored device concepts, such as transistors and solar cells,

*Semiconductors and Semimetals*, Volume 93
ISSN 0080-8784
http://dx.doi.org/10.1016/bs.semsem.2015.07.006

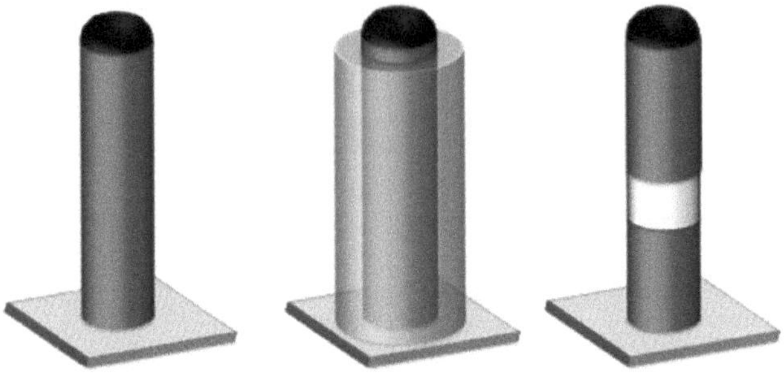

**Figure 1** Schematic of the different Si/III–V nanowire geometries: III–V nanowire on Si, III–V/Si core–shell nanowire, and III–V/Si axial nanowire.

but may also open new applications, such as in quantum information technology. After a discussion on the different challenges related to the combination of Si and III–V semiconductors, we will discuss the growth of III–V nanowires on group IV substrates (Borg et al., 2014; Borgström et al., 2007; Cassell et al., 1999; Dayeh et al., 2007; Hiraki et al., 1972; Huang et al., 2003; Mohan et al., 2005; Motohisa et al., 2004; Mourik et al., 2012), and then focus on the growth of heterostructures within nanowires in the radial (Tomioka et al., 2008, 2012) and axial (Gangloff et al., 2004; Givargizov, 1973; Hiruma et al., 1991; Roest et al., 2006) directions (Fig. 1).

## 2. CHALLENGES IN EPITAXIAL SI/III–V INTERFACES

Silicon is the material platform of microelectronics. Single-crystalline Si wafers, up to 450 mm in diameter, offer a low-cost substrate for advanced electronic and micromechanical components. Silicon has a high thermal conductivity, is mechanically robust, and has a stable oxide. However, silicon has an indirect bandgap (at 1.1 eV) and therefore, it does not efficiently emit light. This significantly limits the applications of silicon in photonic technology. Its isotope $^{28}$Si has zero nuclear spin, which is particularly attractive for quantum information technology, where a single electron spin must be isolated from other spins to be used as a quantum bit.

In contrast, most III–V semiconductors have a direct bandgap and therefore are perfectly suited for optoelectronic devices such as light-emitting diodes (LEDs) and solid-state lasers. Importantly, solar cells of the highest efficiency are made of III–V semiconductors because of the high light absorption. Some III–V's, especially those based on the heavier elements, such as indium antimonide, exhibit extremely high electron mobility, and very strong spin–orbit interaction, enabling fundamental breakthroughs in quantum science and technology such as NW spin–orbit quantum bits (qubits) and Majorana fermions (Mourik et al., 2012). Importantly, the electronic properties, such as the band gap energy and the carrier mobility, of III–V semiconductors can be tuned by varying the chemical composition (e.g., by alloying to form ternary or even quaternary compounds such as $In_xGa_{1-x}As_ySb_{1-y}$). However, III–V semiconductors have poor mechanical integrity (they are fragile) and some of the constituents are in short supply, like indium, which makes them expensive. For a number of important applications that will be discussed here, it is valuable to combine the best properties of Si and III–V semiconductors, but the success depends on the quality of heterojunctions between the two semiconductor classes.

"Epitaxy" is derived from the Greek word meaning "ordered upon" and is the crystalline deposition of material on a substrate with identical lattice structure and orientation. For heteroepitaxial growth, materials with different lattice parameters are combined. If the lattice mismatch between the deposited film and the substrate is large, typically a few percent, misfit dislocations can be incorporated near the interface. Alternatively, adding layers may form three-dimensional nuclei to release the strain in order to minimize the energy in the system, at the cost of creating more surface. The heterostructure growth method that Kroemer and Alferov used was to stack two-dimensional layers of atoms using materials of similar lattice constant and surface energies (to favor layer-by-layer growth rather than island growth) by molecular beam epitaxy (MBE). This growth technique made it possible to produce defect-free bulk $GaAs/Al_xGa_{1-x}As$ heterojunctions. In contrast, semiconductors with higher lattice mismatch suffer from the creation of threading dislocations, which release the strain accumulated during the formation of a new material layer. Dislocations and crystalline defects are the main challenge in crystal growth. They trap carriers and behave as recombination centers for photons, which makes them harmful for device performance and limits practical semiconductor combinations in bulk two-dimensional heterostructures.

In addition to strain, defect formation at the heterointerface can be caused by differences in polarity of the two semiconductors, as is the case for instance for silicon and gallium phosphide (a group IV/III–V interface). The essence of the problem is that GaP can nucleate on Si by either forming Si–P or Si–Ga bonds. Any nonuniformity during growth can lead to the formation of nuclei having opposite polarities, and when these nuclei merge a so-called antiphase boundary (APB) is formed. An APB is a planar defect, in which the regular atomic ordering is interrupted; at the boundary, the atomic ordering is for instance "Ga–P–P–Ga" instead of the regular

"Ga–P–Ga–P". Importantly, when the surface is uniformly covered with one of the precursor elements, APBs can still be formed at monoatomic steps on the substrate surface. Along with twins and dislocations, APBs are a major limitation that specifically reduces the potential of structures that combine Si with III–V semiconductors in the bulk.

Nanowires are one-dimensional crystals with a large aspect ratio. They have a diameter of several nanometers and their length can vary from a few micrometers to several millimeters. From a materials engineering point of view, the very small cross section of NWs is highly advantageous because strain can be elastically released at the interface between semiconductors with large lattice mismatch. Rather than producing crystalline defects, the atoms adjust their spacing to minimize strain because they are free to expand towards the sidewalls of a NW. Moreover, as it will be described below, nanowires grow via a layer-by-layer mechanism which involves a single nucleus per layer. There are therefore no APBs in nanowires. Thanks to their ability to accommodate strain and the absence of APB, nanowires are therefore perfect candidates for III–V/Si heterostructures.

## 3. III–V NANOWIRES ON SI SUBSTRATES

III–V Nanowires can be grown on Si substrates by using different growth mechanisms, such as selective area growth (SAG; Mohan et al., 2005; Motohisa et al., 2004) or by the vapor–liquid–solid (VLS) method. Nanowires grown by the VLS mechanism are nucleated from a nanometer-sized metal seed or catalyst particle that collects the precursor material from the vapor phase, establishing a local supersaturation. There are a few examples showing that the metal particle acts as a catalyst for the decomposition of the precursor molecules (Borgström et al., 2007; Dayeh et al., 2007). A vapor pressure of the precursor components can be created by means of MBE, metalorganic vapor phase epitaxy (MOVPE) and related techniques, pulsed laser ablation, or by simple evaporation of the precursor material.

When the supersaturation has been reached, crystalline material precipitates underneath the metal particle. In order to enable epitaxial growth of a nanowire, it is of crucial importance to have a clean and crystalline substrate surface. Methods to clean Si substrates are well established and an etching step with hydrofluoric acid is mostly included. The catalytic metal particles are deposited directly after substrate cleaning. So far, Au has mainly been used as the active particle, but also alternative metals, oxides, and silicides, being compatible to standard silicon processing, have been used. Importantly, any exposure to air affects the epitaxy, since Au catalyzes the oxidation of Si and tens of nm thick $SiO_2$ layer forms on top of the Au particle within days already at room temperature (Hiraki et al., 1972).

Another method to grow one-dimensional structures is by using SAG. Holes are defined in an amorphous dielectric layer, which cover on epitaxial substrate. Taking into account that growth only takes place from the opened areas pillar structures will be formed. Since no catalyst particle is used with this method, no impurity atoms from the catalyst will be incorporated in the semiconductor material. Another advantage is that the III–V/Si wire/substrate heterointerface will not be affected by alloying of the catalyst particle with the substrate. Such strategy was extended to InAs nanowires selectively grown in vertical $SiO_2$ nanotube templates fabricated on Si substrates. After removal of the template, the InAs nanowires show an epitaxial relationship with Si and are single-crystalline. Misfit dislocations are however present at the InAs/Si interface (Borg et al., 2014).

Scanning electron microscopy (SEM) can give a first indication of nanowire epitaxy. Epitaxial growth results in alignment of the nanowires in specific directions determined by the crystal symmetry of the substrate. However, orientation of one-dimensional structures can also be obtained during growth by other mechanisms, such as interaction with external fields (Cassell et al., 1999) or with a gas flow (Gangloff et al., 2004; Huang et al., 2003). In general, the III–V wires tend to grow in the [111]B direction (Givargizov, 1973; Hiruma et al., 1991), and vertical growth can be induced on [111]B-oriented III–V substrates. We should note, however, that the elemental semiconductors, such as silicon, are not polar, and all $\{111\}$ facets are chemically equivalent. This means that the III–V wires can grow in the four $\langle 111 \rangle$ directions on a [111]-oriented Si substrate; one orientation perpendicular to the surface and three orientations having a 19° angle with the surface, and in-plane components at 120° from each other. This has, for instance, been observed for the growth of InP on (111)-oriented silicon substrates (Roest et al., 2006) having a mismatch of 8.1%. There are three clearly noticeable orientations with in-plane components parallel to the sides of an equilateral triangle. Some wires are oriented perpendicular to the surface, and in this top view they appear as small bright spots.

It is possible to terminate a group IV substrate by either a group III or group V element using surface reconstructions as has been shown by Fukui (Tomioka et al., 2008) using SAG. In this way a quasi-polar surface can be formed. This approach has been successful in obtaining high yield of vertically oriented III–V nanowires on (111) Si substrates. The SEM image in Fig. 2A shows the high yield obtained with this method.

The crystallographic relation between the substrate and an ensemble of nanowires can be substantiated by X-ray diffraction pole figure measurements (Bakkers et al., 2004, 2007; Roest et al., 2006). To record a pole figure, the detector is set at a $2\theta$ angle corresponding to one of the lattice spacings, and the substrate is rotated continuously around $\varphi$, and stepped around $\psi$. Pole figures are measured for the (111) reflections of the substrate and the wires. In the reference pattern of the silicon substrate, plotted in Fig. 3C, (111) spacings were found at four orientations typical for a (111)-oriented single crystal (these orientations are reflected by the wire orientations as observed with SEM). One set of reflections is in the center of the pole figure, corresponding to the substrate normal, and three sets with $\psi = 19°$ and the in-plane angle of 120° with respect to each other. The pole figure for InP wires grown by MOVPE on Si(111) is shown in Fig. 3C. The signals associated with the majority of the InP wires, labeled A (i.e., about

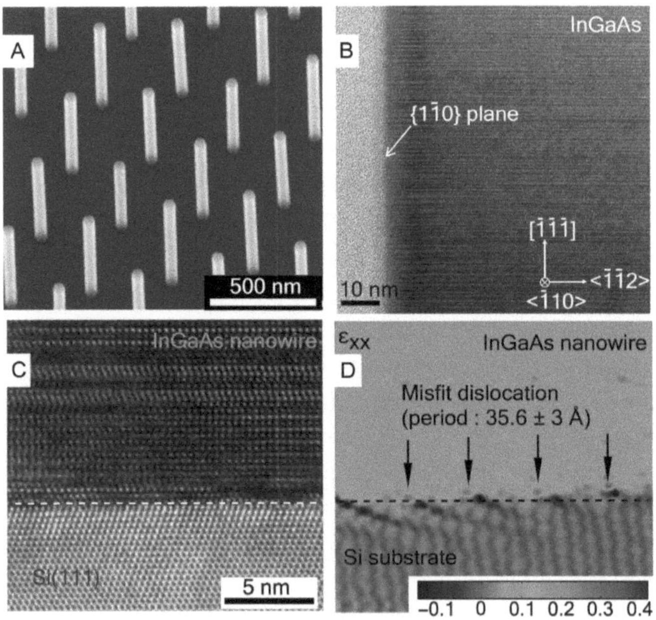


**Figure 2** (A) SEM image of InGaAs nanowires grown on a Si(111) substrate by using selective area growth (SAG). (B) TEM image of a InGaAs showing the crystal structure and the planar stacking faults. (C) HRTEM image of the InGaAs wire/Si substrate interface. (D) Strain map obtained from the image in (C) showing the periodic misfit dislocations (Tomioka et al., 2012).

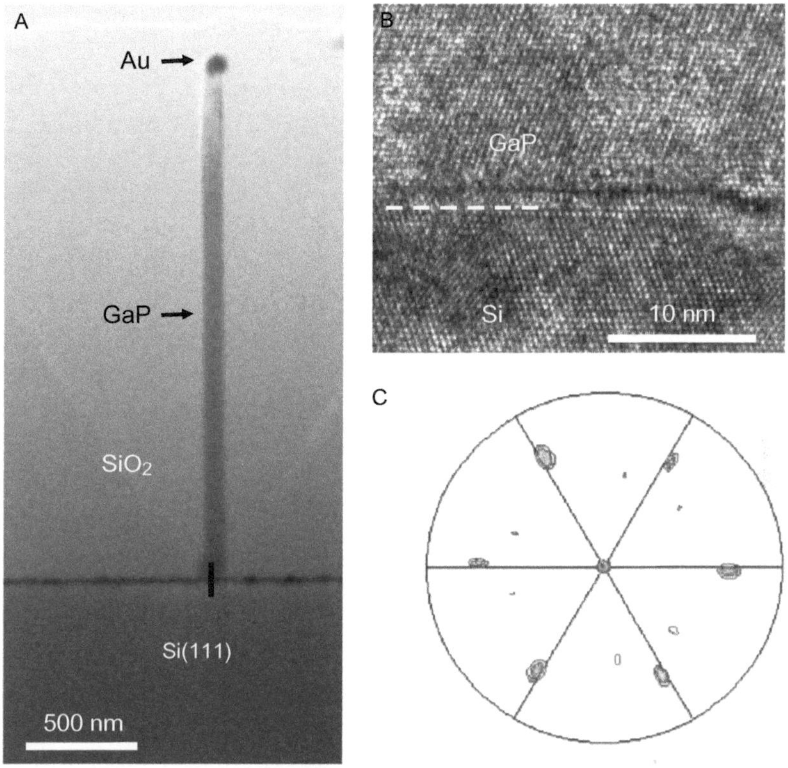


**Figure 3** (A) Cross-sectional TEM image of a GaP nanowire grown on a Si(111) substrate by using the VLS mechanism. (B) HRTEM image of the GaP/Si interface showing the roughness, which is induced by alloying of the catalyst particle with the Si substrate prior to growth. (C) Pole figure of InP wires grown on Si(111). The threefold symmetry demonstrates the epitaxial growth. The three other intense peaks show that there are twin planes present in the nanowires (Bakkers et al., 2007).

two-thirds of the total signal), match the pole pattern of the substrate, providing an unambiguous signature of heteroepitaxial growth. With such pole figures, the epitaxial relation between a range of III–V nanowires, such as GaAs, InP, and InAs, with the Si(111) substrate has been confirmed (Roest et al., 2006). The difference in lattice spacings for the GaP/Si system was too small to be resolved in the pole figures.

The wire/substrate interface of individual wires can be studied in great detail by transmission electron microscopy (TEM). For this study, vertical cross sections were sliced with a focused ion beam and thinned by argon ion milling. To provide mechanical support during this process, the wires were embedded in a microns-thick silicon oxide layer, deposited by spin-on process or by plasma-enhanced chemical vapor deposition (PECVD). Figure 3A shows an overview of a cross-sectional TEM image of a GaP wire grown on a Si(111) substrate. In Fig. 3B, the wire/substrate interface was imaged at high resolution. The lattice planes continue from the substrate into the nanowire showing the epitaxial relation. It can also been seen that the substrate wire interface is not flat. This is probably due to alloying of the Au catalyst particle with Si from the substrate, roughening the surface.

Such alloying can be avoided by using SAG. In Fig. 2C, the flat heterointerface between InGaAs/Si has been imaged at high resolution (Tomioka et al., 2012). The heterointerface was apparently free of threading dislocations and APBs, but had periodical misfit dislocations (see Fig. 2D), indicating that the lattice mismatch has been plastically relaxed. These dislocation networks were formed only at the heterointerface. However, from TEM imaging, it is also clear that these wires grown by SAG contain many planar stacking faults (Fig. 2B). Recently, defect-free wurtzite InP wires have been grown by SAG on corresponding substrates at very high temperatures, which exhibit very high optical quality.

III–V Nanowires grown on Si have been explored for many different device applications. Recently, high-quality field-effect transistors (FET) have been demonstrated exploiting the high electron mobility of the III–V material and the vertical wire geometry to form a gate-around structure as shown in Fig. 4A (Tomioka et al., 2012). The gate-around structure minimizes short channel effects such as drain-induced barrier lowering as is clear

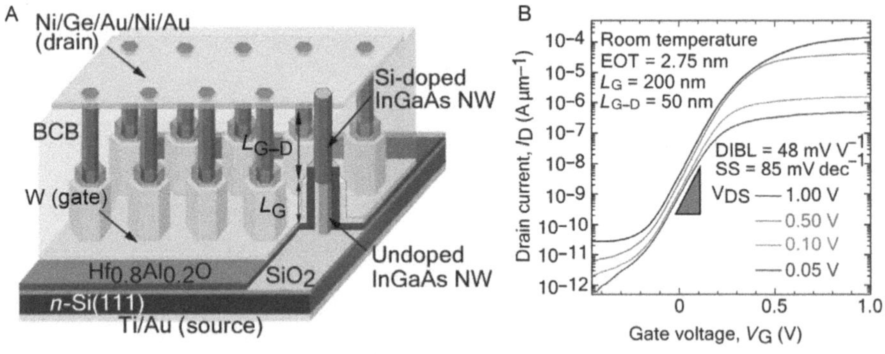


**Figure 4** (A) Schematic image of a field-effect transistor device based on an array of vertical InGaAs nanowire grown on Si(111). Note that the gate is all around the nanowire channels. (B) Transconductance characteristics of the device in (A), showing good performance (Tomioka et al., 2012).

from Fig. 4B. The III–V/Si heterointerface also allows improving alternative FET principles, such as the tunnel FET (Borg et al., 2014; Ionescu and Riel, 2011). Less power is needed to switch a tunnel FET compared to a "normal" FET. By using heterostructures, there is more freedom in tuning the band alignments between the different semiconductors and with this the performance can be optimized. Epitaxial integration of III–V with silicon also introduces direct band gap semiconductors into electronics. Various optoelectronic applications have been studied including LEDs, lasers, and photovoltaic cells. Considering the optical quality of state-of-the-art nanowires, it can be expected that this approach will result in devices, which can be competitive with standard III–V technologies, but then epitaxially integrated in silicon.

## 4. III–V/IV RADIAL CORE–SHELL NANOWIRES

A number of significant steps in the controlled synthesis of core/shell nanowire heterostructures have been recently achieved by various groups. In this review, we illustrate some of these results in the cases in which core and shell of the nanowire are grown using materials of the groups III–V and IV, as we did in the previous section. To begin with, Algra et al. (2011) were able to synthetize the first III–V/IV core–shell system consisting of the combination of GaP and Si semiconductor materials. This group reported a method for transferring the desired crystalline structure from a GaP nanowire to the Si shell. Interestingly, such approach allows tailoring the crystalline structure of the shell material by suitably growing the core material in specific crystalline phases.

In their work, the growth of the GaP nanowires was performed using an MOVPE reactor using trimethylgallium and phosphine ($PH_3$) as precursors by adding diethylzinc to the gas phase during growth. In particular, a designed twinning superlattice in zinc blende crystal structure present in the GaP nanowires was epitaxially transferred to the Si shell at 550 °C using $SiH_6$ as a precursor with a partial pressure of $3.4 \times 10^{-3}$ mbar. This important property allows designing new materials, like hexagonal silicon, which are more difficult to achieve in other systems, opening the path to a number of important applications.

Conesa-Boj et al. synthesized GaAs/Si core–shell nanowires by combining MBE and PECVD (Conesa-Boj et al., 2013). The GaAs nanowires were first grown in an ultra-high vacuum (UHV) MBE chamber. After that, the wafers containing the original GaAs nanowires were transferred to the

PECVD system. Before the deposition of the Si, the wafers were subjected to a $H_2$ plasma treatment in order to reduce the oxide present on the sidewalls of the GaAs nanowires. The concentration of silane radicals and reactive radicals in the plasma determine the nature of the Si shell.

In Fig. 5A and B, two representative GaAs/Si and GaP/Si nanowires are shown. It is remarkable that in this growth approach, the shell adopts the same crystalline structure as the core material, in these two cases in particular Si adopts the cubic and the hexagonal crystalline phases, respectively. The insets in Fig. 5 also show the energy-dispersive X-ray spectroscopy (EDX) compositional maps that illustrate the presence of the GaAs/Si and GaP/Si core–shell heterostructures, respectively.

The structural quality of the junction in core–shell nanowires is one of the essential requirements that must be satisfied in order to implement such nanostructures in realistic nanodevices. It has been extensively demonstrated that one of the main sources of defect formation in core–shell systems is due to the lattice mismatch between the shell and the core leading to the appearance of misfit dislocations at the core and the shell interface. However, both experimental studies and theoretical models of defect formation have been mostly restricted to mismatched materials, where the appearance of defects is driven mostly by the difference in geometry and crystalline structure of the core and the shell.

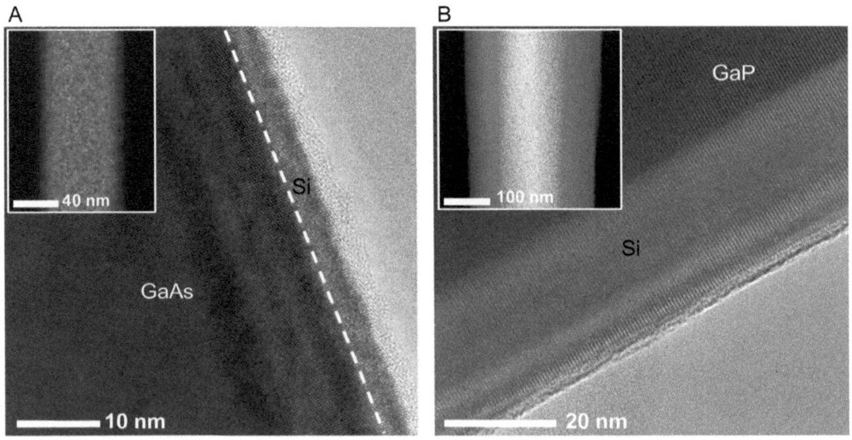


**Figure 5** (A and B) High-resolution TEM images of a representative GaAs/Si and GaP/Si core/shell nanowires, respectively. The insets correspond to the EDX compositional map confirming the presence of the GaAs/Si and GaP/Si core–shell heterostructures. In these maps Si signal is marked in red (dark gray in the print version).

As an illustration of the potentialities of structural characterization in core–shell systems, Conesa-Boj et al. (2014) combined high-resolution transmission electron microscopy (HRTEM) with geometrical phase analysis (GPA) and finite element (FE) methods to determine and quantify the origin of the lattice distortion in GaAs/Si core–shell nanowires. The presence of edge dislocations in such system produces a localized elastic distortion, in particular a tensile line crossing the diameter of the nanowire, which turns out to be around 4%, see the strain map along the axial direction shown in Fig. 6A.

As a useful baseline to understand the essential features of the strain field observed experimentally, and to identify the three-dimensional structure of the defect, FE simulations were performed. In particular, the effects of an edge dislocation loop were simulated, assuming the dislocation loop lying in the $(1\bar{1}1)$ plane at the core–shell interface with Burgers vector $b = a/3[1\bar{1}1]$, as described the sketched in Fig. 6B. Given that the FE simulation results were in agreement with the experimental results, this study suggested that the two edge dislocations observed in the GPA image are the fingerprints of an edge dislocation loop traveling around the nanowire core at the GaAs/Si interface and that the localized tensile strain is induced by a loop-like dislocation configuration.

In addition to highly mismatched materials, defect formation can also be ubiquitous when combining materials with similar lattice parameters. In these cases, the mechanisms of defect formation can be tightly related to

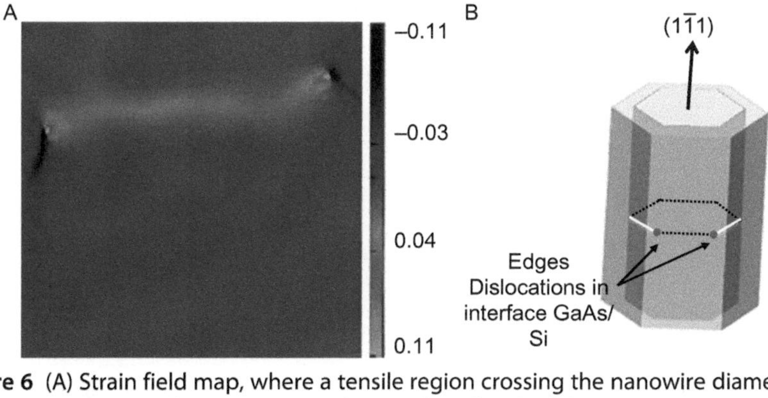


**Figure 6** (A) Strain field map, where a tensile region crossing the nanowire diameter is visible (green (gray in the print version) line) and (B) sketch of the nanowire assuming an edge dislocation loop, lying in a $(1\bar{1}1)$ plane at the core–shell interface with Burgers vector $b = a/3[1\bar{1}1]$.

specific growth process. This phenomenon has been explored for instance in the work of Conesa-Boj et al. (2015), who reported it for GaP/Si core–shell nanowires. Despite the similar lattice parameters, the presence of defects, in particular "crack" defects, was observed. The crack defect can be modeled as a local fracture, that is, a region where the separation between two adjacent atomic layers is increased, as shown in Fig. 7.

In order to reduce the formation of these types of defects, it is important to understand how their appearance is related to the details of the growth processes. In this case, it has been reported in the literature that the formation of such defects and their development into other type of defects is dependent on the brittle-to-ductile transition behavior. In the work we are discussing, different temperatures for the Si shell deposition were tested.

The main point to understand the mechanism for crack formation was that during Si shell deposition, two different steps in terms of temperature were used. The temperature difference was enough to induce local changes of the dilatation coefficient in the crystalline structure of the Si shell and thereby allowing the formation of cracks and other kind of defects such as stacking faults and Frank-type dislocations. Such studies have led to propose an alternative growth strategy to achieve defect-free hexagonal Si.

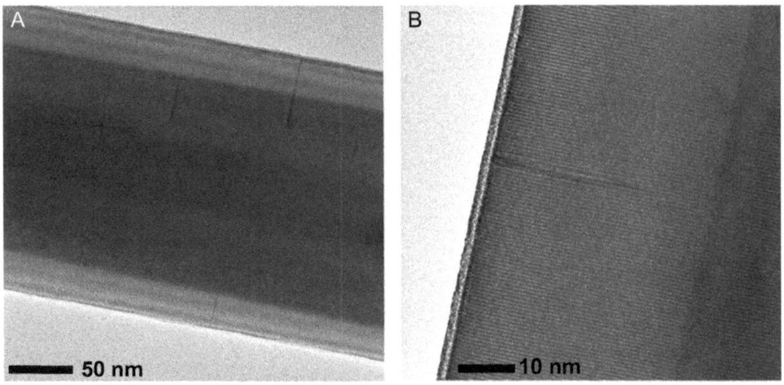


**Figure 7** (A) Low-magnification TEM image of a GaP/Si core–shell nanowire, where the stripes with different contrast indicate the presence of cracks defects; (B) high-resolution TEM image of a lateral of a GaP/Si core–shell nanowire exhibiting a crack defect in the Si shell.

## 5. AXIAL III–V/IV NANOWIRE HETEROSTRUCTURES

Most of axial nanowire heterostructures realized so far involve different materials with the same crystal structure or the same material with different crystal structures. Multifamily semiconductor heterostructures involving different materials with different structures started to appear only recently, increasing material combinations possibilities. Such novel material combinations implicate lattice mismatches ranging from almost 0% up to 10%, adding complexity to the system. Increasing mismatches lead to dramatic consequences in two-dimensional heterostructures when plastic relaxation causes misfit dislocations. Glas (2006) studied the effect of the stress relaxation on axial structures regardless of their chemistry. He theoretically observed that the critical thickness increases with the reduction of the nanowire heterostructure diameter. In other words, the nanowire geometry favors elastic relaxation and allows the combination of materials of large lattice mismatch in contrast with two-dimensional heterostructures. In the particular case of Si/III–V heterostructures, geometrical parameters exist in which crystal defects due to strain relaxation do not appear; therefore, it is theoretically possible to create misfit dislocation-free Si/III–V axial heterostructures. However, the nature of each semiconductor is a major hindrance to axial Si/III–V's, in particular their chemical bonds and surface energy which have a direct influence on the growth mechanisms.

The first experiments on axial Si and III–V combinations were presented by Dick et al. in 2007. They focused their investigations on Si/GaP, GaP/Si, GaAs/Si, and Si/GaAs heterostructure nanowires. The nanowires were grown by MOCVD using standard growth parameters for the respective segments. A purging step under $H_2$ of several minutes was used to remove precursors from the gas phase. This step also allowed to increase/decrease the growth temperature for the next segment. Straight segment were found for Si/GaP and Si/GaAs axial nanowires whereas the reverse stacking showed kinks. A thermodynamic model was developed and suggested that in analogy to 2D growth, the nanowires' morphology (straight or kinked) strongly depends on the relative surface energies between materials.

Later in 2012, Hocevar et al. (2012) published a complete work on multiple axial GaP/Si nanowires. The nanowires were grown in a MOCVD reactor at a constant temperature of 540 °C. A growth interruption step was used between GaP and Si growth sequences. After the GaP segment growth, the group V source was shut down for several seconds, keeping the group III source on. The group III source was then shut down while the Si source was opened to grow the Si segment.

When no growth interrupt was performed between the GaP segment and Si growth sequences, all of the Si segments kinked at the GaP/Si interface. The authors suggested that phosphorous atoms prevent Si entering the gold catalyst. As a consequence, Si does not uniformly alloy with the liquid gold particle and three-dimensional growth occurs. In contrast, when a growth interrupt takes place between GaP and Si growth sequences, all the Si segments are straight at the GaP/Si interface (Fig. 8A). Phosphorous atoms are removed from the surface of the gold either by desorption or by the Ga atoms to form GaP. Consequently, Si absorbs uniformly on the Au surface, which results in a uniform Si concentration throughout the droplet, and layer-by-layer growth occurs.

By engineering the growth interrupts between growth sequences, Hocevar et al. (2012) achieved Si/GaP (Figs. 8B and 9) and multiple GaP/Si heterostructures along the nanowire axis. These nanowires exhibit interesting feature: (1) both materials grow from a AuGa alloy and (2) the nanowires' diameter is smaller along the Si segment diameter than along the GaP segment. This diameter variation is mainly explained by the difference in contact angles (and therefore surface energies) between the AuGa liquid on a GaP surface and on a Si surface.

Inspired by the GaP/Si heterostructures, Hocevar et al. integrated an optical emitter, a GaAs segment between GaP barriers along a Si nanowire. The GaAs segments emitted at 1.495 eV with photoexcited carrier lifetimes over 1 ns, demonstrating a high-crystal quality.

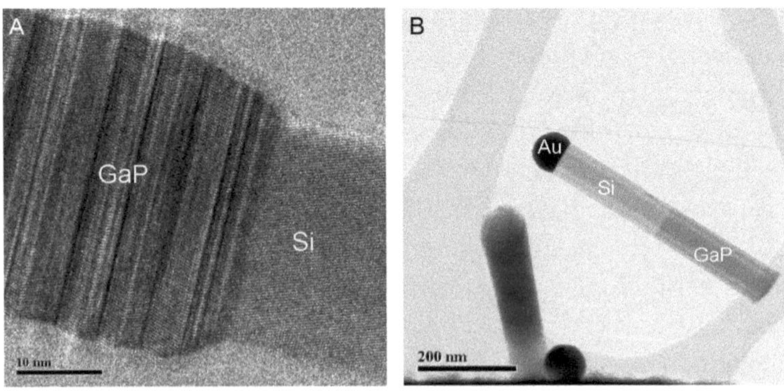


**Figure 8** (A) TEM image of GaP on Si heterostructure nanowire grown by Au-assisted VLS. (B) TEM image of Si on GaP heterostructure nanowire grown by Au-assisted VLS.

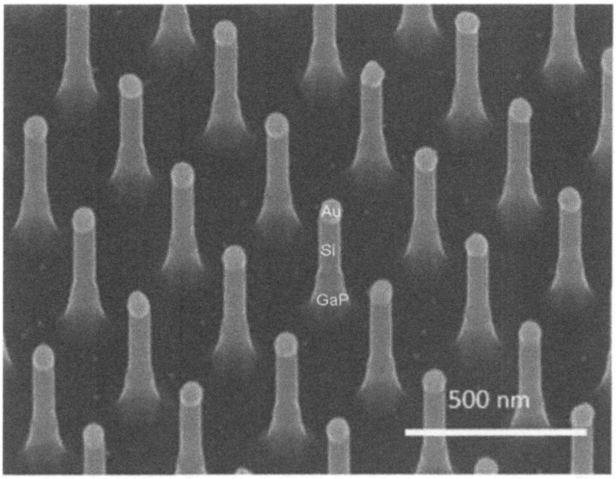


**Figure 9** SEM image of an array of Si on GaP heterostructure nanowires grown by Au-assisted VLS.

Hillerich et al. (2013) grew axial GaP/Si and GaAs/Ge nanowires using various equipments such as MOCVD, CVD, and *in situ* UHV TEM. Between two growth sequences (between GaP and Si for example), the samples were cooled down to room temperature. High-crystalline quality GaP/Si and GaAs/Ge segments were grown and crystalline defects formation some distance after the interface was investigated. From these experiments, it appeared that the droplet geometry drives the growth morphology: if the mismatch between the droplet diameter and the nanowire diameter is too high, crystal defects form in the newly grown segment. Therefore, by optimizing the growth parameters to minimize the gold droplet diameter change during the new growth sequence, straight GaP/Si and GaAs/Ge nanowire heterojunctions can be grown. Another strategy to create defect-free III–V/Si nanowires proposed by Hillerich et al. (2013) is to use an interlayer. They demonstrated that the use of Ge layers between GaAs and Si allowed the creation of defect-free hybrid GaAs/Si nanowire heterostuctures. Conesa-Boj et al., 2013 achieved axial Si/GaAs heterostructures using a gallium-catalyzed VLS mechanism with a combination of MBE and PECVD techniques. They showed crystalline axial growth of Si on GaAs at 250 °C substrate temperature and low silane flow (Fig. 10). This preliminary work is a major step forward as gold-free nanowires have strong potential for future integration of nanowires in CMOS technology.

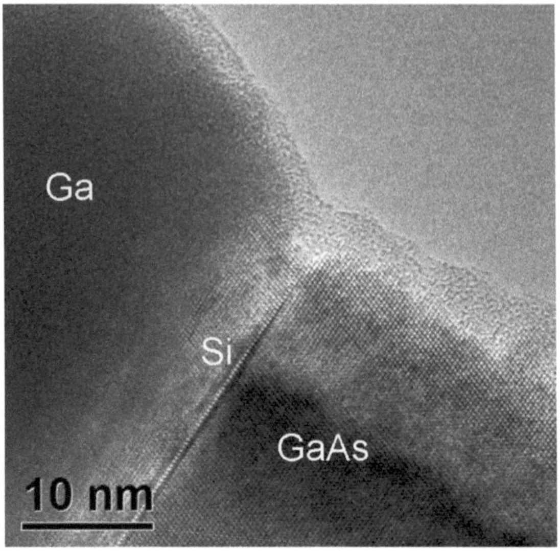


**Figure 10** TEM image of Si on GaAs heterostructure nanowire grown by Ga-assisted VLS

Prucnal et al. (2014) developed a route to create Si/InAs heterostructure nanowires using a combination of sequential ion beam implantation and flash lamp annealing. First Si nanowires were grown by Au-catalyzed VLS mechanism using a low-pressure CVD system. The nanowires were then implanted at selected positions with As and In ions with a fluences of the order of $10^{16}$ ions/cm$^2$ for each element. The chosen ion implantation energies for In+ and As+ allowed the depth distribution of In and As atoms to overlap within the Si nanowires. The samples were annealed with a flash lamp annealing system to recrystallize the nanowires. The obtained InAs quantum dot/Si nanowires showed atomically sharp interfaces between Si and zinc blende InAs segments.

## 6. CONCLUSION

A long-standing vision of the semiconductor industry is to integrate III–V materials on a Silicon platform. Some fundamental issues, associated with this integration, can be solved or at least relieved when using the nanowire geometry. In this chapter, we have reviewed recent progress in the epitaxial growth of a variety of III–V nanowire material on silicon. The vertical geometry allows for new device geometries and device performance, which are up to state-of-the-art devices.

The wire geometry also allows for an even more intimate integration of group III–V and group IV semiconductors in the radial and the axial direction. This opens the possibility to fabricate unprecedented heterostructures, and maybe more importantly, to get access to new crystal structures, with yet unexplored properties. These features create much more freedom in band structure engineering, which has always been the driving force behind the electronics industry.

## REFERENCES


Algra, R.E., et al., 2011. Crystal structure transfer in core/shell nanowires. Nano Lett. 11, 1690–1694.

Bakkers, E.P.A.M., et al., 2004. Epitaxial growth of InP nanowires on germanium. Nat. Mater. 3, 769–773.

Bakkers, E.P.A.M., Borgström, M.T., Verheijen, M.A., 2007. Epitaxial growth of III-V nanowires on group IV substrates. MRS Bull. 32, 117–122.

Borg, M., et al., 2014. Vertical III-V nanowire device integration on Si(100). Nano Lett. 14, 1914–1920.

Borgström, M.T., Immink, G., Ketelaars, B., Algra, R., Bakkers, E.P.A.M., 2007. Synergetic nanowire growth. Nat. Nanotechnol. 2, 541–544.

Cassell, A.M., Raymakers, J.A., Kong, J., Dai, H., 1999. Large scale CVD synthesis of single-walled carbon nanotubes. J. Phys. Chem. B 103, 6484–6492.

Conesa-Boj, S., et al., 2013. Hybrid axial and radial Si-GaAs heterostructures in nanowires. Nanoscale 5, 9633–9639.

Conesa-Boj, S., et al., 2014. Plastic and elastic strain fields in GaAs/Si core-shell nanowires. Nano Lett. 14, 1859–1864.

Conesa-Boj, S., et al., 2015. Cracking the Si Shell growth in hexagonal GaP-Si core–shell nanowires. Nano Lett. http://dx.doi.org/10.1021/nl504813e. 150504094140004.

Dayeh, S.A., Yu, E.T., Wang, D., 2007. Growth of InAs nanowires on $SiO_2$ substrates: nucleation, evolution, and the role of Au nanoparticles. J. Phys. Chem. C 111, 13331–13336.

Dick, K.A., et al., 2007. The morphology of axial and branched nanowire heterostructures. Nano Lett. 7, 1817–1822.

Gangloff, L., et al., 2004. Self-aligned, gated arrays of individual nanotube and nanowire emitters. Nano Lett. 4, 1575–1579.

Givargizov, E.I., 1973. Periodic instability in whisker growth. J. Cryst. Growth 20, 217–226.

Glas, F., 2006. Critical dimensions for the plastic relaxation of strained axial heterostructures in free-standing nanowires. Phys. Rev. B Condens. Matter Mater. Phys. 74, 2–5.

Hillerich, K., et al., 2013. Strategies to control morphology in hybrid group III-V/group IV heterostructure nanowires. Nano Lett. 13, 903–908.

Hiraki, A., Lugujjo, E., Mayer, J.W., 1972. Formation of silicon oxide over gold layers on silicon substrates. J. Appl. Phys. 43, 3643–3649.

Hiruma, K., et al., 1991. Quantum size microcrystals phase epitaxy grown using organometallic Grown crystals. Appl. Phys. Lett. 59, 431–433.

Hocevar, M., et al., 2012. Growth and optical properties of axial hybrid III-V/silicon nanowires. Nat. Commun. 3, 1266.

Huang, S., Cai, X., Liu, J., 2003. Growth of millimeter-long and horizontally aligned single-walled carbon nanotubes on flat substrates. J. Am. Chem. Soc. 125, 5636–5637.

Ionescu, A.M., Riel, H., 2011. Tunnel field-effect transistors as energy-efficient electronic switches. Nature 479, 329–337.

Mohan, P., Motohisa, J., Fukui, T., 2005. Controlled growth of highly uniform, axial/radial direction-defined, individually addressable InP nanowire arrays. Nanotechnology 16, 2903–2907.

Motohisa, J., Noborisaka, J., Takeda, J., Inari, M., Fukui, T., 2004. Catalyst-free selective-area MOVPE of semiconductor nanowires on (111)B oriented substrates. J. Cryst. Growth 272, 180–185.

Mourik, V., et al., 2012. Signatures of Majorana fermions in hybrid superconductor-semiconductor nanowire devices. Science 336, 1003–1007.

Prucnal, S., et al., 2014. III–V Semiconductor nanocrystal formation in silicon nanowires via liquid-phase epitaxy. Nano Res. 7, 1769–1776.

Roest, A.L., et al., 2006. Position-controlled epitaxial {III-V} nanowires on silicon. Nanotechnology 17, S271.

Tomioka, K., Motohisa, J., Hara, S., Fukui, T., 2008. Control of InAs nanowire growth directions on Si. Nano Lett. 8, 3475–3480.

Tomioka, K., Yoshimura, M., Fukui, T., 2012. A III–V nanowire channel on silicon for high-performance vertical transistors. Nature 488, 189–192.

CHAPTER SIX

# Atom Probe Tomography of Nanowires

**Nari Jeon, Lincoln J. Lauhon[1]**
Department of Materials Science and Engineering, Northwestern University, Evanston, Illinois, USA
[1]Corresponding author: e-mail address: lauhon@northwestern.edu

## Contents



## 1. INTRODUCTION

Semiconductor nanowires are highly anisotropic materials whose physical properties vary strongly with size and/or shape. In nanowire heterostructures, the confinement potentials of electrons and photons can therefore be engineered in interesting and useful ways (Hyun et al., 2013). It follows that slight variations in chemical composition can significantly alter electronic and optical properties, requiring analytical characterization tools of high chemical sensitivity and spatial resolution to establish

*Semiconductors and Semimetals*, Volume 93
ISSN 0080-8784
http://dx.doi.org/10.1016/bs.semsem.2015.07.003

relationships between processing, structure, and properties. Because the composition of nanowire heterostructures can be independently modulated along both the axial and radial dimensions, one is often interested in characterizing buried nonplanar heterointerfaces or other composition modulations that exist along multiple axes. In this context, many conventional characterization techniques developed for bulk or thin-film semiconductor materials may be insufficient. In contrast, atom probe tomography (APT) can characterize composition in three dimensions with subnanometer resolution and 1–10 ppm sensitivity. Over the last decade, APT has begun to be utilized (Allen et al., 2008; Perea et al., 2005, 2006, 2008, 2009) in nanowire characterization efforts to address important scientific questions regarding nanowire growth processes and physical properties. For example, APT has been uniquely useful in analyzing the spatial distribution of dilute species such as dopants and impurities, which cannot otherwise be detected directly (Allen et al., 2008; Chen et al., 2012; Connell et al., 2013; Eichfeld et al., 2012; El Kousseifi et al., 2014; Moutanabbir et al., 2013; Murthy et al., 2011; Perea et al., 2005, 2006, 2009; Roussel et al., 2011; Xu et al., 2008; Zeiner et al., 2011). The combination of atom probe studies with other complementary characterization methods enables the nanoscale origins of electronic and optical properties to be established on a quantitative basis. In addition, the three-dimensional composition distributions generated by atom probe studies provide a natural basis for modeling and simulations (Morkötter et al., 2015; Schlitz et al., 2009; Yoon et al., 2013; Zhang et al., 2009).

Though APT has existed for many decades, its contributions to material science have been limited primarily to structural metals until relatively recently. The commercialization of local electrode APT was one significant advance toward the analysis of semiconductor nanowires in terms of relaxing specimen requirements and increasing the field of view, speed of data acquisition, and sensitivity. As described below, the local electrode enables isolation of a single nanostructure for analysis, among other benefits. In 2005, the first commercial pulsed laser APT became available, enabling the reliable analysis of less electrically conductive materials including intrinsic semiconductors. Commercial instruments have advanced rapidly in the last decade to the point where sample preparation has become the limiting factor in data acquisition from semiconductor nanowires. Hence, any discussion of the application of APT to nanowires should also include consideration of how materials are prepared for analysis.

The primary goal of this chapter is to provide a brief introduction to the application of APT to semiconductor nanowires for readers who are familiar with the materials but not with the analytical technique. The principles of APT are discussed at a level that may facilitate understanding of more technical aspects of research papers and are additionally intended to help beginning users understand key considerations and experimental design. More advanced users will benefit from consulting the many excellent reviews of APT referenced within. We first describe the underlying principles of APT and the nature of atom probe data in order to help readers to understand later sections. We then introduce sample requirements for local electrode APT and describe various nanowires sample preparation methods. The subsequent section describes relationships among different experimental variables under user control (such as laser pulse energy, voltage, and temperature) and their influences on field evaporation behaviors and data quality. Approaches to reliable reconstruction of atom probe data into a tomographic model of the specimen are then delineated based on what prior knowledge of the specimen is available. The final section discusses examples of the characterization of semiconductor nanowires by APT including analysis of (1) dopant and impurity distributions, (2) alloy fluctuations and clustering, and (3) buried interfaces. For each analysis topic, we summarize key challenges relating to the principles of the evaporation and reconstruction process, current technical limitations of the instrument, and the geometry of nanowires. Approaches to overcome or circumvent the challenges are also discussed.

## 2. PRINCIPLES OF APT

In APT, a bias voltage of a few thousand volts is applied between a needle-shaped specimen, or tip, and a counter electrode to facilitate electric field-induced evaporation of ions. A laser or voltage pulse induces the evaporation of single ions, which are subsequently captured by the detector (Fig. 1; Kelly and Miller, 2007; Seidman, 2007; Seidman and Stiller, 2009). The primary data of APT consist of *X*- and *Y*-positions of hit events, the time of flight, the voltage at the time of evaporation, and the evaporation sequence. The time of flight is converted to a mass-to-charge state ratio ($m/n$ ratio) by considering the ion flight path, and a histogram of the observed mass-to-charge state ratios is the mass spectrum. In the mass spectrum, ranges of mass-to-charge state ratio are assigned to identified ions or ion complexes. The 2D position of the hit event is converted to the 3D position of the

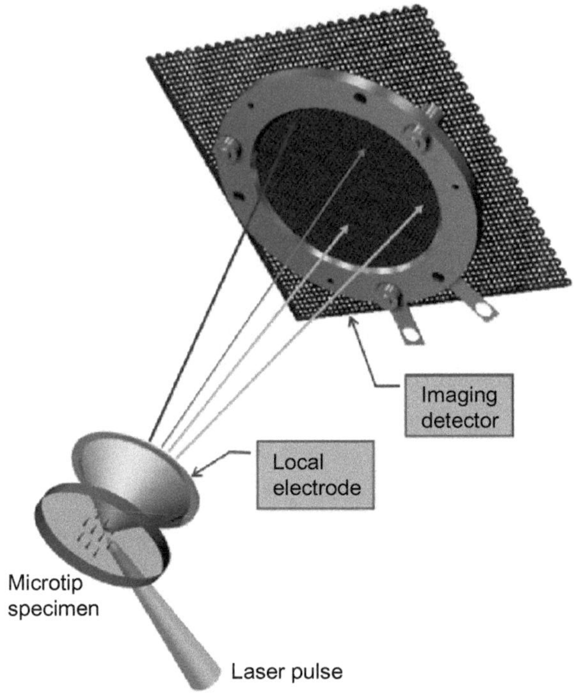


**Figure 1** Schematic of key elements of local electrode atom probe tomograph. *With kind permission from Springer Science and Business Media (Larson et al., 2013).*

species identified in the mass spectrum by taking into account the flight path and the time of the hit within the entire sequence of hit events. Under optimal experimental conditions, individual ions field-evaporate uniformly from the topmost surface of the specimen only during laser or voltage pulses with an equal probability per pulse, and at a typical detection rate of one ion per 100 laser pulses. Uncorrelated evaporation between laser pulses produces background noise and should therefore be minimized. Experimental conditions are adjusted empirically to optimize the quality of data as judged by the following criteria. (1) A mass spectrum should have high signal-to-noise ratio with a minimal number of unidentified peaks. (2) Hit events should be spatially homogeneous without significant influence from processes that obscure actual positions of ions, such as surface migration and clustering. (3) Ion count rate variations with run conditions should be similar for the elements of interest for the measured stoichiometry to the accurate. While the mass spectrum of a specimen is valuable, one is most often interested in the spatial distribution of various elements. Hence, a reconstruction of the data into a representation of the original sample is the starting point for most analysis. With high-quality data and reliable reconstruction parameters, APT can display the distribution of elements with ~1–10 ppm detection sensitivity and subnanometer spatial resolution (lateral: 0.2–0.3 nm, in-depth: 0.1 nm). Furthermore, APT detects all elements with equal efficiency and does not require *a priori* knowledge of compositions. Various quantitative and statistical methods have been developed to analyze and represent data from reconstructions, including frequency distribution analysis, proximity histograms, and isoconcentration surfaces. These methods are introduced along with specific examples below.

## 3. SAMPLE REQUIREMENTS AND PREPARATION METHODS

Successful atom probe analysis requires that an electric field high enough for ion evaporation be achieved at the tip using a voltage that is low enough to avoid sample fracture. Hence, smaller tip diameters facilitate analysis at a cost of reducing the diameter of the analysis region. Though there are variations between materials, a tip diameter of 100 nm or less is a typical target. In addition, the spacing between tips should be larger than ~10 μm to ensure that evaporation and ion detection originate only from the nanowire of interest. The nanowires should extend above the sample plane by ~50 μm or more from the plane to minimize aberration of the electric field by the nearby substrate. The geometry of nanowires, i.e., the small diameter and high aspect ratio, makes them good specimens for APT as they may require less complicated preparation procedures compared to other structures such as bulk and planar samples. Below, we discuss basic sample preparation methods ranging from simple in-place nanowire growth to complicated lift-out/sharpening method using focused ion beam (FIB). The main goal of sample preparation and analysis is to maximize the yield of useful data. Various sample preparation methods for nanowires have been developed with the following related goals in mind: (1) facile alignment of a sample with respect to a local electrode, (2) simplification of processes, (3) protection of a sample from damage during the preparation processes, (4) selection of a target nanowire from a nanowire growth batch, and (5) increased field of view (Figs. 2 and 3).

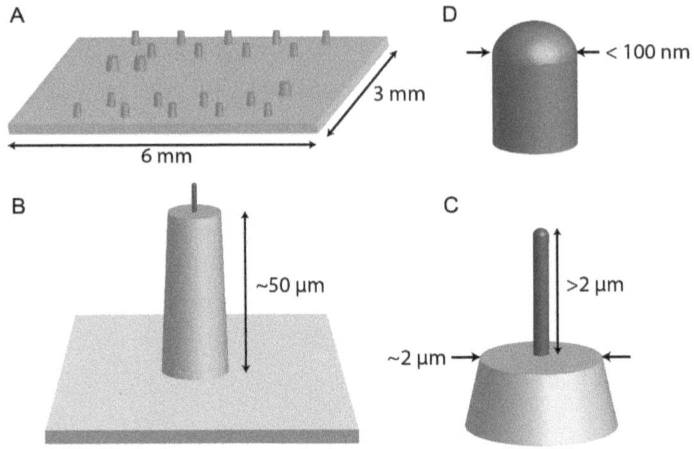


**Figure 2** Schematics of (A) Si micropost array substrate, (B and C) nanowire on a micropost at different length scales, and (D) nanowire tip structure.

## 3.1 Grow in Place

Nanowires grown perpendicular to a planar substrate are usable as an atom probe specimen without further processing provided that a nanowire is tens of microns long and can be properly aligned with a local electrode. In the first such example, epitaxially grown InAs nanowires of 100–140 µm height spaced 500 µm apart were analyzed by voltage-pulsed APT (Perea et al., 2005). To catalyze epitaxial growth in selected locations, Au disk arrays were patterned on a GaAs (111)B substrate by electron beam lithography and lift-off. A patterned nanowire array was also used as a fiducial marker to facilitate the alignment of the target nanowire with respect to a local electrode. Nanowires can also be grown on micropost arrays, as first demonstrated for Si and Ge nanowires (Perea et al., 2006; Xu et al., 2008). Because the microposts are tens of microns tall, the nanowires can be much shorter (a few to ~10 µm) compared to those grown on a planar substrate. Furthermore, micropost arrays with posts ~10 µm in diameter and 60–100 µm height are easily resolved by cameras typically installed in the analysis chamber. While commercial microposts are available for Si(100) substrates, micropost arrays can also be fabricated on Si(111) substrates by reactive ion etching or wafer dicing to facilitate epitaxial growth (Perea et al., 2006, 2008). Thorough etching and cleaning processes are needed to remove oxide layers and clean other contaminants, as described in detail elsewhere (Perea et al., 2006, 2008).

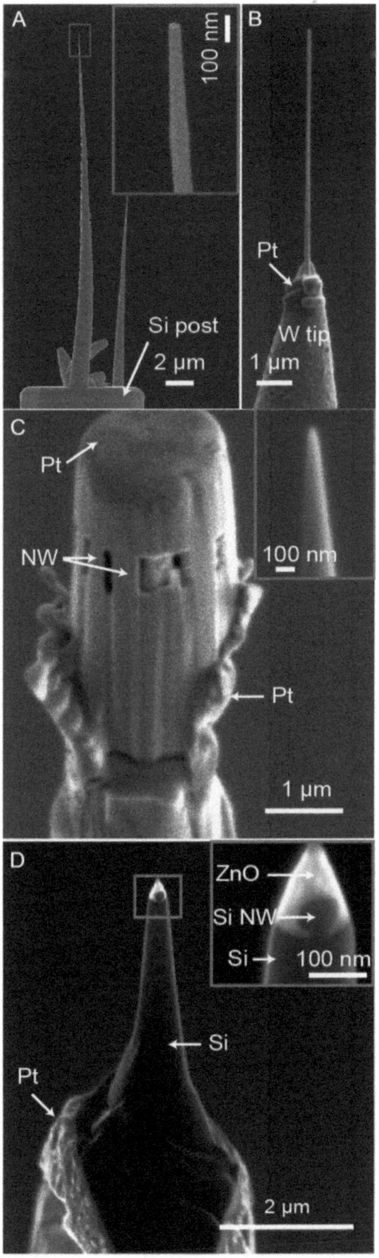


**Figure 3** Various approaches to nanowire sample preparation for APT analysis. (A) P-doped Ge nanowire epitaxially grown on a Si post, (B) GaAs–AlGaAs core–shell nanowire attached to a W probe tip, (C) Pt-capped GaN–InGaN nanowire array mounted on a Si micropost prior to sharpening. Inset: sharpened tip, and (D) ZnO-capped horizontally oriented Si nanowire lying on a Si micropost.

## 3.2 Transfer

Nanowires can be transferred from the growth substrates to other platforms such as a TEM half-grid (Dawahre et al., 2012) or micropost (Eichfeld et al., 2012; Isheim et al., 2012; Sanford et al., 2014). The transfer method utilizes the three-axis movement capability of a nanomanipulator that is mounted in a scanning electron microscope with a gas injection system to enable electron beam-induced deposition. A very sharp probe tip is mounted on the nanomanipulator to make contact with the nanowire of interest. The method encompasses three basic steps—picking up a nanowire, translating it to the platform, and mounting the nanowire to the platform. The probe tip to which the nanowire is attached may itself serve as the mounting platform (Riley et al., 2012). When fixing a nanowire to a probe tip, metal deposition (such as Pt) induced by an electron beam can provide sufficient mechanical stability. In this process, the delicate manipulation of a single nanowire is enabled by the small diameter (<0.1 μm) of the probe tip and fine movement control of the nanomanipulator. The direct pickup and transfer method have the advantage of enabling selection of the nanowire of interest. When direct pickup is not possible, nanowires can be mechanically transferred en masse from the growth substrate to a TEM grid or half-grid and then subsequently transferred by nanomanipulation to a platform. When a properly aligned nanowire is attached to a TEM half-grid, correlated TEM–APT studies can be conducted using specimen holders compatible with both TEM and APT (Gorman et al., 2008).

## 3.3 Lift-Out

Standard lift-out and sharpening procedures (Kelly et al., 2007; Padalkar et al., 2014) for planar samples can be applied to nanowire arrays when they are embedded in a solid matrix. Lift-out typically includes the following steps: (1) deposition of capping layers to protect the sample surface from FIB damage, (2) lift-out of a wedge-shaped section from a planar sample by FIB milling, (3) mounting the wedge onto a micropost, and (4) annular milling of the mounted sample to form a sharp tip of a diameter typically less than 100 nm. The lift-out wedge can be rotated using an axial rotation manipulator so that the direction of the sample with respect to the atom probe analysis direction can be selected arbitrarily. To apply lift-out to nanowire arrays, the arrays should first be embedded in an appropriate material to fill the gaps between nanowires before the lift-out step (Blumtritt et al., 2014; Riley et al., 2013). Selection of materials and deposition methods for the embedding and capping processes must take into account several factors. For example, when the embedding or capping materials will be present in region of field evaporation, dissimilar field evaporation thresholds will introduce complicated field evaporation behaviors. Additionally, poor mechanical adhesion between the materials in a multicomponent system often results in tip fracture. Alternatively, the capping materials can be removed by chemical wet etching after the annular sharpening step (Prosa et al., 2010) or by low-energy FIB milling at the end of the annular sharpening step (Blumtritt et al., 2014). A nanowire lying down on a substrate can be also made into an atom probe specimen via the similar lift-out and sharpening processes (El Kousseifi et al., 2014). In this case, the analysis direction is perpendicular to the nanowire axis, allowing reconstruction across an entire cross-section of a single nanowire. However, the analysis volume will be smaller than that of a nanowire analyzed along its growth direction (Fig. 4).

## 3.4 Shell Deposition

As described above, most sample preparation methods have been developed for analysis along the nanowire axis. This parallel geometry is, in most cases, preferable to the perpendicular geometry because it allows measurement of composition variations along an axial direction while providing a large volume of analysis. However, the entire diameter of the nanowire will not be in the field of view of the detector, and therefore, the reconstruction will not provide data for the entire cross-section. The deposition of a conformal shell

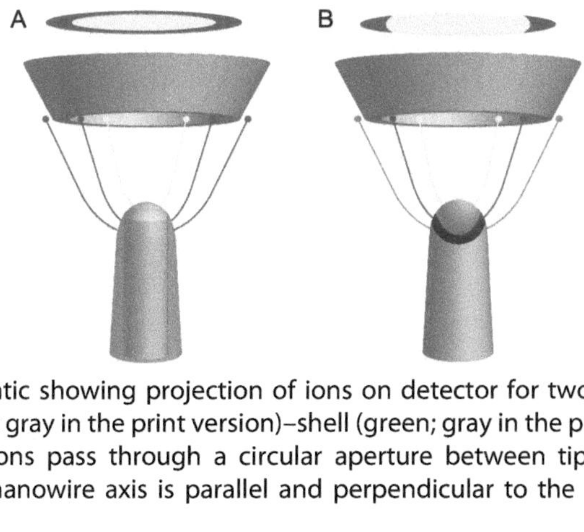


**Figure 4** Schematic showing projection of ions on detector for two orientations of a core (yellow; light gray in the print version)–shell (green; gray in the print version) nanowire specimen. Ions pass through a circular aperture between tip and detector. In (A) and (B), the nanowire axis is parallel and perpendicular to the analysis direction, respectively.

on the nanowire surface either during or after growth effectively decreases the magnification by increasing the diameter, enabling the entire cross-section of the original nanowire to be within the field of view (Connell et al., 2013; Perea et al., 2008). The shell can be deposited *in situ* following nanowire growth (Connell et al., 2013; Perea et al., 2008) or deposited after growth in another deposition such as a sputtering (Connell et al., 2013) or atomic layer deposition system. The details of shell deposition methods for full-diameter imaging are discussed further below.

# 4. ATOM PROBE OPERATION

## 4.1 Operating Guidelines

There are two pulsing modes to induce field evaporation. In the voltage pulsing mode, a voltage pulse triggers field evaporation of the ions by lowering the energy barrier to desorption, whereas in the laser pulsing mode, a laser pulse heats the specimen, providing sufficient thermal energy for ions to field-evaporate. The goal of adjusting the operating conditions is to make it equally probable that any ion will evaporate for a given pulse, and that no ions will evaporate in the absence of the pulse. Furthermore, the pulse magnitude should be small enough that multiple evaporation events in a given pulse are minimized. Laser pulsing is generally preferred to voltage pulsing for atom probe analysis of semiconductors for many reasons including a reduced standing voltage and associated reduced sample fracture rate, higher resolution mass spectra, and more facile optimization of evaporation conditions (Cerezo et al., 1985; Dawahre et al., 2012; Hashizume et al., 1986; Kellogg and Tsong, 1980; Perea et al., 2006). The main user-selectable operating conditions are the energy and repetition rate of laser pulse, the sample base temperature, and the target detection rate. The operating conditions are often reported in terms of laser pulse energy and voltage since they are intimately connected to each other for a given detection rate. To maintain a fixed detection rate, the DC standing voltage is automatically adjusted, while the laser pulse energy is held constant. That is, the voltage is adjusted as a response to any sample changes that influence the real-time detection rate. For example, an increase in the tip diameter will lead to an increase in the standing voltage to keep the average electric field constant at the tip, thereby maintaining the target detection rate. If the standing voltage becomes too high, however, there will be significant ion evaporation that is uncorrelated with the laser pulsing, leading to noise (uncorrelated hits) throughout the mass spectrum. Just below such a threshold, one is likely to observe preferential evaporation of a given species for nonelemental materials. When the laser pulse energy and standing voltage are ideally balanced, evaporation of any species is equally probable. While this is imperfectly realized in practice, reliable elemental analysis is still possible (Riley et al., 2014). On the other hand, excessive laser pulse energy can induce atomic migration on the tip surface, leading to the formation of complex ions and spatially inhomogeneous detection hit maps. The optimal rate of laser pulsing considers trade-offs between the data acquisition rate and the mass range detected. Large ion clusters will have a long time of flight, which imposes an upper limit on the pulse rate if such clusters need to be analyzed. Furthermore, materials with poor thermal conductivity may need to be analyzed at lower pulse rates to allow sufficient time for residual thermal energy to be removed from the tip in between pulses. Because evaporation and diffusion can occur while the tip is cooling down, it is generally desirable to choose the lowest base temperature to maximize the tip cooling rate and minimize uncorrelated evaporation. As a caveat to the above guidelines, field evaporation is a complex process that depends nonlinearly on the temperature and the local electric field. The local electric field is influenced by the tip shape, which also evolves in time and depends on the history of the operating conditions. Therefore, many of the operating parameters exhibit complex interdependencies during atom probe analysis, precluding a simple prescription of operating conditions for a given material. However, successful user experience with a particular material and typical tip shape is generally sufficient to reproduce reliable evaporation behavior.

## 4.2 Metrics of Run Quality

Just as there is no single set of optimum parameters for atom probe analysis, there are no universal metrics for assessing the quality of data being produced in real time. However, it is possible to describe general guidelines for assessing whether the data being produced with particular operating conditions might be improved by adjusting the conditions, in real time. Due to the complexity of field evaporation behaviors, these guidelines are necessarily simplified.

### *4.2.1 Mass Spectrum*

The simplest indicator of run quality is the signal-to-noise ratio in the mass spectrum. Significant DC field evaporation, which is uncorrelated with laser pulsing, introduces a nearly constant background count rate, degrading the signal-to-noise ratio. The uncorrelated evaporation decreases with standing voltage and temperature. Excessive pulse energy and/or slow tip cooling will produce peak broadening in the form of "thermal tails"; when a thermal tail of an abundant species overlaps the peak of another species, the sensitivity is degraded.

### 4.2.2 Homogeneity

Ideally, hit events should be spatially homogeneous for elemental materials except for local inhomogeneities near crystallographic poles and along zone lines, which can be in fact a sign of well-controlled field evaporation. Poles and zone lines are imaged as a result of trajectory aberrations, that is, ion trajectories that are influenced by the local surface crystallography. A detailed explanation of the generation of poles and zone lines may be found elsewhere (Vurpillot et al., 1999a,b). Other inhomogeneous hit events may indicate undesirable phenomena such as surface migration, preferential loss of a certain species, or even tip fracture. Special care must be taken for analysis of compound materials, where field evaporation behaviors can be different for different species. Spatial inhomogeneity in the evaporation of compound cannot be completely avoided and can vary with operation conditions as well as the crystal structure and orientation. Hence, the spatial distribution of the hits of the different ions and ion complexes should be reviewed to determine how inherent inhomogeneities impact the reconstruction and interpretation thereof.

### 4.2.3 Charge State Ratio

For a given element, the propensity to form ions with a particular charge state depends on both materials and operating conditions. However, the charge state ratio between ions of the same element (e.g., $Si^{2+}/Si^{+}$) can be observed during atom probe operation as a measure of the local electric field at the tip surface. Hence, the charge state ratio may provide insight into how various species are being impacted by local variations in curvature. Furthermore, the charge state ratio is often used as an empirical measure of whether ideal operating conditions have been realized following correlation with other metrics of run quality.

### 4.2.4 Golden Hit Percentage

When a single ion species is detected within the detection window following a laser pulse, it is referred to as a single hit; multiple detection events associated with a single laser pulse are referred to as multiple hits. The golden hit percentage is the fraction of detection events that are single hits. Some of the ions detected in the multiple hit events may experience a time delay with respect to the laser pulse and may not be properly identified in a mass spectrum. However, a ~100% golden hit percentage does not always indicate the optimum acquisition conditions, and moreover, a certain portion of multiple hit events may be unavoidable. The optimum range of golden hit percentage can also vary with purposes of analysis because different species have different propensities for single hit events. For III–V semiconductor compound materials such as GaN and GaSb, group V species tend to diffuse on the tip surface and form cluster ions, which are more likely to field-evaporate in multiple hit events (Müller et al., 2011; Riley et al., 2012). For the detailed analysis, one can retrieve identities of elements involved in the multiple hit events and related physical mechanisms in correlation histograms.

### 4.2.5 Count Rate

Detection rate is the number of hit events per laser pulse, which is typically on the order of one per 100–1000 laser pulses, whereas count rate refers to the number of a particular species detected in a given time (Riley et al., 2012, 2014). While detection rate is directly observable during atom probe operation, the count rate can only be determined in post atom probe analysis following calibration of the mass spectrum and reconstruction. For materials in which the elemental constituents evaporate quite differently, analysis of count rates is necessary to analyze trends in run conditions and to determine whether an accurate measurement of composition has been achieved.

## 5. RECONSTRUCTION

In an atom probe, atoms ionize and desorb from the surface of the tip and then accelerate toward the detector following electric field lines. The initial 3D positions of atoms in the sample are established by reconstructing the flight path back from the 2D positions of hit events on the detector and the sequence of evaporation. The ion flight paths are reconstructed by considering the 3D electric field distribution between the tip surface and the detector. The simplest model to reconstruct the flight path assumes that the tip surface is a perfectly smooth half-sphere. In such a model, there are four primary input parameters: the image compression factor ($\xi$), the field factor ($k_f$), the evaporation field ($F_e$), and the detection efficiency ($\eta$). Reconstruction model parameters are derived from the geometry of the

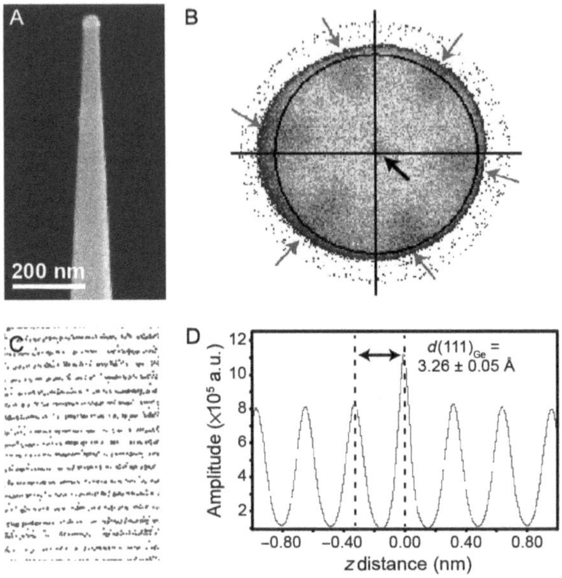


**Figure 5** (A) SEM micrograph showing Ge nanowire tip for APT analysis. (B) Detector hit map showing (111) pole indicated by a black arrow. Blue (gray in the print version) and yellow (light gray in the print version) regions represent regions of low (high) hit density. (C) Reconstruction of the (111) pole region showing (111) planes. (D) Spatial distribution map measured in the (111) pole region. *Derived from Connell et al. (2013).*

sample including the initial tip diameter, the half-shank angle, and the sphere to cone (Fig. 5).

The reconstruction process can be imagined as the process of building up of a series of thin hemispherical shells with a varied diameter. The evolution of the diameter during evaporation impacts ion trajectories, so it is important to specify this evolution for the purposes of reconstruction. There are three primary approaches to doing so: voltage, shank angle, and tip profiling. In voltage profiling, the particular tip diameter at which ion evaporated is calculated by using the standing voltage assuming a constant field evaporation threshold (constant $V/r$). In constant shank angle mode, the diameter evolves from a user-defined initial diameter while assuming a constant shank angle. In a tip profile mode, the user manually inputs tip diameters at different depths along the analysis direction. Users select an appropriate diameter evolution method based on prior knowledge of the tip shape. For example, tip profiles can be obtained by electron microscopy prior to atom probe analysis (Rigutti et al., 2014). When samples have internal structure of known dimension, such as in superlattice samples measured by correlated methods, the known spacing between different layers can be used to gauge the accuracy of the reconstruction, and the reconstruction parameters can be iteratively adjusted (Galtrey et al., 2008; Riley et al., 2014). When an identified crystallographic pole is observed in the detector hit map as a local region of a low hit density, the reconstruction parameters can often be calibrated using the known spacing of crystallographic planes that are resolved within the pole region. For this purpose, a spatial distribution map (SDM) is used. The SDM is a plot of the pair distribution function in a specified direction, and lattice planes are often resolved as peak to peak distances in a 1D SDM generated along a pole axis. For example, detector hit maps of ⟨111⟩-oriented Si and Ge nanowires typically display a (111) pole, and a 1D SDM of the (111) pole can then be used for the calibration of reconstruction parameters (Connell et al., 2013; Perea et al., 2009).

# 6. EXAMPLES OF ATOM PROBE ANALYSIS OF NANOWIRES

Many factors determine the suitability of characterization tools for composition analysis, including the detection limit, spatial resolution, detectability for all elements, analysis volume, and dimensionality of data. In this regard, APT is a unique characterization tool displaying several advantages such as single-atom sensitivity, subnanometer-scale resolution, and equal detectability for all elements. Furthermore, a statistical analysis of composition variations is possible with sufficient analysis volume or sufficient number of atoms detected, and the 3D nature of reconstruction data allows isolation and an analysis of various features. This section reviews up-to-date examples of APT for the characterization dopants and impurities, composition fluctuations and clusters, and interfaces and heterostructures.

## 6.1 Dopants and Impurities

APT is uniquely well suited to the analysis of dopant and impurity distribution in nanowires. Since the first measurements of nanowire impurity concentrations (Perea et al., 2005, 2006), there have been a number of studies reporting inhomogeneous distribution of intentional dopants along an axial or radial direction (Chen et al., 2012; Connell et al., 2013; Murthy et al., 2011; Perea et al., 2009; Schlitz et al., 2009; Yoon et al., 2013). Quantitative mapping of dopant distributions has led to important insights into nanowire growth kinetics and related mechanisms. Perea et al. first reported different dopant incorporation rates in the core and shell regions of P-doped Ge nanowires grown by a vapor–liquid–solid (VLS) mechanism and identified distinct VLS and vapor–solid (VS) incorporation pathways (Perea et al., 2009). Later, Connell et al. reported 100-fold enhancements in the dopant concentration in region near VLS-trijunction in B-doped Si nanowires and P-doped Ge nanowires, proposing that a multiply faceted growth interface is responsible (Connell et al., 2013; Fig. 6).

APT doping studies naturally contribute to greater understanding of the electrical properties of nanowires (Yoon et al., 2013; Zeiner et al., 2011; Zhang et al., 2009). The nature of the conductivity in undoped and P-doped Ge nanowires was explored by combining dopant and impurity concentration measurements with temperature-dependent conductivity

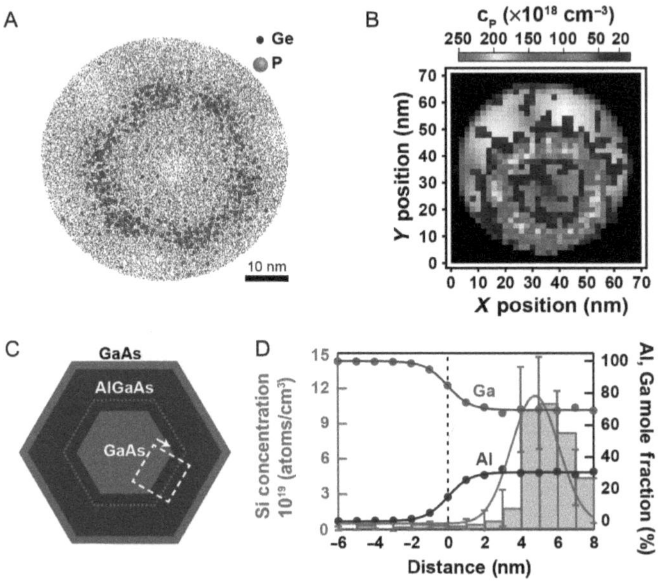


**Figure 6** (A) End view of 3D atom map of P-doped Ge nanowire. Blue (light gray in the print version) and gray dots represent Ge and P atoms, respectively. (B) End view of P-dopant concentration map in the region shown in (A). (C) Schematic of GaAs–AlGaAs core–shell nanowire with Si δ-doping layer embedded in the shell region (red (gray in the print version) dotted line). White dashed box indicates approximate location of the region where the proxigram shown in (D) is generated along the direction of white arrow. (D) Proxigram showing variation of Al and Ga mole fractions and Si-dopant concentration with respect to the GaAs–AlGaAs core–shell interface. *Panel (A): Derived from Connell et al. (2013). Panel (B): Reprinted (adapted) with permission from Connell et al. (2013). Copyright (2013) American Chemical Society. Panel (D): Derived from Morkötter et al. (2015).*

measurements (Zhang et al., 2009). Utilizing a simple conductivity model, contributions from surface states, dopants, and impurities were verified (Zhang et al., 2009). Subsequently, the radial distributions of P dopants and $P_2$ clusters measured from P-doped Si nanowires were utilized as input to a core–shell model for distribution of active dopants, showing that enhanced near-surface doping strongly influences barrier height at metal contact (Yoon et al., 2013). The conductivity increase in Ga-implanted Ge nanowires upon annealing was explained in terms of activation of Ga dopants by correlating Ga concentrations and temperature-dependent conductivity (Zeiner et al., 2011). More recently, the distribution and concentration of a Si δ-doping layer were measured in GaAs–AlGaAs core–shell nanowires, and they were used to simulate the concentration of 2D electron gas (Morkötter et al., 2015).

It is also important to investigate the concentration of impurities in nanowires, especially for VLS-grown nanowires in which the incorporation of catalyst atoms is a concern; catalyst atoms can act as trap sites and recombination centers, degrading electrical and optical properties. Due in part to this concern, many atom probe studies have addressed impurity atom concentrations (Allen et al., 2008; Eichfeld et al., 2012; Moutanabbir et al., 2013; Murthy et al., 2011; Roussel et al., 2011). Allen et al. established an upper bound of Au atoms in Si nanowires by APT and revealed the relative influence of Au atoms and surface states on minority carrier lifetime in Si nanowires (Allen et al., 2008; Kayes et al., 2008). Quantitative analysis of catalyst incorporation has also helped unveil nanowire growth mechanisms. Al catalyst atoms were shown to be incorporated into Si nanowires during VLS growth at a concentration exceeding the equilibrium solid solubility by orders of magnitude (Eichfeld et al., 2012; Moutanabbir et al., 2013). This interesting phenomenon was explained in terms of solute trapping at the edge of fast-growing steps at the growth interface (Moutanabbir et al., 2013). Analysis of environmentally abundant impurities such as O, H, C, and N can be challenging because such species are typically present in the background even under the ultrahigh vacuum conditions ($<10^{-10}$ Torr) of the APT chamber. Background gas atoms and molecules adsorbed on the shank of the tip may readily field-evaporate as $H_2$, CO, and $CO_2$ in response to or even in the absence of a laser pulse. Also, H atoms from the background gases can form hydrides (XH and $XH_2$) with elements (X) from the nanowire. Moreover, the relative contribution can depend on the operation conditions, so control experiments are necessary (Dawahre et al., 2012).

Approaches for dopant and impurity analysis in nanowires, including techniques beyond APT, have been reviewed recently (Holmberg et al., 2013; Wallentin and Borgström, 2011). Here, we provide a few examples of the other techniques that have been used to investigate the issues discussed above. HAADF-STEM (high-angle annular dark field-scanning transmission electron microscopy) can be used to detect individual dopants or impurity atoms by detecting the scattering of the focused electron beam from atoms with high atomic number (the scattering is proportional to atomic number $\sim Z^{1.7}$). Through-focal series HAADF-STEM imaging coupled with simulations has been used to identify distinct atomic positions of Au atoms in Si nanowires (Allen et al., 2008; Oh et al., 2008) and Au atoms in GaAs nanowires (Bar-Sadan et al., 2012). Under appropriate imaging conditions, it is possible to distinguish between substitutional and interstitial defect configurations, which is not generally realized with APT (Oh et al., 2008). When the difference in the atomic numbers between the impurity and the matrix is not sufficient to identify elements by scattering alone, HAADF-STEM can be combined with spectroscopies such as EDS (energy-dispersive X-ray spectroscopy) (Cohen-Karni et al., 2012) and EELS (electron energy loss spectroscopy) (Wang et al., 2011). However, many factors influence the "visibility" of the elements of interest, including the crystal structure, crystal orientation, composition, thickness, and location of the atoms along the beam direction. Moreover, the detection limit of EDS (typically $\sim$1 at.%) is not sufficient for single impurity detection, and quantitative EELS is very sensitive to background subtraction for detection of very dilute species. Due to the varying sensitivity to dilute species, it is not generally possible to obtain concentration profiles of dilute species using S/TEM-based techniques. It is worth noting, however, that off-axis electron holography in a transmission electron microscopy can be used to assess the impact of *active* dopants in semiconductor nanowires (Chung and Rabenberg, 2006; den Hertog et al., 2009; Table 1).

### *6.1.1 Challenges in Measuring Dilute Species*

Accurate quantification of dilute species in nanowires requires optimization of operation conditions, often without the availability of a relevant standard sample, as well as analysis of mass spectra limited by counting statistics. When distinct peaks of dilute species can be identified, the concentration is calculated by dividing the background-subtracted counts of the species by the entire counts of host atoms. However, it may be the case that no distinct peaks of a dilute species are observed in a mass spectrum. Even if a species

**Table 1** Previous Atom Probe Studies on Concentrations of Dopants and Impurities in Semiconductor Nanowires

| Semiconductor | Dopants/ Impurities | Concentration | References |
|---|---|---|---|
| Ge | P | $1.0 \pm 0.2 \times 10^{18}$ cm$^3$ | Perea et al. (2008) |
| Si | B | $1.1 \pm 0.2 \times 10^{19}$ cm$^3$ | Perea et al. (2008) |
| Ge, 25 nm core, 20 nm shell | P | $2 \times 10^{18}$, $2 \times 10^{18}$, 1.5–20 $\times 10^{19}$ | Zhang et al. (2009) |
| Ge | Au | Upper bound $2 \times 10^{17}$ cm$^3$ | Zhang et al. (2009) |
| Si | Au | Upper bound $\sim 5 \times 10^{17}$–$1.5 \times 10^{18}$ | Allen et al. (2008) |
| i-Ge, P:Ge | O | Upper bound $3 \times 10^{17}$, (3–12) $\times 10^{17}$ | Zhang et al. (2009) |
| Ge | Ga | $2.3 \pm 0.2 \times 10^{18}$ | Zeiner et al. (2011) |
| Si core<br>i-Ge shell | B | $8 \pm 2 \times 10^{18}$<br>$2 \pm 0.4 \times 10^{19}$ | Schlitz et al. (2009) |
| Si | B | $1.1 \pm 0.1 \times 10^{19}$ | Chen et al. (2012) |
| Si | B | $1.3 \pm 0.3 \times 10^{19}$ | Xu et al. (2008) |
| Si | P | $8 \times 10^{19}$ | El Kousseifi et al. (2014) |
| InGaN | Mg | $2.7 \times 10^{19}$ | Riley et al. (2013) |
| Si | B | $1.4 \pm 0.1 \times 10^{20}$ | Roussel et al. (2011) |
| Si | P | 2–5 $\times 10^{19}$ | Murthy et al. (2011) |
| Si | Al | $\sim 2 \times 10^{20}$ | Eichfeld et al. (2012) |
| Ge | P<br>O<br>Au | $1.3 \pm 0.2 \times 10^{18}$,<br>$2.1 \pm 0.2 \times 10^{18}$,<br>$2.3 \pm 0.3 \times 10^{18}$,<br>$0.5 \pm 0.2 \times 10^{18}$,<br>$0.3 \pm 0.1 \times 10^{18}$,<br>$1.5 \pm 0.1 \times 10^{18}$,<br>$<0.2 \times 10^{18}$ | Perea et al. (2009) |
| Si | Al | $2.0 \pm 0.5 \times 10^{20}$ | Moutanabbir et al. (2013) |

is present, related peaks may be eclipsed by background counts or neighboring peaks, so the absence of resolvable peaks does not imply that a species is absent. In such a case, a global upper bound on concentration can be calculated by comparing the root mean squared fluctuation in the mass ranges expected for the dilute species with the total count of host atoms (Allen et al., 2008). Clearly, both intrinsic and extrinsic factors determine the detection limit. Intrinsically, counting statistics impose an uncertainty in concentration. For example, even in the absence of uncorrelated background counts, detection of 100 dilute species atoms out of a total of 1 million atoms corresponds to a concentration of $100 \pm 10$ ppm. Extrinsic influences on the detection limit include the background level of uncorrelated evaporation events, thermal tails of identified peaks, and the relative mass-to-charge-state ratios of host and impurity ions. Uncorrelated field evaporation of background gas ions as well as specimen ions contributes to the background by DC field evaporation, which has a larger impact on the measurement of dilute species at lower mass-to-charge-state ratios. Detection of dilute species can be also complicated by thermal tails when the mass ranges of dilute species occur immediately after those of host atoms, such as for phosphorus dopants (mass 31) in a silicon (mass 28, 29, and 30) nanowire host. The thermal tail can be diminished by increasing the taper angle of the nanowires and optimizing focusing of the laser spot to facilitate rapid heat dissipation (Perea et al., 2008). The hybrid geometry of an untapered nanowire grown on a tapered nanowire can also help facilitate tip cooling (Perea et al., 2008).

## 6.2 Alloys: Composition Fluctuations and Clusters

The 3D nature, sensitivity, and resolution of APT offer unique opportunities for quantitative analysis of alloy fluctuations and clustering on the nanometer scale. APT is particularly well suited to determine whether composition fluctuations exceed those of a random alloy and can be used to characterize the distribution of nanoscale clusters in terms of size, composition, and shape. Before introducing specific examples of the alloy analysis, we first briefly explain two basic analyses that have been widely utilized—frequency distribution analysis (FDA) and isoconcentration surfaces. For FDA, the reconstruction is divided up into volume elements (voxels), each of which contains a certain number of atoms. The composition in each voxel is calculated, and a histogram of voxel compositions is then generated. The histogram is compared with the distribution expected for a random alloy, and deviations from a random distribution are quantified with a $\chi^2$-test. Isoconcentration surfaces, or isosurfaces, refer to the surfaces of constant concentration and can be presented as 3D contour plots (Fig. 7).

The example of GaAs–AlGaAs core–shell nanowires (Jeon et al., 2015; Mancini et al., 2014b) highlights many of the unique attributes of APT. Quantitative analysis of the alloy distribution in the AlGaAs shell region by S/TEM alone is difficult because of the complexity of alloy segregation in radial heterostructures, the limited volume of analysis in cross-sectional samples, and the 2D nature of projected TEM images (Heiss et al., 2013; Rudolph et al., 2013; Zheng et al., 2013). While traces of alloy fluctuations in bulk-like AlGaAs regions have been detected in TEM images (Rudolph et al., 2013; Zheng et al., 2013), the nature of the alloy composition fluctuations was clearly identified by APT (Jeon et al., 2015; Mancini et al., 2014b). Moreover, by utilizing isosurfaces, proxigrams, and 2D composition maps,

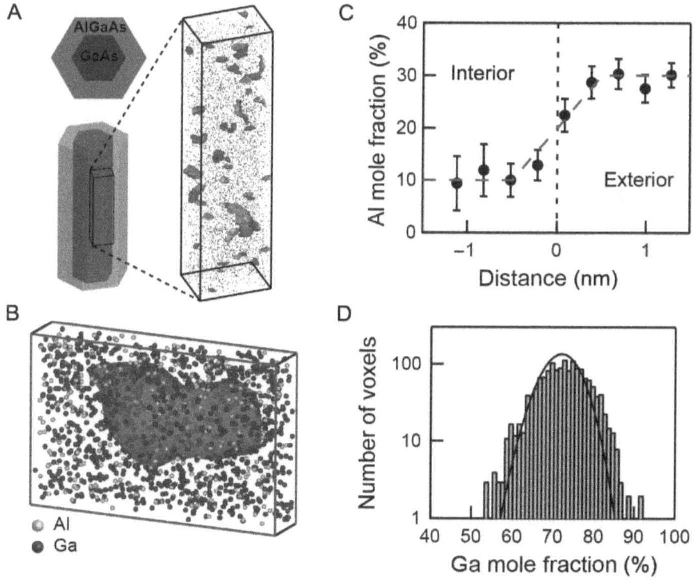


**Figure 7** (A) 3D atom map showing presence of Ga-rich clusters in shell of GaAs–AlGaAs core–shell nanowire (Derived from Jeon et al., 2015). Schematics show analysis location in the shell. (B) 3D atom map showing Ga-rich cluster enclosed by 80% Ga isosurface (Derived from Jeon et al., 2015). (C) Proxigram of Ga-rich cluster showing variation of Al mole fraction across the cluster interface (black dashed line) (Derived from Jeon et al., 2015). (D) Frequency distribution analysis for Al and Ga atoms in the AlGaAs shell region. Bars and solid lines represent observed distribution and pseudo-binomial (random) distribution (Derived from Jeon et al., 2015).

Ga-enriched regions were identified and their size distribution, composition, and shape were statistically studied (Jeon et al., 2015). Although APT provides powerful analytical capabilities for alloy analysis, it is important to quantitatively evaluate the possible influence of artifacts. In particular, artifacts can originate from the dissimilar field evaporation behaviors of different elements, leading to biased composition measurements, as discussed further below (6.2.1).

### *6.2.1 Challenges in Measuring Stoichiometry*

Individual elements in compounds have different thresholds for field evaporation and varying tendencies to diffuse on the surface and form ion clusters. Hence, it is important to examine the count rates of particular ions as well as related groups of ions to assess the accuracy of measurements of stoichiometry. The influence of count rate variations on stoichiometry is well studied in III–V compound semiconductor materials (Agrawal et al., 2011; Mancini et al., 2014a; Riley et al., 2012). It is well known, for example, that group III species tend to have lower field evaporation thresholds and experience preferential loss, particularly when they directly field-evaporate in between laser pulses. Such undercounting can be partially rectified by decreasing the standing voltage and increasing the laser energy. On the other hand, a higher laser energy or decreased DC voltage for a given detection rate may retard the field evaporation of group V species of higher field evaporation threshold and facilitate the formation of cluster ions. Cluster ions may be difficult to identify based on a mass spectrum because of peak overlap; for example, $N_2^{2+}$ ions have the same mass-to-charge state ratio as $N^+$. Count rates also vary with intrinsic sample features such as the orientation and polarity of the facet from which the evaporation takes place. The presence of artifacts arising from variations in ion count rate between species and across the evaporation surface can be empirically evaluated by measuring variations in count rates with respect to changes in operating conditions and deviations in composition observed in reference samples evaluated by correlated methods (Riley et al., 2012, 2014). Based on the knowledge of count rate variations, the optimum operating conditions can be selected based on the information one is most interested in extracting from the sample. For example, to measure the relative amounts of A and B species that share same sublattice in $A_xB_{1-x}C$ alloys, the operation conditions should be optimized to maximize and equalize the detection efficiency for A and B.

## 6.3 Interfaces and Heterostructures

Semiconductor nanowire growth methods enable new types of heterostructures to be realized (Hyun et al., 2013), but the small interface area and nonplanar nature present significant challenges to characterization. APT can be used to analyze such interfaces, which can be visualized as isoconcentration surfaces. Subsequently, various analyses can be performed to describe changes in composition along or across interfaces. Riley et al. analyzed InGaN–GaN multiquantum well (QW) nanowire arrays capped by a Mg-doped InGaN canopy layer (Riley et al., 2013). In each nanowire, there exist QWs of two different crystal polarities—semipolar and nonpolar (Riley et al., 2013). Rendering of the quantum wells using isosurfaces enabled isolated assessments of the In composition in the distinct quantum well regions, including the uniformity of the composition as described above. Moreover, the spatial variation in In composition was correlated with variations in the wavelength of cathodoluminescence. Mg clusters in Mg-doped InGaN canopy layer were additionally identified by FDA and isosurfaces (Fig. 8).

Analysis of the catalyst–nanowire interface can help explore atomic-scale mechanisms of nanowire growth or more specifically, advance of growth fronts (Dayeh et al., 2011; Gamalski et al., 2011; Wen et al., 2011). Quantitative analysis of the chemical abruptness and interfacial roughness is particularly important in semiconductor heterostructures to establish the band profiles that determine electronic properties (Chen et al., 2013; Perea et al., 2011). To evaluate abruptness of interfaces, one can generate a 1D composition profile along a particular axis from a series of thin 2D slices generated at different positions with respect to the interface (Perea et al., 2005). However, a curved but chemically abrupt interface will appear diffuse if the profile is generated in this manner. A proximity histogram, or proxigram, provides a quantitative and often better defined measure of the interface composition profile that is valid for curved interfaces as well. A proxigram is a 1D composition profile along the path of steepest ascent/descent defined by the gradient along the local normal of an isoconcentration surface. A proxigram can be generated from open or closed surfaces. It has been shown using proxigrams that catalyst–nanowire interfaces are very abrupt, with the widths of less than 0.5 nm both for Au-catalyzed InAs nanowires and for Au-catalyzed Si nanowires (Allen et al., 2008; Perea et al., 2005). While S/TEM-based methods can in principle provide a higher resolution, it is necessary to align the specimen such

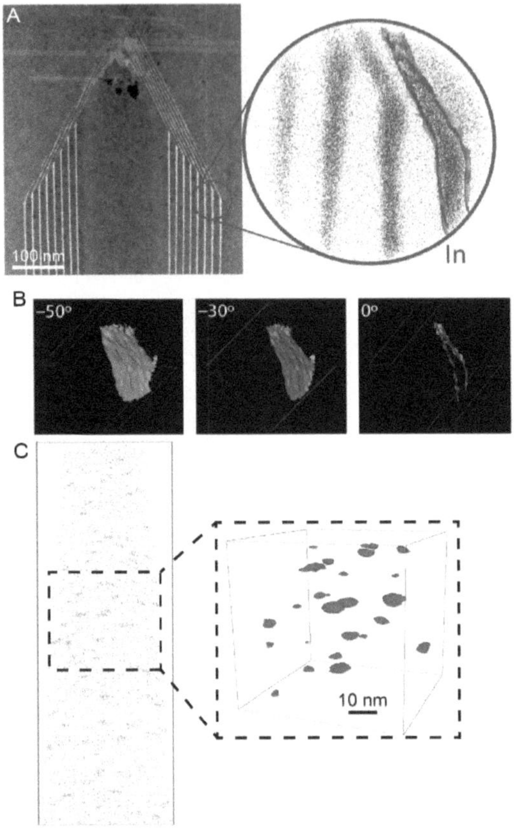


**Figure 8** (A) Cross-sectional HAADF-STEM image of GaN–InGaN quantum well nanowire near the nanowire apex. The enlarged region (red (dark gray in the print version) circle) shows the 3D atom map of In atoms with isosurfaces depicting quantum well interfaces. (B) Isosurfaces with 3.5 atomic % of In at different angles. (C) 3D atom map of Mg-doped InGaN canopy layer showing clustering of Mg atoms (blue (gray in the print version) dots). Right panel shows the isosurfaces of 1.0 at.% Mg for the dashed region in the left panel. *Derived from Riley et al. (2013).*

that the interface is parallel to the electron beam; 1D composition profiles measured by EELS or EDX will overestimate the width of the interface when a specimen is not properly aligned or the interface is not flat. Proxigrams generated from isosurfaces' width are also well suited to analyzing fluctuations in interface width, as demonstrated in planar heterostructures (Müller et al., 2012; Prosa et al., 2011).

### 6.3.1 Challenges in Interface Analysis: Full Diameter Imaging

Due to the limited field of view in APT, the entire diameter of a nanowire specimen is not analyzed under conventional experimental conditions, so the outer surface will not appear in the reconstruction. Tilting of the specimen has been suggested to enable imaging of the peripheral regions of a nanowire (Agrawal et al., 2011). However, it is difficult to finely adjust the degree of specimen tilt. Furthermore, nanowires tend to bend to align with the electric field such that the nanowire axis approaches the analysis direction. An alternative approach, discussed above, involves the deposition of a conformal shell on the nanowire to increase the total specimen diameter, thereby decreasing magnification and enabling the entire peripheral region to come into the field of view (Connell et al., 2013). However, there are many detailed considerations in the selection of shell materials and deposition methods to ensure that they evaporate smoothly and that the interface is stable. First, the shell material should have an evaporation threshold that is equal to or lower than that of the core. Second, peaks from shell materials should not overlap with those of the core. Third, the shell should be deposited conformally with good mechanical stability. Finally, other factors including the differential absorption of the laser pulse will influence the evaporation characteristics. As an alternative approach, one can use a substrate, above which an apex of atom probe tip is positioned only few micrometers, thereby introducing a high degree of image compression (i.e., compression of ion trajectories) and enhancing the field of view (Du et al., 2013).

## 7. SUMMARY

By analyzing nanoscale variations in composition in three dimensions, APT can play a unique role in elucidating processing–structure–property relationships in semiconductor nanowires. Advances in commercial instrumentation and software over the last decade, particularly in pulsed laser local electrode APT, have enabled the technique to be applied to semiconductor nanowires and nanowire heterostructures of both doped and insulating materials. Atom probe studies have provided important insights into atomic-scale mechanisms that govern nanowire growth by providing a three-dimensional perspective on dopant and impurity distributions that can be related to processing conditions. Furthermore, three-dimensional composition data provide a foundation for modeling of materials and device properties that can be correlated with other complementary characterization methods. Significant hurdles in sample preparation have been overcome in recent years and have been described here to educate readers who are familiar with nanowires but not with APT. Considering the known artifacts of atom probe analysis that arise from the field evaporation process, careful optimization of operating conditions as well as thorough analysis of the data generated are necessary to support reliable scientific claims. This chapter has attempted to provide a sufficient foundation for the reader to understand these issues and investigate further as necessary. Finally, we have surveyed the literature for applications of APT to the analysis of semiconductor nanowires and have highlighted important examples that have advanced scientific understanding. Complementary and correlated analysis by more widely available and established methods is often essential to making the best use of the unique information that APT provides.

## ACKNOWLEDGMENTS

This chapter draws on advances made by many students and postdocs in the author's (L.J.L.) research group, including Daniel Perea, Eric Hemesath, Justin Connell, Sonal Padalkar, James Riley, Sonal Padalkar, Zhiyuan Sun, and Nari Jeon. The work described herein was supported by the National Science Foundation under DMR-1306854, DMR-1006069, DMR-0449933; the Office of Naval Research under N00014-09-0182; and the US DOE Office of Basic Energy Sciences through the Energy Frontier Research Center on Solid State Lighting Science. Atom probe tomography was performed at the Northwestern University Center for Atom-Probe Tomography (NUCAPT). NUCAPT is a Shared Facility at the Materials Research Center of Northwestern University, supported by the National Science Foundation's MRSEC program (DMR-1121262).

## REFERENCES

Agrawal, R., Bernal, R.A., Isheim, D., Espinosa, H.D., 2011. Characterizing atomic composition and dopant distribution in wide band gap semiconductor nanowires using laser-assisted atom probe tomography. J. Phys. Chem. C 115, 17688–17694.

Allen, J.E., Hemesath, E.R., Perea, D.E., Lensch-Falk, J.L., Y., Liz, Yin, F., Gass, M.H., Wang, P., Bleloch, A.L., Palmer, R.E., Lauhon, L.J., 2008. High-resolution detection of Au catalyst atoms in Si nanowires. Nat. Nanotechnol. 3, 168–173.

Bar-Sadan, M., Barthel, J., Shtrikman, H., Houben, L., 2012. Direct imaging of single Au atoms within GaAs nanowires. Nano Lett. 12, 2352–2356.

Blumtritt, H., Isheim, D., Senz, S., Seidman, D.N., Moutanabbir, O., 2014. Preparation of nanowire specimens for laser-assisted atom probe tomography. Nanotechnology 25, 435704.

Cerezo, A., Grovenor, C.R.M., Smith, G.D.W., 1985. Pulsed laser atom probe analysis of GaAs and InAs. Appl. Phys. Lett. 46, 567–569.

Chen, W., Dubrovskii, V.G., Liu, X., Xu, T., Lardé, R., Philippe Nys, J., Grandidier, B., Stiévenard, D., Patriarche, G., Pareige, P., 2012. Boron distribution in the core of Si nanowire grown by chemical vapor deposition. J. Appl. Phys. 111, 094909.
Chen, L., Fung, W.Y., Lu, W., 2013. Vertical nanowire heterojunction devices based on a clean Si/Ge interface. Nano Lett. 13, 5521–5527.
Chung, J., Rabenberg, L., 2006. Mapping of electrostatic potentials within core-shell nanowires by electron holography. Appl. Phys. Lett. 88, 013106.
Cohen-Karni, T., Casanova, D., Cahoon, J.F., Qing, Q., Bell, D.C., Lieber, C.M., 2012. Synthetically encoded ultrashort-channel nanowire transistors for fast, pointlike cellular signal detection. Nano Lett. 12, 2639–2644.
Connell, J.G., Yoon, K., Perea, D.E., Schwalbach, E.J., Voorhees, P.W., Lauhon, L.J., 2013. Identification of an intrinsic source of doping inhomogeneity in vapor–liquid–solid-grown nanowires. Nano Lett. 13, 199–206.
Dawahre, N., Shen, G., Balci, S., Baughman, W., Wilbert, D., Harris, N., Butler, L., Martens, R., Kim, S., Kung, P., 2012. Atom probe tomography of zinc oxide nanowires. J. Electron. Mater. 41, 801–808.
Dayeh, S.A., Wang, J., Li, N., Huang, J.Y., Gin, A.V., Picraux, S.T., 2011. Growth, defect formation, and morphology control of germanium–silicon semiconductor nanowire heterostructures. Nano Lett. 11, 4200–4206.
Den Hertog, M.I., Schmid, H., Cooper, D., Rouviere, J.-L., Björk, M.T., Riel, H., Rivallin, P., Karg, S., Riess, W., 2009. Mapping active dopants in single silicon nanowires using off-axis electron holography. Nano Lett. 9, 3837–3843.
Du, S., Burgess, T., Tjing Loi, S., Gault, B., Gao, Q., Bao, P., Li, L., Cui, X., Kong Yeoh, W., Hoe Tan, H., Jagadish, C., Ringer, S.P., Zheng, R., 2013. Full tip imaging in atom probe tomography. Ultramicroscopy 124, 96–101.
Eichfeld, C.M., Gerstl, S.S.A., Prosa, T., Ke, Y., Redwing, J.M., Mohney, S.E., 2012. Local electrode atom probe analysis of silicon nanowires grown with an aluminum catalyst. Nanotechnology 23, 215205.
El Kousseifi, M., Panciera, F., Hoummada, K., Descoins, M., Baron, T., Mangelinck, D., 2014. Ni silicide nanowires analysis by atom probe tomography. Microelectron. Eng. 120, 47–51.
Galtrey, M.J., Oliver, R.A., Kappers, M.J., Humphreys, C.J., Clifton, P.H., Larson, D., Saxey, D.W., Cerezo, A., 2008. Three-dimensional atom probe analysis of green- and blue-emitting $InxGa_{1-x}N/GaN$ multiple quantum well structures. J. Appl. Phys. 104, 013524.
Gamalski, A.D., Ducati, C., Hofmann, S., 2011. Cyclic supersaturation and triple phase boundary dynamics in germanium nanowire growth. J. Phys. Chem. C 115, 4413–4417.
Gorman, B.P., Diercks, D., Salmon, N., Stach, E., Amador, G., Hartfield, C., 2008. Hardware and techniques for cross-correlative TEM and atom probe analysis. Micros. Today 16, 42–47.
Hashizume, T., Hasegawa, Y., Kobayashi, A., Sakurai, T., 1986. Atom-probe investigation of III–V semiconductors: comparison of voltage-pulse and laser-pulse modes. Rev. Sci. Instrum. 57, 1378–1380.
Heiss, M., Fontana, Y., Gustafsson, A., Wust, G., Magen, C., O'regan, D.D., Luo, J.W., Ketterer, B., Conesa-Boj, S., Kuhlmann, A.V., Houel, J., Russo-Averchi, E., Morante, J.R., Cantoni, M., Marzari, N., Arbiol, J., Zunger, A., Warburton, R.J., Fontcuberta, I., Morral, A., 2013. Self-assembled quantum dots in a nanowire system for quantum photonics. Nat. Mater. 12, 439–444.
Holmberg, V.C., Helps, J.R., Mkhoyan, K.A., Norris, D.J., 2013. Imaging impurities in semiconductor nanostructures. Chem. Mater. 25, 1332–1350.
Hyun, J.K., Zhang, S., Lauhon, L.J., 2013. Nanowire heterostructures. Annu. Rev. Mater. Res. 43, 451–479.

Isheim, D., Kaszpurenko, J., Yu, D., Mao, Z., Seidman, D.N., Arslan, I., 2012. 3-D atomic-scale mapping of manganese dopants in lead sulfide nanowires. J. Phys. Chem. C 116, 6595–6600.

Jeon, N., Loitsch, B., Morkötter, S., Abstreiter, G., Finley, J.J, Krenner,, H.J., Koblmüller, G., Lauhon, L., 2015. Alloy fluctuations act as quantum dot-like emitters in GaAs-AlGaAs core-shell nanowires. ACS Nano. http://dx.doi.org/10.1021/acsnano.5b04070, http://pubs.acs.org/doi/abs/10.1021/acsnano.5b04070.

Kayes, B.M., Filler, M.A., Henry, M.D., Maiolo Iii, J.R., Kelzenberg, M.D., Putnam, M.C., Spurgeon, J.M., Plass, K.E., Scherer, A., Lewis, N.S., Atwater, H.A., 2008. Radial PN junction, wire array solar cells. In: IEEE Phot. Spec. Conf., 11–16 May, pp. 1–5.

Kellogg, G.L., Tsong, T.T., 1980. Pulsed-laser atom-probe field-ion microscopy. J. Appl. Phys. 51, 1184–1193.

Kelly, T.F., Miller, M.K., 2007. Atom probe tomography. Rev. Sci. Instrum. 78, 031101.

Kelly, T.F., Larson, D.J., Thompson, K., Alvis, R.L., Bunton, J.H., Olson, J.D., Gorman, B.P., 2007. Atom probe tomography of electronic materials. Annu. Rev. Mater. Res. 37, 681–727.

Larson, D.J., Prosa, T.J., Ulfig, R.M., Geiser, B.P., Kelly, T.F., 2013. Local Electrode Atom Probe Tomography, first ed. New York, Springer.

Mancini, L., Amirifar, N., Shinde, D., Blum, I., Gilbert, M., Vella, A., Vurpillot, F., Lefebvre, W., Lardé, R., Talbot, E., Pareige, P., Portier, X., Ziani, A., Davesnne, C., Durand, C., Eymery, J., Butté, R., Carlin, J.-F., Grandjean, N., Rigutti, L., 2014a. Composition of wide bandgap semiconductor materials and nanostructures measured by atom probe tomography and its dependence on the surface electric field. J. Phys. Chem. C 118, 24136–24151.

Mancini, L., Fontana, Y., Conesa-Boj, S., Blum, I., Vurpillot, F., Francaviglia, L., Russo-Averchi, E., Heiss, M., Arbiol, J., Morral, A.F.I., Rigutti, L., 2014b. Three-dimensional nanoscale study of Al segregation and quantum dot formation in GaAs/AlGaAs core-shell nanowires. Appl. Phys. Lett. 105, 243106.

Morkötter, S., Jeon, N., Rudolph, D., Loitsch, B., Spirkoska, D., Hoffmann, E., Döblinger, M., Matich, S., Finley, J.J., Lauhon, L.J., Abstreiter, G., Koblmüller, G., 2015. Demonstration of confined electron gas and steep-slope behavior in delta-doped GaAs-AlGaAs core–shell nanowire transistors. Nano Lett. 15, 3295–3302.

Moutanabbir, O., Isheim, D., Blumtritt, H., Senz, S., Pippel, E., Seidman, D.N., 2013. Colossal injection of catalyst atoms into silicon nanowires. Nature 496, 78–82.

Müller, M., Gault, B., Smith, G.D.W., Grovenor, C.R.M., 2011. Accuracy of pulsed laser atom probe tomography for compound semiconductor analysis. J. Phys. Conf. Ser. 326, 012031.

Müller, M., Gault, B., Field, M., Sullivan, G.J., Smith, G.D.W., Grovenor, C.R.M., 2012. Interfacial chemistry in an InAs/GaSb superlattice studied by pulsed laser atom probe tomography. Appl. Phys. Lett. 100, 083109.

Murthy, D.H.K., Xu, T., Chen, W.H., Houtepen, A.J., Savenije, T.J., Siebbeles, L.D.A., Nys, J.P., Krzeminski, C., Grandidier, B., Stiévenard, D., Pareige, P., Jomard, F., Patriarche, G., Lebedev, O.I., 2011. Efficient photogeneration of charge carriers in silicon nanowires with a radial doping gradient. Nanotechnology 22, 315710.

Oh, S.H., Benthem, K.V., Molina, S.I., Borisevich, A.Y., Luo, W., Werner, P., Zakharov, N.D., Kumar, D., Pantelides, S.T., Pennycook, S.J., 2008. Point defect configurations of supersaturated Au atoms inside Si nanowires. Nano Lett. 8, 1016–1019.

Padalkar, S., Riley, J.R., Li, Q., Wang, G.T., Lauhon, L.J., 2014. Lift-out procedures for atom probe tomography targeting nanoscale features in core-shell nanowire heterostructures. Phys. Status Solidi C 11, 656–661.

Perea, D.E., Allen, J.E., May, S.J., Wessels, B.W., Seidman, D.N., Lauhon, L.J., 2005. Three-dimensional nanoscale composition mapping of semiconductor nanowires. Nano Lett. 6, 181–185.

Perea, D.E., Lensch, J.L., May, S.J., Wessels, B.W., Lauhon, L.J., 2006. Composition analysis of single semiconductor nanowires using pulsed-laser atom probe tomography. Appl. Phys. A Mater. Sci. Process. 85, 271–275.

Perea, D.E., Wijaya, E., Lensch-Falk, J.L., Hemesath, E.R., Lauhon, L.J., 2008. Tomographic analysis of dilute impurities in semiconductor nanostructures. J. Solid State Chem. 181, 1642–1649.

Perea, D.E., Hemesath, E.R., Schwalbach, E.J., Lensch-Falk, J.L., Voorhees, P.W., Lauhon, L.J., 2009. Direct measurement of dopant distribution in an individual vapour-liquid–solid nanowire. Nat. Nanotechnol. 4, 315–319.

Perea, D.E., Li, N., Dickerson, R.M., Misra, A., Picraux, S.T., 2011. Controlling heterojunction abruptness in VLS-grown semiconductor nanowires via in situ catalyst alloying. Nano Lett. 11, 3117–3122.

Prosa, T.J., Alvis, R., Tsakalakos, L., Smentkowski, V.S., 2010. Characterization of dilute species within CVD-grown silicon nanowires doped using trimethylboron: protected lift-out specimen preparation for atom probe tomography. J. Microsc. 239, 92–98.

Prosa, T.J., Clifton, P.H., Zhong, H., Tyagi, A., Shivaraman, R., Denbaars, S.P., Nakamura, S., Speck, J.S., 2011. Atom probe analysis of interfacial abruptness and clustering within a single InxGa1–xN quantum well device on semipolar (101¯1¯) GaN substrate. Appl. Phys. Lett. 98, 191903.

Rigutti, L., Blum, I., Shinde, D., Hernández-Maldonado, D., Lefebvre, W., Houard, J., Vurpillot, F., Vella, A., Tchernycheva, M., Durand, C., Eymery, J., Deconihout, B., 2014. Correlation of microphotoluminescence spectroscopy, scanning transmission electron microscopy, and atom probe tomography on a single nano-object containing an InGaN/GaN multiquantum well system. Nano Lett. 14, 107–114.

Riley, J.R., Bernal, R.A., Li, Q., Espinosa, H.D., Wang, G.T., Lauhon, L.J., 2012. Atom probe tomography of a-axis GaN nanowires: analysis of nonstoichiometric evaporation behavior. ACS Nano 6, 3898–3906.

Riley, J.R., Padalkar, S., Li, Q., Lu, P., Koleske, D.D., Wierer, J.J., Wang, G.T., Lauhon, L.-J., 2013. Three-dimensional mapping of quantum wells in a GaN/InGaN core–shell nanowire light-emitting diode array. Nano Lett. 13, 4317–4325.

Riley, J.R., Detchprohm, T., Wetzel, C., Lauhon, L.J., 2014. On the reliable analysis of indium mole fraction within $InxGa_{1-x}N$ quantum wells using atom probe tomography. Appl. Phys. Lett. 104, 152102.

Roussel, M., Chen, W., Talbot, E., Larde, R., Cadel, E., Gourbilleau, F., Grandidier, B., Stievenard, D., Pareige, P., 2011. Atomic scale investigation of silicon nanowires and nanoclusters. Nanoscale Res. Lett. 6, 271.

Rudolph, D., Funk, S., Döblinger, M., Morkötter, S., Hertenberger, S., Schweickert, L., Becker, J., Matich, S., Bichler, M., Spirkoska, D., Zardo, I., Finley, J.J., Abstreiter, G., Koblmüller, G., 2013. Spontaneous alloy composition ordering in GaAs-AlGaAs core–shell nanowires. Nano Lett. 13, 1522–1527.

Sanford, N.A., Blanchard, P.T., Brubaker, M., Bertness, K.A., Roshko, A., Schlager, J.B., Kirchhofer, R., Diercks, D.R., Gorman, B., 2014. Laser-assisted atom probe tomography of MBE grown GaN nanowire heterostructures. Phys. Status Solidi C 11, 608–612.

Schlitz, R.A., Perea, D.E., Lensch-Falk, J.L., Hemesath, E.R., Lauhon, L.J., 2009. Correlating dopant distributions and electrical properties of boron-doped silicon nanowires. Appl. Phys. Lett. 95, 162101.

Seidman, D.N., 2007. Three-dimensional atom-probe tomography: advances and applications. Annu. Rev. Mater. Res. 37, 127–158.

Seidman, D.N., Stiller, K., 2009. An atom-probe tomography primer. MRS Bull. 34, 717–724.

Vurpillot, F., Bostel, A., Blavette, D., 1999a. The shape of field emitters and the ion trajectories in three-dimensional atom probes. J. Microsc. 196, 332–336.

Vurpillot, F., Bostel, A., Menand, A., Blavette, D., 1999b. Trajectories of field emitted ions in 3D atom-probe. Eur. Phys. J. Appl. Phys. 6, 217–221.

Wallentin, J., Borgström, M.T., 2011. Doping of semiconductor nanowires. J. Mater. Res. 26, 2142–2156.

Wang, C.-Y., Kuo, C.-L., Liu, C.-P., Wang, T.-Y., Zheng, R.-K., Ringer, S.P., 2011. Valence excitations and dopant distribution of Al doped ZnO nanowires analyzed by electron energy loss spectroscopy. J. Nanosci. Nanotechnol. 11, 10182–10186.

Wen, C.Y., Tersoff, J., Hillerich, K., Reuter, M.C., Park, J.H., Kodambaka, S., Stach, E.A., Ross, F.M., 2011. Periodically changing morphology of the growth interface in Si, Ge, and GaP nanowires. Phys. Rev. Lett. 107, 025503.

Xu, T., Nys, J.P., Grandidier, B., Stiévenard, D., Coffinier, Y., Boukherroub, R., Larde, R., Cadel, E., Pareige, P., 2008. Growth of Si nanowires on micropillars for the study of their dopant distribution by atom probe tomography. J. Vac. Sci. Technol. B 26, 1960–1963.

Yoon, K., Hyun, J.K., Connell, J.G., Amit, I., Rosenwaks, Y., Lauhon, L.J., 2013. Barrier height measurement of metal contacts to Si nanowires using internal photoemission of hot carriers. Nano Lett. 13, 6183–6188.

Zeiner, C., Lugstein, A., Burchhart, T., Pongratz, P., Connell, J.G., Lauhon, L.J., Bertagnolli, E., 2011. Atypical self-activation of Ga dopant for Ge nanowire devices. Nano Lett. 11, 3108–3112.

Zhang, S., Hemesath, E.R., Perea, D.E., Wijaya, E., Lensch-Falk, J.L., Lauhon, L.J., 2009. Relative influence of surface states and bulk impurities on the electrical properties of Ge nanowires. Nano Lett. 9, 3268–3274.

Zheng, C., Wong-Leung, J., Gao, Q., Tan, H.H., Jagadish, C., Etheridge, J., 2013. Polarity-driven 3-fold symmetry of GaAs/AlGaAs core multishell nanowires. Nano Lett. 13, 3742–3748.

# INDEX

Note: Page numbers followed by "*f*" indicate figures and "*t*" indicate tables.

## T



## U



## V

## W



## Z

# CONTENTS OF VOLUMES IN THIS SERIES

## Volume 1 **Physics of III–V Compounds**

*C. Hilsum*, Some Key Features of III–V Compounds
*F. Bassani*, Methods of Band Calculations Applicable to III–V Compounds
*E. O. Kane*, The *k-p* Method
*V. L. Bonch–Bruevich*, Effect of Heavy Doping on the Semiconductor Band Structure
*D. Long*, Energy Band Structures of Mixed Crystals of III–V Compounds
*L. M. Roth and P. N. Argyres*, Magnetic Quantum Effects
*S. M. Puri and T. H. Geballe*, Thermomagnetic Effects in the Quantum Region
*W. M. Becker*, Band Characteristics near Principal Minima from Magnetoresistance
*E. H. Putley*, Freeze-Out Effects, Hot Electron Effects, and Submillimeter Photoconductivity in InSb
*H. Weiss*, Magnetoresistance
*B. Ancker-Johnson*, Plasma in Semiconductors and Semimetals

## Volume 2 **Physics of III–V Compounds**

*M. G. Holland*, Thermal Conductivity
*S. I. Novkova*, Thermal Expansion
*U. Piesbergen*, Heat Capacity and Debye Temperatures
*G. Giesecke*, Lattice Constants
*J. R. Drabble*, Elastic Properties
*A. U. Mac Rae and G. W. Gobeli*, Low Energy Electron Diffraction Studies
*R. Lee Mieher*, Nuclear Magnetic Resonance
*B. Goldstein*, Electron Paramagnetic Resonance
*T. S. Moss*, Photoconduction in III–V Compounds
*E. Antoncik and J. Tauc*, Quantum Efficiency of the Internal Photoelectric Effect in InSb
*G. W. Gobeli and I. G. Allen*, Photoelectric Threshold and Work Function
*P. S. Pershan*, Nonlinear Optics in III–V Compounds
*M. Gershenzon*, Radiative Recombination in the III–V Compounds
*F. Stern*, Stimulated Emission in Semiconductors

## Volume 3 **Optical Properties of III–V Compounds**

*M. Hass*, Lattice Reflection
*W. G. Spitzer*, Multiphonon Lattice Absorption
*D. L. Stierwalt and R. F. Potter*, Emittance Studies
*H. R. Philipp and H. Ehrenveich*, Ultraviolet Optical Properties
*M. Cardona*, Optical Absorption Above the Fundamental Edge
*E. J. Johnson*, Absorption Near the Fundamental Edge
*J. O. Dimmock*, Introduction to the Theory of Exciton States in Semiconductors

## Volume 4 **Physics of III–V Compounds**



## Volume 5 **Infrared Detectors**



## Volume 6 **Injection Phenomena**

## Volume 7 **Application and Devices**



## Volume 8 **Transport and Optical Phenomena**



## Volume 9 **Modulation Techniques**



## Volume 10 **Transport Phenomena**

## Volume 11 **Solar Cells**

*H. J. Hovel,* Introduction; Carrier Collection, Spectral Response, and Photocurrent; Solar Cell Electrical Characteristics; Efficiency; Thickness; Other Solar Cell Devices; Radiation Effects; Temperature and Intensity; Solar Cell Technology

## Volume 12 **Infrared Detectors (II)**

*W. L. Eiseman, J. D. Merriam, and R. F. Potter,* Operational Characteristics of Infrared Photodetectors
*P. R. Bratt,* Impurity Germanium and Silicon Infrared Detectors
*E. H. Putley,* InSb Submillimeter Photoconductive Detectors
*G. E. Stillman, C. M. Wolfe, and J. O. Dimmock,* Far-Infrared Photoconductivity in High Purity GaAs
*G. E. Stillman and C. M. Wolfe,* Avalanche Photodiodes
*P. L. Richards,* The Josephson Junction as a Detector of Microwave and Far-Infrared Radiation
*E. H. Putley,* The Pyroelectric Detector – An Update

## Volume 13 **Cadmium Telluride**

*K. Zanio,* Materials Preparations; Physics; Defects; Applications

## Volume 14 **Lasers, Junctions, Transport**

*N. Holonyak, Jr., and M. H. Lee,* Photopumped III–V Semiconductor Lasers
*H. Kressel and J. K. Butler,* Heterojunction Laser Diodes
*A. Van der Ziel,* Space-Charge-Limited Solid-State Diodes
*P. J. Price,* Monte Carlo Calculation of Electron Transport in Solids

## Volume 15 **Contacts, Junctions, Emitters**

*B. L. Sharma,* Ohmic Contacts to III–V Compounds Semiconductors
*A. Nussbaum,* The Theory of Semiconducting Junctions
*J. S. Escher,* NEA Semiconductor Photoemitters

## Volume 16 **Defects, (HgCd)Se, (HgCd)Te**

*H. Kressel,* The Effect of Crystal Defects on Optoelectronic Devices
*C. R. Whitsett, J. G. Broerman, and C. J. Summers,* Crystal Growth and Properties of $Hg_{1-x}$ $Cd_x$ Se Alloys
*M. H. Weiler,* Magnetooptical Properties of $Hg_{1-x}$ $Cd_x$ Te Alloys
*P. W. Kruse and J. G. Ready,* Nonlinear Optical Effects in $Hg_{1-x}$ $Cd_x$ Te

## Volume 17 **CW Processing of Silicon and Other Semiconductors**

*J. F. Gibbons,* Beam Processing of Silicon
*A. Lietoila, R. B. Gold, J. F. Gibbons, and L. A. Christel,* Temperature Distributions and Solid Phase Reaction Rates Produced by Scanning CW Beams

## Volume 18 **Mercury Cadmium Telluride**



## Volume 19 **Deep Levels, GaAs, Alloys, Photochemistry**



## Volume 20 **Semi-Insulating GaAs**



## Volume 21 **Hydrogenated Amorphous Silicon**

### Part A

## Part B



## Part C



## Part D

## Volume 22 **Lightwave Communications Technology**

### Part A



### Part B



### Part C



### Part D

## Part E



## Volume 23 **Pulsed Laser Processing of Semiconductors**



## Volume 24 **Applications of Multiquantum Wells, Selective Doping, and Superlattices**



## Volume 25 **Diluted Magnetic Semiconductors**

## Volume 26 **III–V Compound Semiconductors and Semiconductor Properties of Superionic Materials**



## Volume 27 **High Conducting Quasi-One-Dimensional Organic Crystals**



## Volume 28 **Measurement of High-Speed Signals in Solid State Devices**

## Volume 29 **Very High Speed Integrated Circuits: Gallium Arsenide LSI**

*M. Kuzuhara and T. Nazaki*, Active Layer Formation by Ion Implantation
*H. Hasimoto*, Focused Ion Beam Implantation Technology
*T. Nozaki and A. Higashisaka*, Device Fabrication Process Technology
*M. Ino and T. Takada*, GaAs LSI Circuit Design
*M. Hirayama, M. Ohmori, and K. Yamasaki*, GaAs LSI Fabrication and Performance

## Volume 30 **Very High Speed Integrated Circuits: Heterostructure**

*H. Watanabe, T. Mizutani, and A. Usui*, Fundamentals of Epitaxial Growth and Atomic Layer Epitaxy
*S. Hiyamizu*, Characteristics of Two-Dimensional Electron Gas in III–V Compound Heterostructures Grown by MBE
*T. Nakanisi*, Metalorganic Vapor Phase Epitaxy for High-Quality Active Layers
*T. Nimura*, High Electron Mobility Transistor and LSI Applications
*T. Sugeta and T. Ishibashi*, Hetero-Bipolar Transistor and LSI Application
*H. Matsuedo, T. Tanaka, and M. Nakamura*, Optoelectronic Integrated Circuits

## Volume 31 **Indium Phosphide: Crystal Growth and Characterization**

*J. P. Farges*, Growth of Discoloration-Free InP
*M. J. McCollum and G. E. Stillman*, High Purity InP Grown by Hydride Vapor Phase Epitaxy
*I. Inada and T. Fukuda*, Direct Synthesis and Growth of Indium Phosphide by the Liquid Phosphorous Encapsulated Czochralski Method
*O. Oda, K. Katagiri, K. Shinohara, S. Katsura, Y. Takahashi, K. Kainosho, K. Kohiro, and R. Hirano*, InP Crystal Growth, Substrate Preparation and Evaluation
*K. Tada, M. Tatsumi, M. Morioka, T. Araki, and T. Kawase*, InP Substrates: Production and Quality Control
*M. Razeghi*, LP-MOCVD Growth, Characterization, and Application of InP Material
*T. A. Kennedy and P. J. Lin-Chung*, Stoichiometric Defects in InP

## Volume 32 **Strained-Layer Superlattices: Physics**

*T. P. Pearsall*, Strained-Layer Superlattices
*F. H. Pollack*, Effects of Homogeneous Strain on the Electronic and Vibrational Levels in Semiconductors
*J. Y. Marzin, J. M. Gerárd, P. Voisin, and J. A. Brum*, Optical Studies of Strained III–V Heterolayers
*R. People and S. A. Jackson*, Structurally Induced States from Strain and Confinement
*M. Jaros*, Microscopic Phenomena in Ordered Superlattices

## Volume 33 **Strained-Layer Superlattices: Material Science and Technology**

*R. Hull and J. C. Bean*, Principles and Concepts of Strained-Layer Epitaxy

## Volume 34 **Hydrogen in Semiconductors**



## Volume 35 **Nanostructured Systems**



## Volume 36 **The Spectroscopy of Semiconductors**

## Volume 37 **The Mechanical Properties of Semiconductors**



## Volume 38 **Imperfections in III/V Materials**



## Volume 39 **Minority Carriers in III–V Semiconductors: Physics and Applications**



## Volume 40 **Epitaxial Microstructures**

## Volume 41 **High Speed Heterostructure Devices**



## Volume 42 **Oxygen in Silicon**



## Volume 43 **Semiconductors for Room Temperature Nuclear Detector Applications**

## Volume 44 **II–IV Blue/Green Light Emitters: Device Physics and Epitaxial Growth**



## Volume 45 **Effect of Disorder and Defects in Ion-Implanted Semiconductors: Electrical and Physiochemical Characterization**



## Volume 46 **Effect of Disorder and Defects in Ion-Implanted Semiconductors: Optical and Photothermal Characterization**

## Volume 47 **Uncooled Infrared Imaging Arrays and Systems**



## Volume 48 **High Brightness Light Emitting Diodes**



## Volume 49 **Light Emission in Silicon: from Physics to Devices**

## Volume 50 **Gallium Nitride (GaN)**



## Volume 51A **Identification of Defects in Semiconductors**



## Volume 51B **Identification Defects in Semiconductors**

## Volume 52 **SiC Materials and Devices**



## Volume 53 **Cumulative Subjects and Author Index Including Tables of Contents for Volumes 1–50**

## Volume 54 **High Pressure in Semiconductor Physics I**



## Volume 55 **High Pressure in Semiconductor Physics II**

## Volume 56 **Germanium Silicon: Physics and Materials**



## Volume 57 **Gallium Nitride (GaN) II**



## Volume 58 **Nonlinear Optics in Semiconductors I**

## Volume 68 **Isotope Effects in Solid State Physics**



## Volume 69 **Recent Trends in Thermoelectric Materials Research I**



## Volume 70 **Recent Trends in Thermoelectric Materials Research II**

## Volume 71 **Recent Trends in Thermoelectric Materials Research III**



## Volume 72 **Silicon Epitaxy**



## Volume 73 **Processing and Properties of Compound Semiconductors**

## Volume 79 **Semiconducting Chalcogenide Glass II**



## Volume 80 **Semiconducting Chalcogenide Glass III**



## Volume 81 **Conducting Organic Materials and Devices**



## Volume 82 **Semiconductors and Semimetals**

## Volume 83 **Semiconductors and Semimetals**



## Volume 84 **Semiconductors and Semimetals**



## Volume 85 **Semiconductors and Semimetals**

## Volume 86 **Advances in Semiconductor Lasers**



## Volume 87 **Advances in Photovoltaics: Volume 1**



## Volume 88 **Oxide Semiconductors**

## Volume 89 **Advances in Photovoltaics: Part 2**



## Volume 90 **Advances in Photovoltaics: Part 3**



## Volume 91 **Defects in Semiconductors**

## Volume 92 **Advances in Photovoltaics: Part 4**

FSC
www.fsc.org
MIX
Papier | Fördert gute Waldnutzung
FSC® C083411

Kolibri 360 GmbH
Ferdinand-Jühlke-Straße 7
99095 Erfurt, Deutschland
produktsicherheit@kolibri360.de